T0323293

THE THREAD OF ENERGY

The Thread of Energy

MARTIN J. PASQUALETTI

OXFORD
UNIVERSITY PRESS

OXFORD
UNIVERSITY PRESS

Oxford University Press is a department of the University of Oxford. It furthers the University's objective of excellence in research, scholarship, and education by publishing worldwide. Oxford is a registered trade mark of Oxford University Press in the UK and certain other countries.

Published in the United States of America by Oxford University Press
198 Madison Avenue, New York, NY 10016, United States of America.

Library of Congress Control Number: 2021943214
ISBN 978–0–19–939480–7

DOI: 10.1093/oso/9780199394807.001.0001

1 3 5 7 9 8 6 4 2

Printed by Integrated Books International, United States of America

Contents

Contents

Acknowledgments

A conceit of *The Thread of Energy* is that it is the first book to propose a central theme that ties together the varied roles that energy plays in the theater of human activities. While there are many good books on energy, most of them tackle the subject in quite similar and familiar ways. Either they review basic tenets of energy physics and engineering, expose in detail a single topic such as solar energy or oil, or examine energy resources individually with explanations and descriptions that are usually accurate but are not adequately keyed to today's growing social consciousness. I wanted something different. During its long gestation, I have been supported and aided by the encouragement, insights, inspiration, and examples of many people. These include Will Graf, a steadfast friend for over forty years in the competitive and unexpected tumult of academic life; Andy Blowers, whose elegant writings and tireless advocacy are an enduring inspiration to all those with interests in nuclear energy; Jerry Dobson, who joined me in the late 1970s in promoting energy geography, and who then went on to make more meaningful contributions to the breadth of the discipline of geography than just about anyone I can name; Gary Dirks, who brought peerless leadership skills with him when he returned to his alma matter from the corporate energy world in Asia at just the right time to deepen my appreciation of energy as a basic human right; Doug McQueen, whose quiet pestering to finish this book was always accompanied by underlying faith and support for the project; Benjamin Sovacool, who raised the bar just about out of reach for the rest of us with his many insights to the energy/society nexus; Marilyn Brown, whose contributions to our understanding of the relationships between energy and human activity have been exemplary at many levels of applied research and government policy; and Elizabeth Monoian and Robert Ferry, founders of the Land Art Generator Initiative (LAGI), who have managed to marry art, grace, and style with renewable energy in a way that has widened its appeal as an alternative to fossil and nuclear fuels. Thanks are also due to those who aided me with late stages of preparation. These include Jacob Bethem, Ankit Bagga, Barbara Trapido-Lurie, Nate Anbar, Debaleena Majumdar, Prashansa, John Hofmeister, Barry Solomon, and Arina Melkozernova, who created one of the original illustrations of the thread of energy that I have used repeatedly in my energy classes. Last, I wish to acknowledge the thousands of energy students who graced me with their presence, attention, experiences, questions, and curiosity. While I in no way hold any of these people responsible for any gaffes or missteps in what follows, I owe them all my heartfelt thanks.

—Martin J. Pasqualetti, Sun Valley, Idaho, June 2021

1

Discovery

Encountering the Thread of Energy

When you open your eyes each morning, whether your first thought is a hot pot of coffee or a hot planet, whether you mount a bicycle or start a car, whether you eat bread or toast, or whether life is getting easier or is heading in the other direction, there will be one constant, always and forever. It affects your every experience and the experiences of all who share the planet. It weaves its way through all activities, from the most uncomplicated daily chore to the most complex international agreement, and through all ideas, from every memory of the past to every dream of the future. We are talking here about the thread of energy. As simple a concept as it may seem, its role in all human interactions is as fundamental as it is overlooked. I know this truth because I missed the centrality of its importance for many years—until I made the suggestive connection that forms the basis of this book. Allow me to explain.

For as long as I can remember, I harbored a longing to travel, to see the world, to explore just for the surprises I would encounter and the things I would learn. The destination did not matter very much. It was the trip itself I coveted. From an early age, I consumed *National Geographic*, pored over maps for fun, and watched travel shows on television, but nothing satisfied. Nothing slaked my thirst for the "real thing." I just wanted to get out of the house and scratch the wanderlust itch without a predetermined destination or personal restriction. My only goal was to visit places I had read about, explore places that were "off the grid," dive into different cultures, perhaps learn a new language, and see some of the world's remaining wild animals before they were all gathered into zoos or extinct.

Eventually, reluctantly, I came to accept that such trips were nothing more than fantasies. They would never happen. They would cost too much and take me away from the job that would be given to others if I left for an extended period. In short, my fantasies were utterly impractical and highly unlikely.

Then one day I received a letter from an attorney informing me that an elderly relative had just passed away, leaving me a hefty inheritance. It was the windfall that we all secretly hope for but never really expect to receive. It would permit me the means to fulfill my lifelong dream to wander the world, unfettered by the banalities of employment, commuting, housekeeping, or money.

The Thread of Energy. Martin J. Pasqualetti, Oxford University Press. © Oxford University Press 2021. DOI: 10.1093/oso/9780199394807.003.0001

In no time I had stored my belongings, sublet my apartment, resigned my job, and booked a seat on the first available flight. The destination was of no interest; I was off on the grand adventure of my dreams, with high expectations of what I would find, the lessons I would learn, and the questions that I could ponder and seek to answer.

Weeks quickly melted into months, and my experiences accumulated. I was learning more than I could have imagined about how to live an unassuming life. Traveling fast made me more flexible, more alert, more open to new ideas and alternative ways of thinking. One of the first things I noticed was that people were pretty similar everywhere, in that they desired pretty much the same things to sustain themselves: water, food, clothing, shelter, good health, security for their family, and the occasional dash of joy.

I also noticed an interlocking natural balance. This balance was especially apparent in the biosphere—the part of the planet holding all living things. I realized that Nature was continually working to establish equilibrium. When a landslide blocked a river, the river eventually eroded its way through the debris, reestablishing the water's relentless journey to the sea. When a forest burned, the ashes helped fertilize the soil, and the forest eventually regrew. When volcanic ash darkened the sky, it caused catastrophic reductions in crop yields, but the skies eventually cleared and the crop yields returned. But I also noted something else, a common element woven throughout all aspects of human life, from the food we eat to the wars we wage. Furthermore, this common element was indifferent to a person's station in life, age, culture, history, or ethnicity. Soon I found myself asking how I could have overlooked the presence of this connective material for so long. My oversight was even more puzzling given that this entity held together—without exception—all human needs and all social activities. I had uncovered something more essential than water, air, or land. I had discovered the thread of energy.

From that point in time, the thread of energy snapped into focus wherever I looked. I saw it curl endlessly around all life experiences, from personal health to planetary health. I came to appreciate that energy is an essential element in the tapestry of life. I realized that following the thread of energy and understanding the paths it takes helps bring both order and understanding to the messy jumble of human activities that spin around us in an endlessly complicated swirl (Figure 1.1).

I learned that following the thread of energy humanizes all our activities, desires, aspirations, and shortcomings. It highlights the relationships that are broadly global (such as national security and military readiness) as well as those that are predominantly personal (such as quality of life, health and safety, comfort, security, and our role as consumers). In short, following the thread of energy helps us explore how we live, what we covet, the risks we perceive, and the

Figure 1.1 Energy is a social issue with a technical component, not the other way around. Graphic design by Arina Melkozernova.

future that we hope comes to pass. I also learned that while we have gotten pretty good at explaining the physics and chemistry of energy, we have fallen short in identifying and explaining the complex interactions between energy and people (Figure 1.2).

This book digs into the relationships between energy and society by prodding us to go beyond description to expand our vision and make an attempt to explain what energy means to the world community that shares our vulnerable and crowded blue marble. Cradling this ambition, we strive for a fuller understanding of the amalgam of supply, demand, distribution, economics, environmental cost, and political realities that can quickly overtake complacency, multiple distractions, and our lame and fraught attempts at planning for the future.

Try this experiment: While thinking about energy, see if you can list those places in the world that have access to all the energy they can use, and those areas that have little or no access to most of the energy they need to survive. What did you find? If you are like most of us, you noticed that those living in the Global North already have most of what they need to make themselves comfortable, secure, and mobile, with a bit left over for pure entertainment. People in the Global South, on the other hand, must get by on whatever energy resources they can

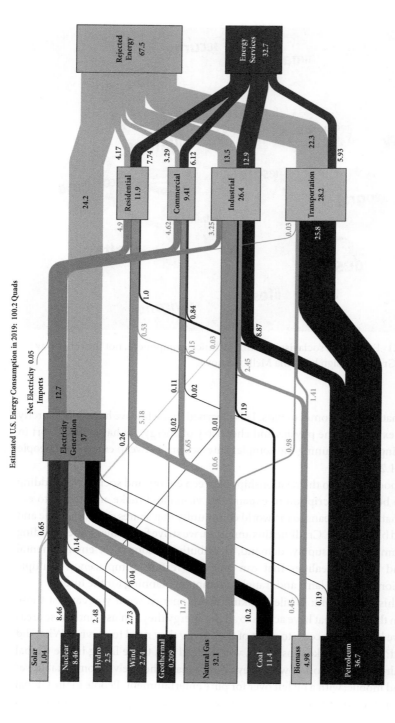

Figure 1.2 Estimated U.S. energy consumption in 2018: 101.1 quads. One quad = one quadrillion Btu. For reference, 1 pound of coal from the Kayenta coal mine in Arizona, when converted to electricity, produces about 1 kilowatt-hour. A typical household in the Phoenix area uses about 13,000 kWh per year. See plates for color version.

Source: Lawrence Livermore Laboratory, 2019, https://flowcharts.llnl.gov.

scrape together. Often the accumulated results of their search are barely enough to prepare meals, remain comfortable, and feel relatively safe and secure. For such folks, no alternative exists to devoting many hours every day to the burdensome task of gathering fuel wherever they can find it—even if it is a great distance from where they live—and then carrying it home. Such fuel is most commonly wood twigs or animal dung. Worse, gathering it is more than an exhausting and lengthy activity for the people who undertake it, who are mostly women: it robs them of time for schooling and exposes them to the possibility of physical assault.

The purpose of this book is to draw back the curtain and reveal the myriad relationships between energy and society—relationships that often attract little public attention. It aims to adjust that dynamic by casting a light on the innumerable ways that energy, in all its forms, weaves in and out of all our lives.

The mushrooming literature on these relationships recognizes the vital importance of acquiring a social vantage point when considering the role energy plays in every human activity, challenge, and puzzle. As one data point, consider the growing relevance of a new journal, *Energy Research & Social Science (ER&SS)*. While many journals must go begging for manuscripts, *ER&SS*—launched only in 2014—today receives an average of 1,200 submissions yearly. Such quick success highlights the pent-up need for such a journal, and its manuscript rejection rate of about 80 percent allows it to publish only those works that adhere to a high standard of excellence. Moreover, the number of contributing authors— exceeding 2,000—and the breadth of the topics they address have helped reveal a sea change in attitudes regarding energy studies. Energy is no longer a standalone topic of investigation. The long-standing dominance of technology has expanded to include the societal dimension. It is just such a shift that prompted the creation of this book.

As the contents of *ER&SS* demonstrate (Table 1.1), the social science emphasis on energy has been progressing from description to theory and then to explanation. This evolution not only reflects growing academic maturity but also increasing relevance to the lay public. A short paragraph from a recent article in *ER&SS* illustrates this developing emphasis:[1]

> It is all too common for energy researchers to generally undervalue social science discoveries, ignore possible interdisciplinary awareness, and marginalize diverse perspectives. . . . [S]ecuring our energy future will require that this pattern changes. We must alter infrastructure and technology and support social change if we are to achieve a future energy system that enhances human wellbeing that is sustainable and just. Such an energy future can be realized only by *integrating* insights from the physical and social sciences. [Emphasis added]

Table 1.1 Range of Topics Covered in *Energy Research & Social Sciences*

- Science and energy studies (discussions about the state of the field and the need for social science; *not* a generic catchall for articles)

- Acceptance of energy systems (social attitudes, perceptions, and experiences with energy technology)

- Consumption and behavior (all things connected to how people consume and use energy)

- Politics and national profiles (studies about politics and/or national case studies)

- Institutions and governance (global energy governance, the state of institutions such as OPEC or the IEA, common pool resource problems, polycentrism)

- Transitions (socio-technical transitions, carbon lock-in, historical case studies)

- Equity and justice (fuel poverty, energy poverty, energy access, energy and development, ethics and justice)

- Innovation and research (RD&D [research, design, and development], patents, innovation styles)

- Analyses and media representations (communication studies, framing, mass media)

- Energy and demographics (energy and gender, energy and race, energy and class)

- Energy and the environment (climate change, energy externalities)

- Education and knowledge (indigenous peoples, inclusion, curriculum, training)

For those of you stimulated by this book to pursue energy studies beyond an introductory level, it might be helpful if I introduce you to some other refereed journals that find a place within the interlaced realm of energy and society. They include *Energy Sustainability and Society* (first published in 2011), *Sustainable Cities and Society* (2011), *Energy and Buildings* (1977), *Renewable and Sustainable Energy Reviews* (1997), *Energy Economics* (1979), *Food and Energy Security* (2015), and *Energy Policy* (1973).[2] Of all these, one should note, *ER&SS* is ranked the highest—number 22 on a list of more than 1,000 energy journals. This quick rise in rank not only provides strong evidence that the *raison d'être* for the journal's creation was sound but also suggested to me the conceit of trying to capture, at least in broad strokes, the sweep of the journal in the book you have before you.[3]

In some ways, *The Thread of Energy* rests on the shoulders of the contributions that precede it, including prescient books by Fred Cottrell and Harold Schobert.[4] Both of these books follow a largely historical model, with Schobert mixing in a more generous portion of energy chemistry and physics. For advanced students, at least two additional books warrant notice. *Sparking a Worldwide Energy Revolution: Social Struggles in the Transition to a Post-Petrol World*, edited by

Kolya Abramsky and published in 2010, includes fifty-eight chapters on a wide range of topics.[5] And in 2018, a group of scholars, all of them steeped in the integrative discipline of human geography, published *Energy and Society: A Critical Perspective*, noting in the opening pages the importance of addressing contemporary concerns about energy from a social science perspective.[6]

Sorting through these books—as well as hundreds of journal articles published just in the last decade—has helped me refine the four goals that guide the discussion in *The Thread of Energy*.

Goal 1: To illustrate that the thread of energy weaves through all individual and collective human activities

Goal 2: To explain the operation of these activities from the perspective of social science, thereby setting aside most discussions of physics, chemistry, biology, and engineering

Goal 3: To emphasize that energy challenges are most clearly understood and resolved when they are examined primarily through a social science lens

Goal 4: To broaden readers' employment opportunities once they understand how to apply their newfound appreciation of energy to the workplace

As we weave our way through this book, we touch on how energy uses have changed, from controlling fire to exploiting the power of the atom (Chapter 2, "Transitions"); how you use energy in everyday life, such as in your house, in your car, on vacation, and in every other activity (Chapter 3, "Life"; the importance and significance of energy as seen in artistic expressions (Chapter 4, "Art"); the environmental costs of energy in terms of the quality of our air, water, and land (Chapter 5, "Environment"); the threat our need for energy is posing to the global climate (Chapter 6, "Climate"); how policy decisions guide energy availability and the consequences of those decisions (Chapter 7, "Policy"); the international importance of energy trade, supply and demand, and military conflict (Chapter 8, "Geopolitics"); the concentration, impacts, and opportunities of energy use in urban environments (Chapter 9, "Cities"); the implications of energy demand for disparities in energy development, energy availability, and energy use in different cultures and at different times (Chapter 10, "Justice"); the relationship between how prosperous, comfortable, and secure you feel and the quantity and quality of the energy you have available (Chapter 11, "Lifestyles"); the risks that are a part of developing and using energy (Chapter 12, "Dangers"); the job opportunities that accompany meeting energy demands with energy supplies (Chapter 13, "Business"); and, last, what energy supply and demand might look like in coming years and how we might react to those changes (Chapter 14, "Futures").

This book rises from the foundational idea that energy supply and demand are entwined with every human activity. You think of some of these activities every day, like heating and cooling your house or apartment. Others, such as wars in the Middle East or tensions in the South China Sea, are less obvious but potentially even more critical.

It is increasingly apparent that all of us need energy, and we need it all the time—whether directly, in the food we eat, or indirectly, in the products we use. While we know that the average human body needs roughly 2,400 calories per day, that is only a small slice of the energy pie that each of us consumes every day. If we divide the total amount of energy we use in the United States by the total population of the country, we realize that the annual per capita consumption in the United States is quite large, about 300 million British thermal units (Btu).[7] That is equivalent to an astonishing 30,000 pounds of coal per person per year. It means that the United States, with less than 5 percent of the world's population, consumes about 25 percent of the world's energy. Absorbing the implications of such numbers, we realize that this heavy energy burden cannot be technically or ethically sustained indefinitely. Other countries need energy as well. The United States should not hog energy just because it has the power and the will to do so.

As we continue expanding and deepening our understanding of the relationships between energy and society, we will continue gaining further understanding of our daily behavior, innovations of every description, how they meld with one another, how we interact with them, and how they will influence what the future will bring.[8] Such considerations help humanize energy because they more thoroughly illustrate how it permeates the fabric of our lives. In this way, it underscores a principal lesson of this book: *energy is a social issue with a technical component, not the other way around.*

Our need for energy inspires decisions just as surely as it limits behaviors, often in ways few people appreciate and fewer people control. For these reasons, it is more important than ever that we understand the centrality of energy and elevate its importance within our consciousness and sense of responsibility to one another and our natural environment. Following the thread of energy helps create strategies that will contribute to prosperity, justice, and happiness for all who live on Earth, the only home we have.

It is this thread that creates the fabric that at once traps us but also protects us. We cannot escape its influences, its mandates, its promise, or the threats that would accompany its absence. We must strive for a better understanding of how the thread of energy weaves the tapestry of our lives. That is what I hope reading this book will accomplish for you, and it is what I discovered when I was exploring the world.

2

Transitions

Evolving Uses of Energy

We Need Energy

Energy is now and has always been fundamental to our well-being, our survival, our security, and our happiness. You need energy all the time. You rely on its constant supply, and nowadays you count on its reliable supply, even as you have nothing to do with making those supplies available. In the earliest days of human existence, reliability of supply was an unknown concept. It was more a question of day-to-day survival. You were on a daily search for your next meal. Would you have enough energy to make it to tomorrow? Starvation was a constant possibility. You would eat when you could, store the energy in your muscles, and search for the next installment of food to recharge your body to feed your muscles and maintain body temperature. Until fairly recently in human existence, people were self-reliant for food to eat and woody growth to use for cooking and heating. All that changed for most people with the widespread use of more concentrated forms of energy.

Today, we use these more modern forms of energy to do much more than maintain body temperature and organ function. Today, we use such resources to heat and cool buildings, pump water across deserts, speed cars from city to city, charge phones, power the internet, and hold all kinds of modern-day gadgetry in a "fast-on" mode for when we want to use them without delay.

All the energy resources we use today, with minor exceptions, originated with the sun. They were created and stored in the earth's crust millions of years ago and had been patiently waiting to be discovered, developed, and put to use. Some of it—such as coal—we can use right out of the ground with little to no processing; some of it we use in various refined forms, such as gasoline and diesel; and some of it we cleverly convert to electricity.

Over the past century, our appetite for energy has become thoroughly addictive. The costs accruing to feed our addiction have been streaking upward on a trajectory that is rising parallel to the triple indicators of population growth, rising per-capita energy demand, and competition for diminishing supplies. We now find ourselves on a never-ending and intense quest to expand our resource base while simultaneously minimizing the environmental costs of using those

The Thread of Energy. Martin J. Pasqualetti, Oxford University Press. © Oxford University Press 2021.
DOI: 10.1093/oso/9780199394807.003.0002

resources. We are attempting these naturally contradictory tasks while keeping energy affordable for people of widely varying financial means and energy needs.[1] Being successful at even the most essential of these steps—such as identifying the location and possible use of energy resources—has become relentless, challenging, costly, and sometimes ruthless, as more and more people vie for the same resource in an increasingly competitive world.

Join me as we follow our slow but steady climb up the energy ladder, from one energy transition to another, from a time when berries, nuts, fruits, and wood were the only energy resources we recognized to a time when—as fast as a chain of lightning bolts—we were bounding with dizzying speed from one good energy resource to another that was even better.[2]

Early Hunting and Gathering

Imagine yourself a thousand centuries ago, living in the raw and untamed landscapes of the East Africa Rift Zone. Imagine as well that you are on an unending search for food, a search that is not always unsuccessful. You have few tools, no ability to make fire, limited communication skills, and the mortal fear that at any time a hungry carnivorous animal might pay you an unexpected visit. Now, shift your focus. Imagine yourself standing alone within the cold and trackless permafrost of what is now Siberia, or, the red rocks cliffs of Wadi Rum in southern Jordan. Wherever you are, your needs would be just as minimal, and they would be vital. You would need water, food, protection from the elements, and a watchful eye for a wide assortment of hazards and dangers. In which scenario would you be able to survive the longest? Arguably, you would be best off in East Africa. At least the natural conditions would give you a better chance for survival: the climate would be more moderate, there would be more sources of food, more water to drink, more possibilities to create shelter. Once outside the tropics, you would encounter more challenges: scorching heat, deep cold, progressive aridity, and an increasing threat of starvation.

Keep in mind that even though your bigger brain holds much more knowledge than those of any of your distant ancestors, your body has not changed very much. Physically you are heavier and taller than your ancestors who lived thousands of years ago, but you are on average less fit and certainly less self-sufficient.

Try this mental experiment: strip away the most common accouterment of modern life that have come along as we have come to control more and more energy resources. Discard your cars and trucks, televisions and radios, computers, all synthetic fibers, fertilizers, airplanes and trains, frozen foods, mobile phones, malls, grocery stores, pharmacies, medical testing equipment, aluminum, concrete and steel, permanent shelter, even electricity. Then take a good look at your circumstances. What do you see? How do you feel? Likely, you would be poorly clothed, hungry, thirsty, and afraid. Could you survive? Unless you have an unusual set of skills, or friends who have those skills, you would not last very long.

Now, let's address a fundamental question. What is the origin of almost all the energy you use as a modern-day citizen? Answer: the sun, either directly, in the forms of warmth and vitamin D, or indirectly, in the forms of naturally available fruits, vegetables, nuts, animals, and fossil fuels such as coal. When you use energy, you are using the sun. It has always been this way, and it still is. The only substantive difference between then and now is what we can call "upgrades." That is, houses instead of tents, indoor plumbing instead of outhouses, artificial fibers instead of cotton, surpluses instead of subsistence crops, shoes instead of sandals, furnaces instead of hearths, and so forth. While we are now all used to these upgrades, the period of their existence is but a blink of an eye in the history of human existence.

Several essential inventions preceded these upgrades, the first being when we learned to use fire. At first, no one knew how to *make* fire. Instead, our ancestors knew only how to *use* fire, and then only when and where it became available by circumstances, such as from lightning strikes or spontaneous combustion. Once *Homo erectus* learned to use fire, perhaps as long as 1.5 million years ago, everything changed. Fire provided light and a protective buffer from predators. It was used to harvest protein by driving animals off cliffs. It expanded food sources by making raw vegetables and meats more digestible. Fire allowed people to turn clay into pottery, stone into metal, sand into glass. Most significantly, controlling fire allowed humans to migrate from the warm tropics to just about wherever they wanted to go.

Having fire first requires having fuel to burn. Except in the driest areas, the earliest fuel was biomass, including grasses, shrubs, and trees. Not only was there plenty of it in most areas of the inhabited world, but it was easy to use, assuming you could ignite it. For hundreds of thousands of years, wood was the only concentrated energy resource available for heating, lighting, cooking, and personal protection. Its greatest attribute was that it regenerated itself, and it did so without any help from us. How ironic it is that as we look toward a future based on renewable energy resources today, for most of our time on the planet we used renewable energy resources almost exclusively. It is as if we are reverting to past practices in order to survive into the future.

Plant and Animal Domestication

Cities are a recent creation in human history. Before their invention, people in small groups tended to be always on the move, following game animals, seeking uncontested land, escaping adversaries. Eventually, after many millennia, people began forming larger groups and became more sedentary. No one knows exactly why this occurred or how it could have occurred simultaneously and independently in several areas of the world. What is agreed, however, is that almost inevitably it derived from the ability to produce surplus food.

The beginning of agricultural surpluses—and the more settled and diversified lifestyle such surpluses enabled—appears to have occurred first in the Middle East, the nexus of travel routes crossing from Africa, Europe, and Asia. Trade in goods, ideas, and seeds—coupled with the advent of irrigation on the rich alluvial soils in the Fertile Crescent—spawned pre-industrial cities like Ur in present-day Iraq, Harran in present-day Turkey, and Jericho in present-day Palestine (Figure 2.1). These were not the only places where cities developed—Cambodia, India, China, and Mesoamerica are some of the others—but the Middle East offered an ideal combination of water, soil, climate, trade routes, and a cross-fertilization of ideas, germplasm, and techniques. Wheat and barley were the dominant crops in this location, and eventually it would be here that animals such as sheep, cattle, pigs, and goats were also first domesticated.

The usual connotation of "domestication" is human control, meaning that plants and animals were no longer wild. Eventually, domesticated plants and animals would become so totally dependent on humans and so altered from their

Figure 2.1 The Fertile Crescent, site of some of the earliest examples of plant and animal domestication. Drawing by Carlos Driscoll. Used with permission. See plates for a color version.

original genetic roots that they often could no longer survive without constant attention from us.

From this point on, after literally millions of years of dead-slow progress in human cultural development, things began changing more quickly during the Agricultural Revolution. Plant and animal domestication provided surpluses of food. Grains were fed to animals that produced not only meat but milk, wool, and many other products. Surpluses also freed a portion of the population to pursue non-food-production-related activities such as toolmaking, ceramics, writing, mathematics, engineering, scientific inquiry, and the creation of rules essential for the development of a more civilized lifestyle. Living standards improved, life expectancy increased, and population growth—for the first time unencumbered by the worries of day-to-day survival—started accelerating (Figure 2.2). It marked the beginning of a way of life that was to take on characteristics recognizable today.

Wood, Wind, and Water

Wood, wind, and water have been useful renewable resources for a long time. Given enough time without human interference, more forests will regenerate, and there is abundant evidence of this trait. Think of the forests of New England, in states such as Vermont. These forests succumbed to the ax and saw more than two centuries ago. Nonetheless, people swarm to Vermont in the fall every year to see the colorful tree foliage. The forests have come back. Properly managed, most forests can provide an endless supply of wood for fuel and timber for construction.

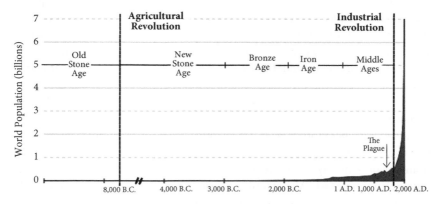

Figure 2.2 The growth of the population accelerated with the invention of agriculture and, subsequently, with the adoption of the use of fossil fuels. *Source:* http://people.eku.edu/ritchisong/RITCHISO/envscinotes3.html.

Between the Tropic of Cancer and the Tropic of Capricorn, we find a land where winter never comes. The tropical rain forests that dominate this swath of the planet support millions of species of plants and animals. No other latitudes on the planet can boast such diversity. There is a lot of rain, abundant sunlight, and no freezing temperatures. It is arguably the most valuable natural region on the planet, but it is vanishing under the plows and machinery of large-scale agriculture. Such conversion is not adaptive to natural conditions. Small-scale farming, however, works well. So-called slash-and-burn agriculture can be the basis of a sustainable form of agriculture. Trees are cut down and then burned, adding organic material to the soil. The replacement growth thrives for a while, but it is a short-lived bounty. In the absence of native plants, primary sources of fertility are lost, the sun bakes the soil into rock-hard laterite, and the spurt of growth of crops quickly gives way to unmanageable weeds, lost productivity, and the need for massive additions of artificial fertilizer. It would be better to abandon the land and allow it to recover naturally. Left unmolested, natural succession will over time create a new forest canopy, and the process can begin again. Put too many people on the land, rotate back to the same site too quickly, or try to continue growing inappropriate crops, and the entire enterprise is doomed. Given enough time, the slash-and-burn approach can be an endless closed loop, making the whole system "renewable" in the best sense of the word.

A slash-and-burn system is poorly adapted for use outside the tropics. Natural conditions in mid-latitude locales such as the United Kingdom foreclose the tropical sequence of clearing, farming, abandonment, and timely regrowth. Although regeneration *can* occur eventually, the necessary time was never available in Britain.

For centuries, forest clearing progressed to strip the land bare. Fuel and building materials came into short supply, and agriculture became the replacement land cover. By the mid-1700s, few forests remained except in the Royal Parks. With a shortage of wood to burn, all but the privileged class were left to suffer in dismal living conditions of rain, mud, and snow. By luck and invention, however, it was at about this time that coal mining came along to provide a substitute fuel and "save the day." Coal was not renewable, but it was abundant, and it had the much higher energy density necessary to support an industrial economy.

Elsewhere, where forests were scarce, other forms of renewable energy came into use, and new levels of work became feasible.[3] Looking back on these times from our present-day perch, we might think that learning how to use these resources is a simple task. Alas, it took centuries—even millennia—for some bright individuals to envision how to use energy resources that are so commonplace today. One of the first of these was the energy of rivers.

Rivers are natural threads of energy. Like so many other energy resources, rivers are partly the result of solar energy. The heat of the sun evaporates water. Then the vapor rises to altitudes where it acquires potential energy, or the "energy of position." Given proper conditions, this vapor condenses into liquid and falls toward earth under the influence of gravity as precipitation. When it reaches the ground, a portion of this water—again under the force of gravity—flows downhill until it reaches a base level of some sort, such as a lake or an ocean. It is during this downhill movement that its kinetic energy is put to use.

The most straightforward work performed by flowing water is also the most common. Anyone who has ever floated down a river has experienced this phenomenon. Anything caught in the current moves downstream with the water. This particular thread of energy carries, drags, and pushes objects from places of higher elevation to positions of lower elevation. People finally recognized the power of moving water; they adapted that power to carry themselves. Someone finally figured out how to create watercraft—from rafts of logs lashed together to canoes hewn out of solid wood. As such craft became more sophisticated, goods and people moved downstream not just by accident or circumstance but intentionally and efficiently. It was hydropower in an intentional application.

Gravity was also useful in irrigation. In arid areas such as Mesopotamia, farmers eventually invented artificial canals to irrigate their crops. While irrigation became a common practice in the Tigris/Euphrates Valley in the Middle East, the Huang Ho in China, and other locations, it was for the longest time based solely on natural downslope movement from gravity. That restriction changed with the invention of the noria, one of the earliest water-driven mechanical devices. The timing of its first use—perhaps in Syria—is uncertain, but it might have been as early as the fifth century BCE.[4]

The noria was the earliest mechanical device propelled by means other than human or animal. It sparked the development of a wide variety of hydraulic and rotating machines to lift water, grind grain, churn butter, card cotton, saw wood, and substitute for human labor in innumerable other jobs. Hydropower was to become a stable renewable energy resource that supported everything from early agricultural life to the early stages of industrial societies.

While hydropower was one of the early renewable energy resources we put to human use, there were others, notably wind. Wind, that invisible energy resource everyone experiences, was first used to propel trade vessels along the Nile River as early as 5000 BCE. Sailing vessels of one design or another were used in many other places as well, and not just for human transport and for purposes of trade. They carried explorers, at first along coastlines in sight of land, but later across vast stretches of ocean.

By the end of the fifteenth century, the northeast trade winds propelled Christopher Columbus and his flotilla of three small ships across the Atlantic Ocean from Spain to the islands of the Bahamas. Columbus's voyages initiated the Age of Discovery. Other expeditions were soon followed by sailors and conquistadors, including Cortés, Vasco da Gama, and Magellan. Although Columbus's discoveries expanded the map of the world, the Chinese may have beaten him into the open oceans. Decades earlier, they perhaps made it to the west coast of North and South America.[5] The Norse made it to Newfoundland centuries before that. So Columbus was probably not first, but his long-distance voyages propelled by the renewable energy of the wind were the most consequential.

Over the years, wind energy propelled larger and faster ships. By the mid-nineteenth century, clipper ships could travel over 400 miles in twenty-four hours. The world was becoming a much smaller place as brave explorers and eager merchants tied one country to another with wind power. Over many more centuries, wind power came into use to drain swamps, pump groundwater, grind grain, and for many other purposes. Today we use it for all those things plus many more, including generating electricity (Figures 2.3 and 2.4).

Figure 2.3 Windmills such as these near Amsterdam were used to drain the wetlands over the centuries.

Figure 2.4 A modern wind turbine near Palm Springs, California.

Fossil Fuels: Coal

You can understand that the cultural evolution from simple gathering and hunting to the control of fire and then the domestication of plants and animals took a long time. Any improvements in the daily lives of people came along slowly as well. It was only when the use of coal became common that lifestyles started improving more quickly. The solid form of fossil fuels, coal, was known for centuries. Much like its sister fossil fuels, it even occasionally ignited naturally and could burn underground for decades or longer. So it was not that people were unaware of coal. Instead, like oil products and cars, its practical and widespread use was tied to other advancements.

For coal, the pertinent advancement was the steam engine, as designed and improved by Thomas Savery, Thomas Newcomen, and James Watt. Not only would their machine pump water out of coal mines, but it would also come to be fueled by the very coal it helped bring to the surface. Once installed, these pumps allowed coal mines to be deeper and more numerous.

Nevertheless, although the social conditions of the past started fading, new social costs came along to replace them. Most of these costs circled the dangers of underground coal mining. Even children were put to work, their smaller size more suitable for the cramped mines, as Charles Dickens noted in his novels. Eventually, following Dickens's intervention, it became illegal for women or for children under the age of ten to work down in the mines (Figure 2.5). Regardless of who was doing the extraction—or what risks they had to accept—coal fueled a quick rise in the many elements that make for more civilized life. So quickly would these changes ripple through society that the period came to be called the Industrial Revolution.

In Europe, especially in the British Isles, the age of coal came just in time. Most forests had been cut down, but coal—a superior (although not renewable) fuel—was becoming commercially available for space heating and industrial processes. Its higher energy density could produce much higher temperatures than wood, especially once Abraham Darby had invented a technique to convert coal to coke (which is more concentrated). By 1840, coal had supplanted wood as an energy source in Britain (Figure 2.6). With higher temperatures came the capabilities for more sophisticated processes, including the manufacturing of steel, a durable material that remains a staple of building today.

The world was changing fast. The widespread acceptance of a coal-based economy helped create the world you see all around you, including its industrial productivity and the resulting creative additions available today, such as buildings, ships, airplanes, cars, and even cities. But because economically recoverable coal reserves exist unevenly within the earth's crust, growing demand stimulated the construction of a complex web of support infrastructure that included railroads and canals.

Canals came first, and they had to be carefully constructed. All parts of each canal—including tunnels through hills, viaducts over valleys, and locks to accommodate elevation changes—had to keep the water levels perfectly flat; otherwise, the water would simply run away. Barges, often 7 feet wide but up to 12 feet in width, could each move 30–40 tons of product, most commonly coal (Figure 2.7). Considering the poor quality of roads, these canals proved invaluable in supporting the burgeoning industrial economy in many countries. "Of over 150 canal acts from 1760 to 1800, 90 were for coal purposes. At the time—before

Figure 2.5 Child labor was common in the coal mines of England and elsewhere. *Source:* Sunday Observer, 2009.

the railways—only canals could have coped with the swiftly rising demand for coal from industries like iron."[6] By 1840, nearly 4,500 miles of canals crisscrossed Britain. Today in Britain, canals still account for a sizable portion of the 2,000 miles of navigable inland waterways (Figure 2.8).

Railroads came later. After 1840, railroads began replacing canals. They could carry more products, including passengers, more efficiently and faster than the walking pace of the canal boats. Although the use of canals to transport bulky

History of energy consumption in the United States, 1775–2009

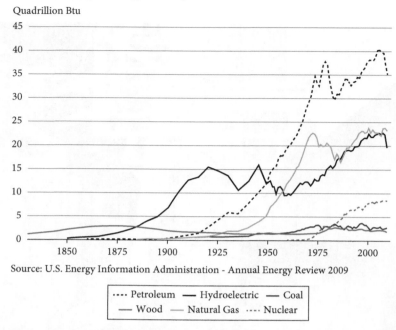

Quadrillion Btu

Source: U.S. Energy Information Administration - Annual Energy Review 2009

···· Petroleum — Hydroelectric — Coal
— Wood — Natural Gas ···· Nuclear

Figure 2.6 U.S. Energy Consumption, Per Capita (1775–2009), showing the rapid decline in the use of wood as an energy source with the rise in coal after 1850 and continuing its decline with the rise in oil use 1900. See plates for color version. *Source:* U.S. Energy Information Administration

Figure 2.7 Barges being towed through Regent's Park, London during the 1930s—the first two carry coal, the third appears to be carrying timber. Photo courtesy of Miss Catherine Bushell.
Source: http://tringhistory.tringlocalhistorymuseum.org.uk/Canal/c_chapter_10.htm.

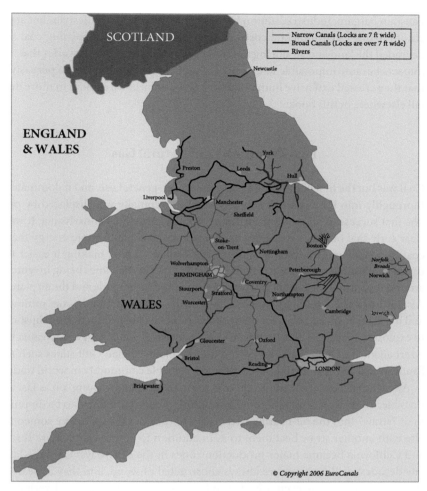

Figure 2.8 Canals, used to move heavy goods such as coal in Great Britain and elsewhere, are used today for recreation. See plates for a color version.
Source: http://www.newandusedboat.co.uk/images/canal-network-map.jpg.

materials ebbed, many of these canals still exist in such places as Britain, central France, the Netherlands, and the northeastern United States. These threads of energy are among the easiest to spot, and some of them are still in use. Many have become tourist attractions as they wind leisurely through rural countryside, often lined with paths for traffic-free jogging, biking, and horseback riding. In some places, visitors can hire canal boats to ply these waters. They have become an appealing example of a repurposed energy landscape.

The age of coal also created several other enduring energy landscapes, most of which are not as appealing as canals. These include mining scars, squalid towns, and buildings soiled black by air pollution produced by the burning of coal. In a

Faustian bargain, industrialization brought not only improved lifestyles but also accumulating environmental costs. Mining, transporting, and burning coal so degraded the land and air during the eighteenth and nineteenth centuries that its impacts became impossible to ignore or avoid. These impacts were so pervasive that they created extensive humanitarian costs, a subject we explore in more detail elsewhere in this book.

Fossil Fuels: Oil and Natural Gas

Coal was but the first of the fossil fuels to come into general use, and it dominated thoroughly into the twentieth century, when it eventually was displaced by oil. The first successful oil well was drilled in 1859 in Titusville, Pennsylvania. It was clear early that oil was a superior energy resource. It contains more energy than coal by weight and by volume, and it is—of course—liquid, making it easier to use as an illuminant and more adaptable to the types of machines being invented around the turn of the twentieth century, such as the automobile and the airplane. Also, it was easier to transport, it avoided the dangers inherent in coal mining, and it was an ideal feedstock for many thousands of useful products, from lipstick to explosives. Early on, this particular thread of energy ran from Pennsylvania to Azerbaijan, the Dutch East Indies, and then to many Persian Gulf states such as Saudi Arabia and Kuwait. It is today the most valuable commodity in world trade.

After the success in Titusville, there was a mad scramble to pump oil as fast as possible. There were few restrictions or controls. Everyone seemed to be dipping their "straw" into the oil-bearing rock, eager to suck out the oil before someone else with another straw beat them to it. In addition to Pennsylvania, East Texas and California became major production zones in the early days. Until the last few decades, drilling rigs were always constructed of wood, and they appeared everywhere there was even a hint of oil. Their proliferation created veritable forests of derricks (Figure 2.9).

Figure 2.9 A forest of oil drilling rigs at Signal Hill, California, 1923, illustrating the principle of the "right of capture," which rewards whoever gets to the oil first. *Source:* Library of Congress, http://www.loc.gov/pictures/item/2007660408/.

For many years, the development of the oil fields was rambunctious, messy, competitive, inefficient, and uncontrolled. Entrepreneurs, such as John D. Rockefeller, saw the potential to make huge profits in the U.S. and foreign oil markets. By 1870, Rockefeller's company, Standard Oil, was refining 90 percent of the country's oil, and not long afterward he became the richest man in the world. The construction of railroads and pipelines followed, both of them snaking from oil fields to refineries, ports, and consumers.[7] Pipelines full of oil and gas spread their tendrils all over the map.

The Nuclear Age

Nuclear fuel came into commercial use in the 1950s, and at the time, many believed it would become the preferred fuel for the generation of electricity. Nuclear power plants were touted for their many advantages: no emissions or ash, virtually no noise during generation, fuel with a high energy density, and the likelihood that such plants would become innocuous good neighbors perfect for boosting tax revenues and employment opportunities. All this and the electricity would be too cheap to bother metering. Advocates promoted nuclear power as clean, safe, inexpensive, quiet, and an ideal alternative to dirty coal. What was not to like? As it turned out, plenty.

Do a bit of research on nuclear power and see what you discover. Would you find that it had lived up its publicity? Had it become a popular source of electricity? Would commercial reactors, scattered in every state, produce cheap electricity with high reliability? Here is the present status: more than sixty years after the first commercial nuclear power plant went critical, the jury is still out as to whether the power of the atom will be the basis of a new transition away from fossil fuel generation. What has been the problem? At its most basic level, nuclear power has been slowed by its most fundamental characteristic: the fact that it is, after all, nuclear.

World War II taught us that nuclear weapons could quickly and efficiently destroy cities. It was a prospect that continues to worry us. Whether by accident or intent, such weapons have the potential to seriously threaten civilization. No other energy resource has attracted so much opposition or been burdened with such a severe array of unsolvable safety challenges. It is ironic that the very characteristic that makes nuclear fuels so valuable—their immense energy density—can also do great harm if an accident occurs. Yet, even with its inherent dangers, nuclear power will continue to entice us with its advantages.

Nuclear power is especially appealing in areas of high and rising demand, places with insufficient reserves of fossil fuels, countries where there are growing concerns about greenhouse gas emissions, and nations with interest in nuclear weapon development. Combine some or all of these characteristics with a centralized governing system and you find China. Thus, it should not seem remarkable that China is planning to build at least three dozen nuclear power plants in the next few years. Despite devastating accidents at Fukushima, Japan, and Chernobyl, Ukraine, to China the advantages of nuclear power more than counterbalance any possible concerns.

The concerns, however, are worth examining. Among them is radiation, the invisible and dangerous demon that accompanies all the steps of the nuclear fuel chain, especially transport and permanent waste disposal. Public apprehension has crimped the ambitious early plans to make nuclear power the basis of the next energy transition. Such worries have slowed new construction to a snail's pace in many places, notably the United States and numerous countries in Western Europe. At the extreme example, the German government committed to shuttering all its nuclear plants in the wake of the Fukushima accident, deciding instead to take the giant next step toward a sustainable energy future that would be the essence of the ultimate next energy transition.

Final Thoughts

Meaningful cultural developments were slow for thousands of years. Then they accelerated in the past 250 years, starting with the Industrial Revolution. Soon we began going through a series of energy transitions, marked by technical advancements and large-scale enhancement in human lifestyles. Today, in the first quarter of the twenty-first century, we are continuing such transitions, ultimately arriving at the one that will lead us all to a sustainable energy future. Reaching that goal will not be immediate. Given the spectacularly complex, multitrillion-dollar infrastructure we have developed around our current energy resources, it will undoubtedly take us some time to make the shift.

One of the reasons that the "ultimate" energy transition will not be quick is that we have invested trillions of dollars in the existing energy system. Sunk costs, legacy costs, and the hub-and-spoke designs of centralized power systems all carry substantial inertia. In addition, public opinion is commonly slow to adjust to new ideas; we already hear public opposition to renewable energy. While one might hope for a more rapid transition, consider the example of nuclear power. The first commercial reactor came online over sixty years ago, yet today nuclear power still provides less than 10 percent of total energy supply. Even though advocates expected nuclear power to dominate energy portfolios around the world, most countries have no nuclear power plants at all.

As at present, so in the future: most demand will be met by a portfolio of energy resources. Despite impressive inroads, the pathway to a sustainable energy future is quite long. Nevertheless, the transition has started.

3

LIFE

Everyday Threads of Energy

Uses of Energy

Your body relies on energy in every way you can imagine, and many you do not. Inside your body, you need energy to fill your lungs, pump your blood, and spark your brain. Outside, energy provides light so you can see, propels cars so you can travel, and powers vacuums so you can clean your living room carpet. We use energy to power our factories, our airplanes, our ships, our military operations, our entire economies. Without energy, everything stops.

So valuable is energy access to our lives that, as a people, we commonly risk environmental harm, economic coercion, and even armed conflict to ensure that we can do everything we want to do, whenever and wherever the opportunity exists. Energy is the key to how you live today and the life you aspire to for your family and your descendants. Energy is so indispensable to political stability that countries are rarely willing to limit the actions that might be required to keep supplies coming.

From one person and one culture to another, the demand for energy rests somewhere on a continuum that stretches between lives of penury and lives of abundance. Your position on this energy continuum strongly depends upon such factors as your place of birth, your economic station, and the effectiveness of your various representatives in getting you what you need when you need it for a price you can afford.

The motivations to control energy resources are apparent. Throughout history, countries rich in energy often travel a smoother road to prosperity. Others, not as blessed, often find their citizens doomed to scratch out a living with whatever meager supplies they can acquire through their labor and inherent wits. Sometimes, if countries with original abundance find their reserves waning, they are inclined to expand the radius of their search for resources they can bring home to make up the difference. Such activities can take two forms: a willing business arrangement or various types of conflict and competition.

All this may seem either curious or, perhaps, obvious. If the former, it is because you have never faced energy shortages and never had to suffer the inconveniences and risks that can result. If the latter, your life experience has tutored you in the vital importance and stiff penalties that shortfalls of energy

The Thread of Energy. Martin J. Pasqualetti, Oxford University Press. © Oxford University Press 2021.
DOI: 10.1093/oso/9780199394807.003.0003

supplies create. In either case, you must understand and appreciate the ties between energy and the way you live your life. What you may not fully appreciate, however, are some of the details. My intent in this chapter is to get specific, to illustrate how the thread of energy runs through your everyday life, whether or not you give its presence much notice. I concentrate on four themes: food, reliability and safety, comfort and shelter, and transportation. I call them the "everyday threads of energy."

Food

Knowing that a steady supply of energy is critical to every aspect of your well-being entails the question of how to ensure you get all you need whenever you need it. You need such energy to fuel your routine activities. Unless you are a photosynthesizing plant, baking in the sun will not get you the energy you need, regardless of how much time you spend outside. For your body to accept, process, and utilize sunlight—the ultimate origin of all the energy you use—you will need some help. Such help comes in the form of food, the intermediary between the sun and your taste buds.

Gaining access to the right type of food is vital to your long-term health. But how much do you need? According to the Food and Agricultural Organization of the United Nations (FAO), the minimum dietary energy requirement (MDER) is the amount of energy needed for light activity in a person of minimum acceptable weight for attained height.[1] Such a requirement varies by location and time and depends upon the gender and age structure of the population. For an entire population, the MDER is the weighted average of the minimum energy requirements of the different gender-age groups in the population. That requirement ranges between 1,700 and 2,000 kilocalories (kcal) per day.[2] Consuming one cheeseburger with fries and a large cola just about gets you there.

Knowing the amount of energy you need is the demand side of the equation. The other side of the equation is the supply side, the dietary energy supply (DES). The DES is the food available for human consumption, again expressed in kilocalories per person per day (kcal/cap/day). At the country level, it is calculated as the food remaining for human use after subtracting all non-food utilization, including exports, industrial use, animal feed, seed, wastage, and changes in stocks. In 1961 the average global calorie availability was as low as 2,193 kcal/cap/day; by 2011, however, it was much higher, reaching 2,868 kcal/cap/day. It continues to move upward. The source of these calories is a rather narrow base of staple grains as well as meat and dairy products.[3]

If you compare energy needed (MDER of 1,700–2,000 kcal/day) with energy supply (DES of 2,868 kcal/cap/day), you should realize that there are enough calories of food being produced in the world to meet the needs of everyone. Yet people die of starvation and malnutrition every day. One in nine people is not able to regularly satisfy their minimum dietary needs. The FAO reports that more than 800 million people are chronically undernourished. In other words, these people typically do not have access to energy food to keep themselves alive, let alone healthy (Figure 3.1).

Do you know where these 800 million people live? Almost half are in India and China, with the remainder mostly in sub-Saharan Africa. It is ironic to note that even though desperate conditions still exist in some parts of the world, people in many other parts of the world consume more food than they need. Such overindulgence has been leading to what is commonly called an "epidemic of obesity" in places like the United States and other developed countries.

Caloric requirements vary by individual, depending on age, physical condition, and levels of exertion. People in their early twenties are typically at the peak of physical activity and caloric needs. As you age, your energy needs (read: needs for nourishment) drop naturally, partly because you are less active and partly because your metabolic rate declines.

As you have no doubt been told many times, not all calories are created equal. That is another way of saying that they do not all have equivalent nutritional value when it comes to maintaining health. If, for example, you were to drink several gallons of sugary soda or eat a dozen doughnuts each day, you would be meeting your MDER, but gaining scant nutrition. Absent enough proteins, vitamins, and other nutrients, the quality of your health would decline. You would become weaker, find it difficult to marshal stamina, have trouble concentrating, and become more susceptible to illness. Despite its high calorie count, such a diet would eventually result in chronic fatigue, poor health, and even death.

Because grocery stores in OECD countries—the United States, United Kingdom, France, and Sweden as examples—are generally well stocked with food, shoppers in these countries presumably give little thought to the possibility of shortages or to where their food supplies originate.[4] We can get what we want when we want it and even eat "out of season." Our food might start its journey to market just down the street or in a country halfway around the world.

Even if you lived in a hot and remote place such as Phoenix, Arizona, you would have no trouble—regardless of season—buying fruits, vegetables, and other foods that are not grown locally. You can pick up bananas, avocados, kiwis, coconuts, artichokes, asparagus, and just about anything else you desire within a mile or two of home.

Figure 3.1 The World Food Programme's Hunger Map depicts the prevalence of undernourishment in the population of each country in 2016–18. From Africa and Asia to Latin America and the Near East, there are 821 million people—more than one in nine of the world's people—who do not get enough to eat. *Source:* https://www.wfp.org/publications/2019-hunger-map. Courtesy of World Food Programme. Used with permission. See plates for color version.

Embodied Energy

Most people can find a plentiful variety of food at their local grocery store. It is unlikely, however, that many of these same people can pinpoint where that food originates or how it gets to their dinner table. It is a rather complicated and fascinating story. The fascinating part is that very often, the majority of the energy in the food you eat has nothing to do with the calories it contains. We call such energy "embodied energy," and this is where the complexity starts. The embodied energy includes the energy to power the machinery that tills the soil; plants and harvests the crops; pumps the irrigation water; and processes, packages, preserves, and transports the food. While you do not metabolize such embodied energy, it nonetheless contributes to the price you pay and the total energy you must commit to making that food available.

Of all the activities that we should track when we talk about embodied energy, the easiest to imagine is the energy that we call "food miles." These are the miles the food must travel between farm and family. Scan the shelves of the grocery store you patronize, and you should easily find olive oil from Italy, almonds from California, cheese from Wisconsin, water from Fiji, grapes from Chile, apples from New Zealand, wine from Australia, and just about anything else you want, regardless of the season or country of origin. Whichever mode of transport one uses—trucks, trains, ships, or even (for high-value items) airplanes—none can operate without fuel, usually derived from oil. We call these "food miles" and someone—often you—has to pay for them (Figure 3.2). Such food miles can mount up quickly. Sometimes they can be substantial, as when the food is grown on another continent. Other times, such as when they come from a farmer's market, they can be next to nothing.

While shortening distances between producers and consumers can reduce the costs of food miles, other forms of embodied energy can quickly add up. One of these is irrigation, a particularly energy-intensive step, especially when the food is grown in arid areas such as Arizona. Sometimes the embodied energy is associated with pumping well water. Other times the embodied energy is associated with pumping water along canals, through tunnels, and over mountains. The Central Arizona Project (CAP) is a case in point.

In each year from about 1993 until 2019, 24.3 percent of the power from the Navajo Generating Station near Page, Arizona, pumped water along the CAP from Lake Havasu on the Colorado River about 336 miles until the terminus about 13 miles southwest of Tucson. It requires 2.8 million MWh annually to pump and distribute an average of 1.5 million acre-feet of water to users along the way, with the ending point about 1,900 feet higher than where it is first pumped out of Lake Havasu.[5] Those 2.8 million MWh—roughly what is required to meet the yearly demand of about 200,000 homes in the Phoenix area—are embodied within the water.

FOOD MILES
Distance from Farm to Plate

*How many gallons of fuel are used
and pounds of carbon dioxide emitted
to trasnport 1 ton 1,000 miles?*

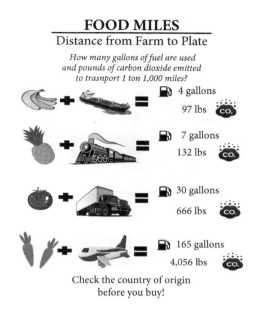

Check the country of origin
before you buy!

Figure 3.2 It is common for food to be transported long distances between farms and consumers. Such transport requires energy. These are called food miles. *Source:* http://teambia.weebly.com/uploads/4/8/9/6/48964919/6267584.jpg?419

Food miles and pumping costs are part of the energy it takes to provide us with food, but there are other energy costs to consider, including processing, refrigeration, packaging, and marketing (Figure 3.3).[6] By one estimate, it takes a total of 12 calories of these types of energy expenditures to produce 1 calorie of food energy.[7] Of these 12 calories, we use 1.6 fuel calories in the agricultural sector and 2.7 calories to process and package food. Distribution (which includes transportation), wholesale and retail outlets, and food service operations such as restaurants and catering services use another 4.3 fuel calories. Finally, food-related household energy use adds another 3.4 calories.

Farms account for about 4 percent of all energy consumed in the United States. In contrast, the food system in the United States accounts for approximately 17 percent of all energy consumed. It takes 3 calories of fossil fuel to produce 1 calorie of food energy on average. This ratio is 35:1 for a beef feedlot.

These amounts have been trending upward. In 2007, the agronomist Eric Garza estimated that it took about 14 fuel calories to deliver 1 calorie of consumed food.[8] When he extrapolated this to 2013, the energy requirement of the U.S. food system had grown to about 15 calories per food calorie. In other words, fifteen times more energy is needed to get the food to your dinner table than you get from eating it. Naturally, if fuel prices rise, the price of food increases to

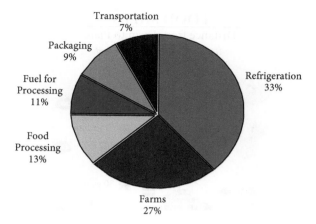

Figure 3.3 Food system energy consumption (percentage of total U.S. consumption). *Source:* http://mcensustainableenergy.pbworks.com/w/page/20638072/Food%20 System.

accommodate these costs. In other words, your food carries a large surcharge, both in energy and in money (Figure 3.4).

The Energy Costs of Meat

Now that you are equipped with some appreciation of the concept of embodied energy, here is a question for you to consider. What category of food is consumed in great abundance in developed countries but rarely found or consumed in the least developed countries? If you answered "meat," you would be right. As we can see in Figure 3.5, per-capita income is strongly correlated with meat supply. It is often considered a surrogate for wealth. The higher the income, the greater the meat supply. Put starkly—and ignoring religious restrictions—rich people have access to more meat than poor people. Put another way, poor people cannot afford meat because of the low conversion efficiencies and high environmental costs that are embodied within each morsel they put in their mouths (Figure 3.6).

Now let's look more deeply into what it takes to put meat on the table. Start by imagining the food chain for a typical steak. At the far front end of the chain is energy from the sun. At the distant other end is the steak sitting on your plate, perfectly grilled and ready to eat. In between these two extremes is an abundance of steps, each one soaking in embodied energy. It starts when the sun shines on green plants. A small portion of that solar energy is converted by photosynthesis into animal food, such as alfalfa. As the animal converts the alfalfa into muscle

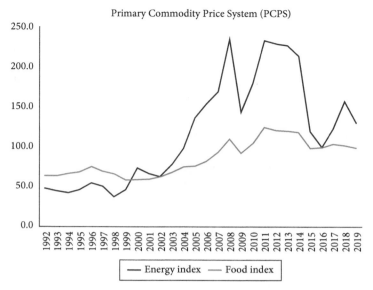

Figure 3.4 Food and fuel prices show a strong correlation, particularly after 2005. This demonstrates the importance of embodied energy in what you pay for your food.

Source: IMF, as reproduced in Eric Garza, 2013, The Energy Return on Energy Invested of US Food Production. https://www.resilience.org/stories/2013-09-09/the-energy-return-of-energy-invested-of-US-food-production/. See plates for color version.

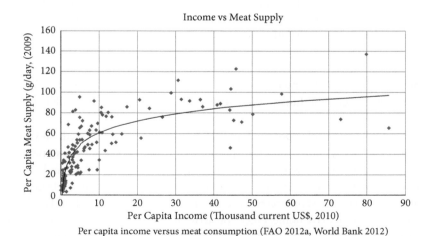

Per capita income versus meat consumption (FAO 2012a, World Bank 2012)

Figure 3.5 Per capita income versus meat consumption (FAO 2012, World Bank, 2012). United Nations Environment Programme, http://na.unep.net/geas/getUNEPPageWithArticleIDScript.php?article_id=92.

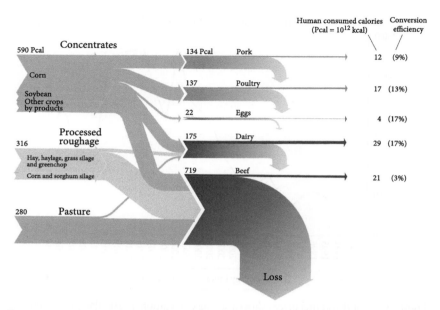

Figure 3.6 A Sankey flow diagram of the U.S. feed-to-food caloric flux from the three feed classes (left) into edible animal products (right). On the right, parenthetical percentages are the food-out/feed-in caloric conversion efficiencies of individual livestock categories. Caloric values are in Pcal (10^{12} kcal). Overall, 1187 Pcal of feed are converted into 83 Pcal edible animal products, reflecting a weighted mean conversion efficiency of approximately 7 percent.
Source: A. Shepon et al., "Energy and Protein Feed-to-Food Conversion Efficiencies in the U.S. and Potential Food Security Gains from Dietary Changes," *Environmental Research Letters* 11, no. 10 (2016): 105002. See plates for color version.

and fat, it gains weight and becomes increasingly valuable to the rancher. While the rancher might not think in these terms, the animal is a form of "solar energy on the hoof."

Eventually, each solar-powered animal is sent off to be processed. Next, the meat is packaged, chilled, and transported to a retail market, perhaps one near you. You burn fuel to drive your car to the market, which is well lighted and cooled or heated twenty-four hours a day. You walk in and find the steak in a refrigerated compartment powered by electricity. You remove the steak, pay for it, and get back inside your car to drive home. Once you arrive home, you place the steak in your refrigerator, which is powered by electricity that the local utility company has generated from one energy resource or another. Eventually, you remove the steak and cook it using natural gas or electricity (or perhaps charcoal heavily processed from cut wood). Each step in this food supply chain adds embodied energy to the steak, far outweighing the calories that are in the steak you consume.

Considering all the preceding, you might ask: What is the total cost—in money and environmental damage—for meat consumption? One way to address that question is to consider a vegetarian diet. Aside from whether or not you adhere to this practice or even think such a diet is healthy, you might wish to consider the following. Eating "higher upstream" on the food chain, vegetarians cannot be held accountable for as much of the damage to the environment as all those who prefer meat in their diet.[9] By one calculation, cattle consume 16 times more calories in the form of grain than they produce as meat.[10] That is a ratio of 16:1. You might wish to consider that ratio when deciding whether you want to switch to a meatless diet.

Knowing about the energy and environmental cost of meat raises the question of whether the planet can continue aspiring to a meat diet. Worse still, this estimate of a 16:1 ratio is probably on the low side; others have estimated that a more accurate ratio is 35:1. This means that the net energy efficiency for meat is about 2.87 percent. In other words, beef requires 35 calories of energy input for every calorie of energy output.[11] Lamb is even worse; Cornell University professor David Pimentel estimates that it takes approximately 57 calories of energy to produce 1 calorie of lamb.

As compelling as all this discussion might be, we should not end our review of the environmental costs of meat without acknowledging that energy is not the only resource embodied in meat. Water is also embodied, sometimes a lot of it. Again, beef leads the pack. While it takes 718 gallons of water per pound of meat to raise pigs and 518 gals/lb to raise chickens, beef cattle require 1,847 gals/lb.[12] If you decided to eliminate beef from your diet, you would save over 300,000 gallons of water a year. This savings is much higher than you could save by never showering, if that might actually be something you were considering.[13]

Yet another cost of beef is in the amount of land cattle need. On 2.5 acres of land, for example, you could satisfy the caloric needs of 22 people if you were growing potatoes, but just one person if you were raising beef, at least on a caloric basis. According to the United Nations, livestock now occupies 30 percent of the earth's entire land surface, mostly permanent pasture. Worse still, 33 percent of the globe's arable land is used to produce feed for livestock. Raising beef is also a significant driver of deforestation; seven football fields of land are bulldozed worldwide every minute to create more room for farm animals. Together, 70 percent of former forests in the Amazon have already been converted to grazing.[14]

Regrettably, we are still not done adding up the costs of beef. There is a cost in terms of greenhouse gases. For example, if you were to go entirely vegetarian, you would avoid emitting 1.5 tons of CO_2 each year, just by yourself.[15] Or, to put this into more digestible terms, if all Americans eliminated just a single ¼ pound serving of beef per week, the reduction in global warming pollution would be equivalent to taking 4 to 6 million cars off the road, according to Natural Resource Defense Council (NRDC) estimates. *Time* magazine puts it another

To produce a 1/4 lb beef patty takes:

–0.3 kWh of energy
–6.7 lbs of grain
–52.8 gallons of water
–74.5 sq ft of land to grow feed crops and graze

Figure 3.7 The resources required for one quarter-pound hamburger. (Not including the animal's waste or the methane emissions from its digestion.)
Source: J. L. Capper, "The Environmental Impact of Beef Production in the United States: 1977 Compared with 2007," *Journal of Animal Science* 89, no. 12 (December 2011): 4249–4261. Producers: Eliza Barclay, Jessica Stoller-Conrad. Designer: Kevin Uhrmacher/NPR.

way: "Giving up . . . meat . . . is one of the greenest lifestyles changes you can make as an individual" (Figure 3.7).[16]

All this brings up a question: if obtaining energy from beef is so costly to pocketbooks and the environment, what is beef's strong attraction? There are many possible factors. First, meat is a staple; it has been a source of protein for millions of years. Humans grew physical and mental capabilities by eating a protein-rich diet. It has been only in recent decades, with increasing population pressure, that we have come to worry about the costs of producing and consuming beef.

Second, raising meat for sale is profitable. That is, meat production is generally a more profitable activity than growing grains such as wheat and canola, especially in years of average or high rainfall.[17]

Third, ignoring any preference for the pleasures one derives from eating meat, being able to afford it projects status and prestige. Why is this so? Because it is expensive, and everyone knows it. One of the reasons it is so costly is because it requires so much energy to make it available to buyers. If people are eating a lot of meat, others will assume they are wealthy. To raise their community standing, they covet meat, even if they cannot afford it.

All this discussion might be a way of explaining why consumers in wealthy countries are beginning to eschew a meat diet. They already know that people envy their comfortable lifestyles. The added prestige that might come from eating more meat is of less importance to them. Now, the next step—at least for these people—is to move past meat back to vegetables and fruit again.

These and many other factors combine to throw light on the fascinating tangle of energy threads that we weave together when we grow the food we need to survive and prosper. But the developing message is becoming clearer by the day. On

May 24, 2020, the *New York Times* published an article entitled "The End of Meat Is Here." As the article put it, "If you care about the working poor, about racial justice, and about climate change, you have to stop eating animals."[18]

Safety and Reliability

If you live in the United States, you have come to expect—and even trust—that your living space is a cocoon of comfort, safety, and security. For the most part, you can enjoy these essentials courtesy of the complex array of energy networks that bring you light at night, warmth in winter, cool air in summer, and the conveniences of appliances and gadgets in growing abundance. Day after day, these services and amenities come to you without much concern or thought on your part. They are just there, at least until the power goes out, everything shuts off, and an unusual quiet takes over. Only then do you realize that you have been taking energy for granted. You immediately wonder what to do and how you can get on with your life if the power stays off for more than a few minutes. Eventually, after what might seem like forever, the lights come back on, the TV cable box resets, the air conditioner resumes its hum, you return to your daily routine, and you forget about the interruption. As long as it does not happen again, you are happy. But, of course, it will.

But not very often. Power interruptions in most parts of the United States are so infrequent that few people prepare ahead of time to deal with them when they occur. You stumble around in the dark, talking to yourself: *Where are the candles? Where are the matches? Do we have a flashlight? Why doesn't it work? Where are the batteries? How long before the steaks in the freezer start to thaw? Why am I not better prepared?*

If the outage drags on, inconvenience quickly escalates to annoyance and then to concern and even danger, depending on when and where it has occurred. Its impacts on you and your family depend on whether you are sleeping, eating, working out, riding an elevator, moving along a subway, manufacturing computer chips, or having open-heart surgery. Outages are rarely welcome, but given the complexity of the entire system, you would be amazed that the "grid" works as well as it does.[19]

The Electric Grid

The electric grid is the largest, if not the most complex machine ever made. In the United States, our electrical network consists of more than 450,000 miles of high-voltage transmission lines and millions of miles of lesser lines. The combined total of the western, eastern, and Texas power grids supplies more than 140 million customers—industries, churches, schools, stores, and residences—with electricity.[20] The shiny network of aluminum and copper conveys a stream of electrons to just about everyone in the country every minute of every hour, day after day, year after year. These electrons flow across landscapes of all types—arid lands, lush farms, jagged mountains, dense forests, and raging rivers—with the single purpose of bringing power to your fingertips with a flip of a switch. Transmission pylons and the wires they support are ubiquitous and rarely out of sight, especially in treeless southwestern deserts (Figure 3.8).

The national grid connects power plants with consumers with better than 99 percent reliability. We always are a bit surprised—and, dare I say, shocked—when something disrupts this service. Yet it does happen, sometimes on a truly epic scale, such as the blackout of November 9, 1965, which affected 30 million people in New York, neighboring states, and Ontario, or the even larger Northeast blackout of 2003 that affected 55 million people. Outages in other countries have been far worse, sometimes affecting hundreds of millions of people, as has occurred in India at least twice since the turn of the twenty-first century.

Utility engineers and government regulatory agencies work diligently to maintain reliability, developing detailed procedures to avoid mishaps, breakages, vandalism, and sabotage. They try to anticipate possible disruptions by developing scenarios for corresponding responses and running unannounced practice drills

Figure 3.8 Long-distance transmission across lands of the Navajo Nation in northern Arizona brings electricity to metropolitan Phoenix. The volcanic San Francisco Peaks, near Flagstaff, are in the background.

as often as possible. Nevertheless, no company can predict or afford to prepare for every imaginable disaster.

The outages that still do occur stem in part from the fact that the entire network of wires and pipe has grown "organically" in both size and complexity. Indeed, there has never been a "grand plan" that guided the pattern, voltages, and rights-of-way for transmission lines for the country. For the most part, expansions have simply followed demand. In response, engineers have to be agile in coping with uncertainty, patterns of growth, and political and public interventions in weaving each mandated addition into an existing fabric. Starting from scratch is rarely an option.

We have become so accustomed to uninterrupted supply that outages can produce severe economic impacts. For example, power outages cost the U.S. economy as much as $52 billion in productivity between 2003 and 2012.[21] Some estimates for that period have been as high as $100 billion, an expense mostly attributable to short interruptions to commercial users.[22] In response, many people are advocating a restructured electricity supply network. Often called the "smart grid," it promises to be more resilient and reliable, allow greater flexibility with alternative energy resources, and permit enhanced two-way communication between utility companies and users (called "demand-response").[23]

Pipelines

As complex as the electric grid is, it is not the only network providing citizens with reliable energy resources. In contrast to transmission lines, which produce a national grid with unavoidable visibility, another intricate energy network lies primarily out of sight, underground. It is a network not of wires but of pipes. These pipes transfer natural gas and petroleum products to customers all over the country, albeit with somewhat less dense coverage, especially in the West and New England. Like the electricity distribution system, the pipeline network also totals hundreds of thousands of miles (Figure 3.9). Significantly, not only are they less noticeable, but they are also less vulnerable to the vagaries of weather and accident, or to intentional disruption.

Although they are underground and out of sight, such pipelines have some disadvantages. For example, they are less convenient to access, maintain, and repair; trenching is necessary; they offer less flexibility for rerouting supplies; flows cannot be stopped as quickly; products move far more slowly; leakage can contaminate soil and water; and the public rarely knows their location. This last circumstance becomes significant if an unsuspecting or careless public is put in harm's way by leaks, breaks, pilferage, and sabotage, especially if combined with the risk of ignition.

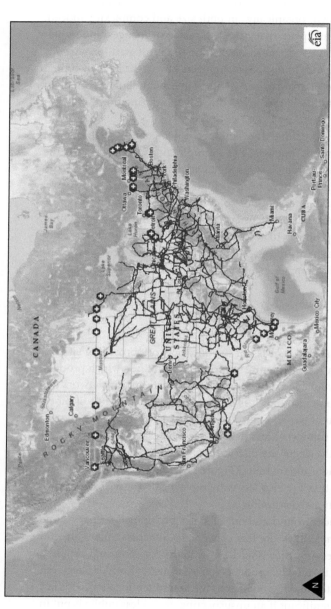

✪ Electricity Border Crossing

— Electric Transmission Line >= 345 kV

Figure 3.9 The natural gas infrastructure. A representation of the more than 210 natural gas pipeline systems, 305,000 miles of interstate and intrastate transmission pipelines, and more than 1,400 compressor stations that maintain pressure on the natural gas pipeline network and ensure continuous forward movement of supplies. There are more than 11,000 delivery points, 5,000 receipt points, and 1,400 interconnection points that provide for the transfer of natural gas throughout the United States. Twenty-four hubs or market centers provide additional interconnections. There are also 400 underground natural gas storage facilities and 49 locations where natural gas can be imported/exported via pipelines. Eight LNG import facilities and 100 LNG peaking facilities are also in place, with more to come.
Source: http://www.eia.gov/state/maps.php

Comfort and Shelter

Aside from your need for water, food, and personal security, few threads of energy are as strong as those concerned with your comfort. The desire for comfort is particularly apparent when you are indoors. Have you ever tried working at a desk when the room temperature is in the sixties or the nineties? It can be uncomfortable and distracting, often sapping productivity. Maintaining a personal bubble of comfort is the epitome of an ordinary thread of energy.

For many years, the standard office temperature in the United States was 72–74°F. More recently, in response to a growing awareness of energy and environmental costs, the standard setting has wandered more widely.[24] Facility managers and private citizens have often lowered the acceptable thermostat in winter to 70°F or lower, especially at night, while raising it in summers to 78°F or higher. You can save as much as 10 percent a year on heating and cooling by simply turning your thermostat back 7–10°F for eight hours a day from its typical setting.[25]

People have been responding to this widening range of indoor temperature settings, often by using personal fans and heaters. People have also begun adjusting their choice of clothing by wearing warmer clothes in winter and more relaxed garments in the summer. On the other hand, some people would reset the thermostat rather than throw on a sweater. These folks might find that heftier energy bills provide incentives to modify their choices and behavior, especially if your house or apartment is a wild "energy hog."

There are several ways to domesticate that hog. The most straightforward and most helpful first step is to arrange for a home energy audit; most utility companies provide that service, often at a subsidized rate. If you do not want to deal with your utility company, you could get some impression of how you are doing by comparing your energy bill with those of your neighbors. If your statement is on the higher end, it will be a strong clue that you are using more energy than necessary to meet your needs. Quite often, there are simple fixes that will save you money; the internet is full of such advice. One place to start is by looking at this list of tips from the publishers of *Consumer Reports* (Table 3.1).[26]

Another approach is to take advantage of time-of-use rates available from most utility companies. Opting for such rates is a painless way to save money (but not energy). Under such rate plans, the utility motivates you to shift as much of your use as possible to off-peak hours (10 p.m.–5 a.m. and holidays, for example), when they have a surplus of generating capacity and are willing to charge you less to buy it. All you have to do is make a few adjustments, such as setting your pool pump timer to off-peak hours. The same is true of when you wash and dry clothes and run your dishwasher. Most of these appliances give you options for delayed operation.[27]

Table 3.1 Energy-Saving Tips

Heating: 43 percent of total energy savings

- If you have a forced-air system, having your ductwork sealed by a pro can save you hundreds each year because 25 to 40 percent of conditioned air (hot and cold) is lost to leaks.
- Plug drafty windows and doors with caulk or weatherstripping.
- Insulate the attic adequately. The typical residence needs 11 inches of fiberglass or rock wool or 8 inches of cellulose insulation.

Water heating: 16 percent of total energy savings

- If your water heater is among the 41 million units in the United States that are more than ten years old, consider an upgrade. Energy Star is working with utilities and retailers to offer rebates to consumers who make the switch. Go to energystar.gov/ waterheaters.
- Wash your clothes in cold water. Many laundry detergents deliver superb cool-water cleaning.
- Install low-flow faucets and showerheads throughout the home. They'll save water as well as energy.

Appliances: 9 percent of total energy savings

- Consider trading in an older refrigerator. A current Energy Star refrigerator uses 50 percent less energy than a refrigerator from 2001. Of course, you should retire the old model rather than keeping it running in the basement or garage.
- Older washing machines are also worth trading in, especially after tougher federal standards that took effect in March 2015. If your old unit is more than ten years old, it costs you about $180 more per year than a new one.
- Run the dishwasher only when it has a full load, and use the "rinse hold" feature sparingly because it uses 3 to 7 gallons of hot water each time.

Cooling: 7 percent of total energy savings

- If your home has a central air-conditioning system that's more than a decade old, a reliable new system could be up to 40 percent more efficient. Work with a reputable contractor who will size the system correctly; you might be able to downsize if you've made other efficiency upgrades, such as new attic insulation.
- Install a programmable thermostat, which can automatically adjust the temperature in your home for maximum savings and comfort (in summer and winter).
- Don't replace windows just to save energy. But if your windows are failing, choose new windows with a low-E coating that reflects heat yet lets light in.

Lighting: 5 percent of total energy savings

- Switch to high-efficiency LEDs, which use up to 80 percent less energy than traditional incandescent bulbs.

Table 3.1 *Continued*

- Place dimmable fixtures on dimmer switches. They'll enable you to save even more energy by maintaining lower light levels.
- For outdoor fixtures, save energy with a motion sensor or a photocell that turns the lights on at dusk and off at dawn.

Electronics: 4 percent of total energy savings

- Ask your cable company to upgrade the set-top boxes in your home to ones that meet the latest Energy Star specifications, making them 35 percent more efficient on average.
- Unplug computers, stereos, and video game consoles when not in use, as they draw power when plugged in even when they're off.
- Trade-in that decade-old flat-screen TV. It costs about $66 per year to run, compared with $25 or so for a new high-efficiency television.

Other: 15 percent of total energy savings

- Plug your laptop's AC adapter into a power strip that can be turned off. That saves energy because the transformer in the adapter draws power even when the laptop isn't attached.
- If you have a stand-alone freezer with manual defrost, still a standard feature, don't let frost build up more than ¼ inch, because that will affect the efficiency of the unit.

Source for the energy-use breakdown: Energy Information Administration. (Total doesn't equal 100 percent because of rounding.)

Note: A version of this article appeared in the October 2015 issue of *Consumer Reports* magazine.

Source: https://www.consumerreports.org/cro/magazine/2015/08/tame-energy-hogs-in-your-home/index.htm.

If you wish to get deeper into the weeds of home energy use, you will find many other factors that affect your energy consumption. When you buy a house or choose an apartment, consider the orientation of your windows and even what type of landscaping is in place. While direct solar gain through exposed windows can be welcome in winters because it can help heat your living space, it is the exact opposite in the summer. If you do have such exposure, consider a roll-down blind outside your windows, keep inside window coverings closed when you are not at home, plant shade trees, install inexpensive caulking around your doors, and so forth. There are many things you can do with little effort or expense.

The most important consideration is the climate. Your single option is to change where you live. For example, while the fundamental factor of an energy efficiency house is similar from place to place, the specifics will shift. You can get a sense of the impact of the demands of climate by knowing the climate zone in which you live. For example, you will find that standards for insulation

Table 3.2 Recommended Insulation Levels for Retrofitting Existing Wood-Framed Buildings

Zone	Add Insulation to Attic		Floor
	Uninsulated Attic	Existing 3–4 Inches of Insulation	
1	R30 to R49	R25 to R30	R13
2	R30 to R60	R25 to R38	R13 to R19
3	R30 to R60	R25 to R38	R19 to R25
4	R38 to R60	R38	R25 to R30
5 to 8	R59 to R60	R38 to R49	R25 to R30

Insulation levels are specified by R-value. R-value is a measure of insulation's ability to resist heat traveling through it. The higher the R-value, the better the thermal performance of the insulation. The table above shows what levels of insulation are cost-effective for different climates and locations in the home.

Source: https://www.energystar.gov/campaign/seal_insulate/identify_problems_you_want_fix/diy_checks_inspections/insulation_r_values.

in northern Maine and Alaska are much higher than they are in Arizona (Table 3.2). Knowing your zone will provide you with basic information about how to make your house less of an energy hog.

Identifying the appropriate amount of insulation is usually a good first step regardless of where you live, but there is more to it than that. You should educate yourself about the energy requirements of your home. You could start with considering an energy "pie chart" (Figure 3.10). You will see that on a national basis, the leading energy users are heating, cooling, water heating, and lighting. Appliances such as refrigerators have become much more energy efficient than they were even twenty years ago. The same is true of dishwashers, clothes dryers, and washing machines. Adding to your energy consumption, however, is the fast-growing segment of electronics. They are gobbling up more energy than ever before.

If you wish to rid yourself of dependence on the utility company completely, there is another option. It has become more popular and more affordable. The choice is to live "off the grid." By that I mean, separated from the utility company. No utility bills. No downed power lines to concern you. Self-sufficiency. As you no doubt realize, it is the dependency you have on your utility companies—as vital as they have been for over a century—that provides their leverage over you; unpaid bills can result in the suspension of service. The threat of this action explains why utility bills are among the first to be paid each month. Going solo rids you of that dependency.

Of course, if you are thinking of going off the grid, you must accept responsibility for your energy reliability. That also assumes you have done some

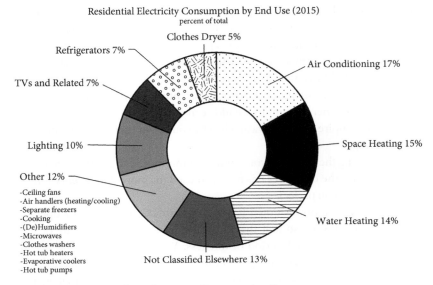

Figure 3.10 2015 Residential Energy Consumption Survey.
Source: Energy Information Administration.

homework in advance to determine the ways you can control your energy use. You could start by creating an Energy Star account: https://www.energystar.gov/campaign/createAccount?srcPage=/assessYourHome.

Space conditioning (i.e., heating and cooling) makes up about one-third of the household energy demand, on average. Even where this is not the case, other costs can mount up too. For example, using average expenditures for U.S. residential customers, lighting costs over $250 per year, assuming about 11 cents per kilowatt-hour, which is on the low end of the electricity costs, albeit typical for Arizona. In Hawaii, for example, 2019 residential prices vary from 31 cents to 42 cents per kWh, with the lowest on Oahu and the highest on Lanai. The lesson here: always consider regional differences in utility costs whenever you are calculating what it will cost to live in one place or another.

Your household energy bills are a total of a lot of somewhat mysterious fees, escalator clauses, and taxes, plus the bill for the energy you use. Whatever actions you take to lower your bills will only influence that last category, which tends to be the largest anyway. There are many actions you can take to reduce your bills.

You can try the following:
- Turning lights off that are not needed, especially if they are incandescent or halogen lights. Somewhere between 10 and 20 percent of total household energy consumption can be trimmed just by taking this action.[28]

- Lowering "phantom energy" expenses. Often also called "vampire energy," this is the energy your appliances use when they are in the "off" position. Such devices include instant-on appliances and anything with a clock. For an average American household, such energy use costs about $100 per household per year. For more on this, take a look at the advice from the U.S. Department of Energy here: https://www.energy.gov/articles/4-ways-slay-energy-vampires-halloween
- A lot of vampire energy is wasted just by your computer. Consider doing the following:[29]
  - Adjusting the setting of your computer monitor so that it turns off the display after the computer has not been used for more than twenty minutes (or less, if you prefer).
  - Employing a power-down feature on your PC through your operating system software, so that the CPU goes into a low-power state or turns off after a specified period of idleness. You must change these settings yourself; the power management features usually are not enabled when a computer is purchased. Learn how to activate the power management features on your computer.

Reducing unneeded lighting and eliminating vampire energy drains will have the additional benefit of lowering your cooling loads. Substantial help is available on the internet, and much of it is free.[30] To begin, here are eight steps recommended by the U.S. Department of Energy:[31]

1. *Install and set a programmable thermostat.* You could save an estimated 10 percent per year on heating and cooling costs by using a programmable thermostat. By resetting it when you are asleep or away from home, you will not have to sacrifice comfort. Nowadays, there are even thermostats you can control from your smartphone.
2. *Use sunlight to your advantage.* While the sun's rays can contribute heat in the winter, they can force air conditioners and fans to work harder—and use more energy—in the summer. During winter months, take advantage of sunlight by opening your curtains during the day to allow the sun to heat your home naturally. During warmer months, use light-colored window shades or blinds to reflect heat outside, keeping your home cooler and more efficient. Using natural lighting effectively will also reduce the need to use artificial light.
3. *When replacing appliances or purchasing electronics, look for Energy Star appliances, fans, and electronics.* Your home's appliances and electronics account for close to 20 percent of your energy bills. Using Energy Star–certified products—which incorporate advanced technologies that use

10–15 percent less energy and water than standard models—throughout your home could save nearly $750 over the lifetime of the products. For example, Energy Star clothes washers use about 40 percent less energy than conventional clothes washers while reducing water bills. Energy Star washers also require less detergent and are gentler on clothes, saving you money on clothing expenses.

4. *Choose energy-saving lighting.* About 10 percent of the energy your home uses goes to lighting costs. By just replacing five of your home's most frequently used lights with energy-efficient Energy Star bulbs, you could save $75 a year in energy costs. Compared to traditional incandescent bulbs, compact fluorescent bulbs can yield as much as 75 percent energy savings and last six times longer. You can get even more energy savings, longer life span, and significantly reduced heat by switching to LED lighting.

5. *Use a power strip for your electronic equipment.* Many electronic devices and equipment continue to consume unnecessary energy, even when not in use. Such energy vampires cost families about $100 a year. Use a power strip for electronic devices and turn the strip off when not in use to eliminate energy vampires. Also, be sure to unplug your chargers—they often draw power whenever they are plugged in.

6. *Reduce energy for water heating.* Water heating is a large energy consumer, accounting for 14–18 percent of your utility bills. By taking low-cost steps, you can reduce your water heating bills. Make sure your water heater is set to no higher than 120°F. Install low-flow showerheads or temperature-sensitive shower valves. Newer water heaters have more insulation than older ones. If you can feel heat by putting your hand on the water heater, it is wasting energy.

7. *Hire a professional to maintain your heating and cooling system.* Arrange for annual maintenance with a qualified technician. Such maintenance should include checking the airflow over the coil, testing for the correct fluid (refrigerant) level, checking that the combustion process and heat exchanger are operating safely, and ensuring proper airflow to each room. In addition, you should clean the air filters in your heating and cooling system once a month and replace them regularly.

8. *Consult a home performance contractor to achieve significant savings.* There is a growing industry of professionals who are qualified to make recommendations to homeowners on how to improve the overall energy efficiency of their homes. These professional energy assessors will do a comprehensive energy audit of your whole house using special tools—such as a blower door test and an infrared camera to locate air leaks—to measure home energy efficiency.

It would be instructive to take a bit of a deeper dive into item 2. Paying closer attention to building orientation can make a big difference in heating and cooling bills. If you have windows facing east or west, direct sunlight will pour into your house mornings and afternoons. In frigid areas, this is not such a bad idea, but in a hot environment, it is going to drive up your energy needs and your energy bills. If you wish to reduce this problem, installing inexpensive solar screens will help significantly.[32] Landscaping, natural ventilation, skylights, shade trees, low-E2 window coatings, double-pane windows, and light roof color are among literally dozens of other things to consider as part of a "whole house" path to lowering energy bills (Figure 3.11).

Your initial strategy to implement home energy savings should be to work in full awareness of local environmental conditions when you are choosing your

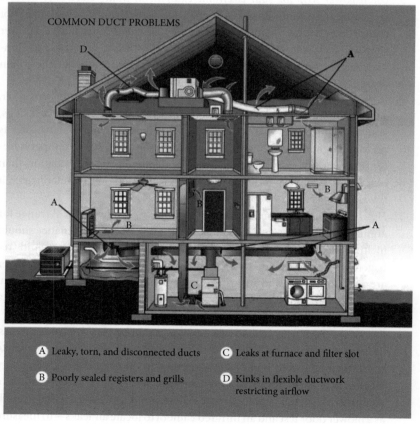

Figure 3.11 The whole-house solution to energy efficiency.
Source: U.S. Department of Energy.

house. This seems simple, but few people consider the energy impacts of natural conditions when they buy a home. For most people, other factors have a higher priority, such as whether the house has granite countertops in the kitchen and a hot tub in the backyard. But if you want your living space to be comfortable, take more care in the orientation of your house to the angle of the sun, whether windows are directly exposed to sunlight, what building materials are used, and on and on. The reason we have gotten away from thinking about such matters is that for many decades, energy was pretty cheap and our demand was pretty low. We did not have all the appliances we have today, including air-conditioning. Conditions have changed. We use more energy nowadays: it is more expensive, and the environmental costs of supplying it have become more apparent.

It pays to be choosy. If you can save 50 percent of the energy that your house uses for twenty years, that savings can help pay for a college education. Plus—and this is the most critical part—there are many better solutions to how we build and operate our homes. Many of these options fall under the heading of "passive solar design," several of which we have mentioned already: building orientation; window placement, size, and coverings; strategic shading; and insulation, to name a few. For inspiration regarding possible passive design, an excellent place to start is by looking back several centuries, well before any modern mechanical devices existed. Ask yourself, *How did the ancients manage to live in hot or cold places without the equipment we now expect in our homes? What was their strategy?* They did not have the option of merely installing more and more machines to solve the problem of personal comfort. If people have been living in every possible climate zone for millennia, including scorching hot deserts, the question is, how did they make life there possible?

The answer is ingenuity. Not having the option of simply plugging in an air conditioner or a heater, they paid more attention to melding lifestyles with local environmental conditions. For example, early residents in central Arizona found that adobe—that is, earth and organic material, or mudbrick—provided a locally available building material that provided valuable insulation from the harsh and hot desert environment (Figure 3.12). Another alternative in the American Southwest was to use natural rock formations to offer themselves protection from both the sun and the rain (Figure 3.13). Elsewhere, people living in present-day Iran developed "windcatchers," which are traditional Persian architectural elements to create natural ventilation in buildings (Figure 3.14). When they were away from homes and other buildings, they adjusted by donning clothing that was appropriate to the climate.

These and many other examples demonstrate impressive resourcefulness that has been accumulated over time, only to be ignored when thought-free technical alternatives became available regardless of their energy costs. Now, with the rising overhead costs of energy, societies around the world are beginning to

Figure 3.12 Casa Grande Ruins National Monument in Coolidge, Arizona, includes a multi-story adobe "Great House," one of the largest known ancient structures in the United States. Adobe provides natural insulation from heat and cold.

accept once again that living more in concert with the environment is a better long-term strategy. We realize the absolute need for and ultimate long-term benefits of alternative approaches to what has become the norm in the global North. In response, many people—especially younger people in grade school and college—favor returning to earlier practices that melded rather than ignored what the environment offered. If you are one of those people, you will take pains to make your house more pleasant to occupy and less of an energy hog.

Transportation

Imagine it is the mid-1800s. You are living along the eastern seaboard of the United States. One day while scanning the morning paper, you come across an account of a rugged mountain man named Jim Bridger. His description opens your eyes to a natural wonderland 2,000 miles to the west, complete with yellow-walled canyons, booming waterfalls, a profusion of wildlife, cauldrons of boiling mud, and hundreds of geysers spewing hot steam into the air everywhere in sight. Initially, you doubt such outlandish stories. It all seems so implausible. Yet

Figure 3.13 Betatakin ruin, Navajo National Monument. Betatakin was built in an enormous alcove, measuring 452 feet high and 370 feet across, between 1250 and 1270 CE. The residential ruins themselves were constructed of sandstone, mud mortar, and wood. Together, they totaled about 120 rooms at the time of abandonment in 1300 CE. Such cliff dwellings are common in the southwestern United States. They provided protection from the high, hot sun of summer, but allowed in the warming low sun of winter. In today's parlance, this is an example of passive solar design because temperatures were controlled without mechanical equipment. *Source:* https://www.nps.gov/nava/index.htm.

his vivid descriptions capture your imagination. Once excitement replaces skepticism, you start yearning for a personal visit to see for yourself this fascinating landscape. But getting to the wildlands of what would later become part of the state of Wyoming would be a long, challenging, and dangerous journey.

Despite the anticipated rewards, and even armed with great determination and personal courage, you are not sure you are capable of making such an arduous trip. Transportation choices are slim, conditions challenging, and maximum speeds sluggishly slow. You can travel part of the way by riverboat. After that, your means of travel would be by walking, on horseback, or possibly by wagon. However you traveled, you will have to follow faint and often muddy paths, somehow cross the many rivers, and skirt Indian lands occupied in many cases by people who are not pleased with your presence on their heritage lands.

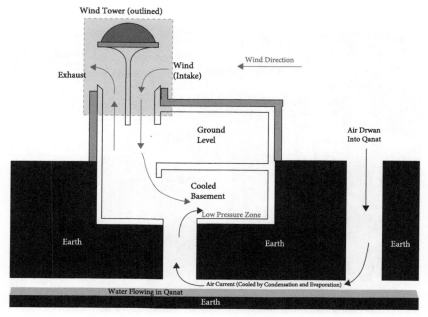

Figure 3.14 An *ab anbar* with double domes and windcatchers in the central desert city of Naeen, near Yazd, Iran, provide an example of architecture can adapt to weather conditions without the use of mechanical equipment. Windcatchers come in various designs: uni-directional, bi-directional, and multi-directional. Windcatchers remain present in many countries and can be found in traditional Persian-influenced architecture throughout the Middle East, including in the small Arab states of the Persian Gulf (including Bahrain and Dubai) as well as Pakistan and Afghanistan. *Source:* Wikipedia.

Nonetheless, you fancy adventure and you are willing to endure whatever inconvenience and discomfort you might encounter in order to experience for yourself the high country that Bridger describes. Finally arriving, you are stunned by what you see. It was worth every hardship and personal sacrifice!

Today we know this piece of land as Yellowstone National Park, and getting there is no longer much hampered by distance, expense, time, danger, or physical restriction.[33] People can travel just about anywhere they want, whenever they want, by any means that they prefer, across the country or around the world. The only consistent limitations are those of one's health and the ability to afford the expense.

Transportation improvements experienced over the century and a half since the establishment of Yellowstone National Park in 1872 rub up against standards of the miraculous. We have progressed from controlling a single horse with a flick of the wrist to controlling many horses with a tap of the foot. We have advanced from being at the mercy of the elements to the freedom to largely ignore their challenges. Visiting Yellowstone has become such a routine undertaking that about 5 million people each year manage to do it with a measure of comfort and convenience unimaginable to tough Jim Bridger.

Among the many modes of travel now available to most of us, automobiles and personal trucks of various sizes and descriptions are the most popular. There are about 290 million registered vehicles in the United States. Their number is almost equivalent to the population of the country itself, including those too young or too old to have a license. They even share some human traits. For example, we have to feed them, take care of their ailments, wash them, admire them, and treat them with affection. We have developed a co-dependent relationship with them. They are symbols of success and status, and we consider them even more important than owning a house. Together, people and their cars have created ubiquitous, enduring, and iconic signatures on the land.

Such impacts are only scarcely connected to where these vehicles are made or sold, but to everything that has spun outward from the core of their creation. To get an idea of what I mean, write down everything you see on your next trip that is tied to the existence of motor vehicles. You will quickly compile a long list: dealers, parking garages, junkyards, vanishing pedestrians, horizontal cities, shrinking sidewalks, suburbs, growing gridlock, blankets of smog, and drive-in establishments of such variety that people in the mid-twentieth century would likely slap their heads in amazement upon seeing them: restaurants, banks, laundries, liquor stores, cinemas, coffee bars, doughnut shops, car washes, wedding chapels, and tunneled giant sequoias (Figure 3.15). No one wants to leave their cars, even to get married, an accommodation that is available in Las Vegas, Nevada.

Nowhere are the threads of energy more abundant and more evident than those created in accommodation to our love of cars. Most obvious is the vast network of roads that crisscross our country and gorge on our land. Environmental historian Martin Melosi reminds us that by 1960 about 59 percent of the ground area in Los Angeles's central business district (CBD) was devoted to streets and parking, with about 35 percent for roads, streets, alleys, and sidewalks, and 24 percent for parking lots and garages that are not an integrated element in buildings having other purposes.[34]

We now find ourselves with millions of vehicles traveling billions of miles along roads that are just about everywhere. Here are some raw numbers, just for the United States: 48,000 miles of interstate highways; 158,000 miles of U.S. highways; hundreds of thousands of miles of paved state highways, secondary highways, and county highways; and over a million miles of unpaved roads. Urban and rural roads in the United States sum to over 4 million miles.[35] There is more than one mile of right-of-way for every square mile of the country.

Figure 3.15 Cars quickly became vehicles for recreational transportation, as pictured here at Wawona Tree in Yosemite National Park in the early twentieth century. *Source:* https://en.wikipedia.org/wiki/Wawona_Tree#/media/File:Wawona_tree1.jpg. Public domain.

We travel these omnipresent threads of energy in cars, pickup trucks, SUVs, semi-trailers, motorcycles, quads, dune buggies, golf carts, snow machines, and anything else we can get our hands on. So much do we drive that each year Americans, just by themselves, complete the equivalent of 4 million trips to the moon and back, literally trillions of miles (Figure 3.16).[36]

If you assume that all this traveling burns boatloads of fuel, you would be correct. In 2019, transportation energy use accounted for 28 percent of all U.S. energy use, which totaled about 100 quadrillion BTUs of energy. In 2019 fuel consumption by light vehicles was over 125 billion gasoline gallon equivalents (GGEs) and another 50 billion GGEs for medium/heavy trucks and buses (Figure 3.17).[37]

One way to appreciate the abundance of energy we use for transportation in the United States is to compare ourselves with other countries. It is an eye-opening exercise. Our energy use in transportation in the United States is

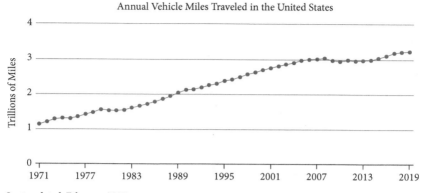

Annual Vehicle Miles Traveled in the United States

Last updated: February 2020
Printed on: July 30

Figure 3.16 The relentless upward trend in total vehicle miles traveled (VMT) in the United States (expressed as a moving twelve-month count) from 1971 through 2019. The long rise in VMT has seen three periods of flattened growth or decline, triggered by the oil price spikes of 1974, 1979, and 2008. The VMT flattening that started in 2008 continued long after oil prices recovered, largely because of an economic recession. Starting in 2012, VMT began rising again largely because the U.S. economy recovered and petroleum prices remained relatively low. The rate of increase has been lessening over the last three years portrayed in the figure. *Source:* http://www.afdc.energy.gov/data/10315.

more than twice the total energy (i.e., all sectors of the economy) used by either Germany or Canada. That is almost equal to the total amount of energy used in all the countries in Central and South America and 1.5 times more than all fifty-seven countries comprising the entire continent of Africa.

Another way to grasp the scale of our transportation energy demand is to run a little calculation. Burning a pound of reasonably good coal in a power plant will generate about 1 kilowatt-hour of electricity. That is enough power to keep one 100-watt incandescent light bulb shining for 10 hours. If we divert the 28 quadrillion BTUs (28 quads) of energy we use for transportation into illumination, we could power 280 billion 100-watt lightbulbs for an entire year. That is more than twice the number of lightbulbs in the whole world.[38] To repeat, the United States uses enough energy in the transportation sector of its economy to power all the lightbulbs in the world twice over.

Not to be outdone, other countries seem intent on catching up with the Americans. As they continue moving in the direction of a cars-everywhere society, these countries are beginning to appreciate the high price they are paying for accepting cars as symbols of status and their rewards for hard work. Traffic

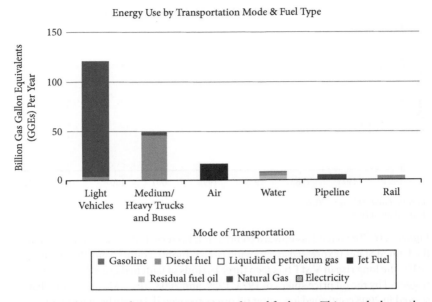

Figure 3.17 Energy use by transportation mode and fuel type. This graph shows the energy consumption (in gasoline gallon equivalents, GGEs) of the transportation sector by mode and fuel type for the year 2016. For the most part, each transportation mode is dominated by a different fuel type. Light-duty vehicles use the most GGEs of fuel per year, followed by medium/heavy trucks and buses. Water transportation relies mainly on residual fuel oil, a by-product of producing the light products that are the primary focus of a refinery. Pipelines are the only mode of transportation that uses non-petroleum fuels predominantly.
Source: Table 2.7, Oak Ridge National Laboratory's *2019 Transportation Energy Data Book*; heat conversion factor is taken from Appendix A3 of EIA's May 2019 *Monthly Energy Review*; https://afdc.energy.gov/data/?q=fuel+consumption++in+transportation. See plates for color version.

jams that we nowadays think of as "normal" are devolving into gridlock. One of the worst incidents occurred in China in 2010. Over 60 miles long, the traffic jam took ten days to clear up, during which time passengers lived in their cars, often moving only half a mile per day.[39]

Private vehicles are notoriously underutilized as a means of transportation: one car tends to transport one person. For this reason, and it might seem obvious, the per-person energy costs of transportation drop substantially when usage density increases. Using a bus provides an excellent example of this principle (Figure 3.18). If you want to maximize the energy efficiency of your transportation, a bicycle is a good option; it is even more efficient than walking. Bikes are 984 person-miles per gallon (PMPG). Walking is 700, while a car is about 36.[40]

Figure 3.18 Using buses for public transportation reduces per capita energy consumption as it reduces highway congestion.
Source: https://transportist.org/2016/02/12/cars-people-buses-bikes/.

Biking is trendy in many cities, and it keeps congestion and air pollution to a minimum. For example, residents of Copenhagen swear by the bike. There are 650,000 bikes in the city, which is more bikes than residents, and 52 percent of Copenhagen residents use their bikes every day to get to their school or place of work, even if this means going farther than the city boundaries.

As you must realize, the amount of fuel your car uses is closely tied to your driving habits, especially to how fast you drive. This relationship has not always been a concern. Over the course of our species's existence, most human travel was accomplished by walking. The speed was only 3 to 4 mph at best, so it is hard to be injured if you accidentally collide with another walker. You also did not have to worry about fueling anything except yourself. By about 5,000 years ago, people started using camels to help carry light loads. They were ideal for deserts, and they walked at about the same pace as you would. Horses were domesticated by about 4,000 years ago, and they had a major advantage over the camel; they were faster, and they were quite good at carrying people over the land. Clipper ships came along centuries later, and they held the obvious advantage of water travel. Plus, you didn't have to quench their thirst or satisfy their hunger. Their energy source was free, but their movement was slow, rarely managing even 200 miles in a day and often much less. These drawbacks were not to decrease until the addition of steam power, a major improvement that added not only consistent speed but also the ability to disregard the direction of predominant winds and currents more easily.

Barges are a specialized form of watercraft. Used on artificial canals and some streams, they were popular in Europe, Asia, and North America for the transport of heavy cargoes such as coal and wood in the centuries leading up to the advent of the railroad. Although they were an effective means of moving cargo, they were no quicker than the pace of the draft animals that pulled them along.

Trains, starting with the first steam locomotive in 1829, could haul heavy loads long distances. Everyone grasped the advantages of railroads, as reflected in the speedy expansion of railroad tracks. In 1831, only 73 miles of track existed in the whole United States. Ten years later, 8,000 miles of track were in place, stretching the length of the Northeast and linking the coast to interior settlements. By 1860, there were more than 30,000 miles of railroad track in the United States.

Trains stitched together the vast North American continent, just as ships pulled the United States closer to the rest of the world. However, there was a cost for the added speed and capacity they afforded. Soon it became important to select the most appropriate modes of transport by weighing energy costs against the desired speed and the value of the product being moved. When speed is less important, low-energy bulk carriers are suitable; the unit cost they charge is the lowest possible. As speed increases, energy requirements rise exponentially. High speed is best suited to lightweight, perishable, critical, and high-value products (Figure 3.19).

For most of a century, no mode of transport surpassed the speed and convenience of trains. Then, two disruptive technologies—automobiles and

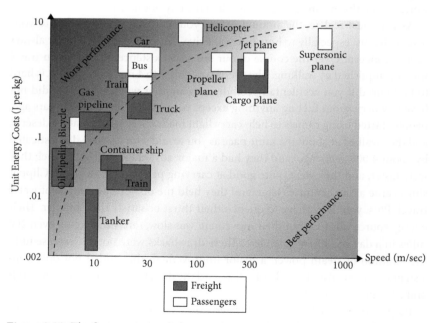

Figure 3.19 The faster you travel, the more energy you use. For that reason, bulk carriers of oil, natural gas, and coal are slow-moving ships.
Source: Adapted from J. D. Chapman, *Geography and Energy: Commercial Energy Systems and National Policies* (New York: Longman Scientific and Technical, 1989).

airplanes—changed everything. They did so in ways that would have challenged the most prophetic nineteenth-century visionary. These two disruptive technologies shrank our world, increased our mobility, and broadened our independence, just as their voracious appetites wolfed down energy to keep both in motion.

The practicability of increased speed has lessened within the past few decades, particularly after the public became more aware of the environmental and economic wisdom of reducing energy use. Several reasons explain this phenomenon. Of course, personal dangers rise with speed, and so too does the energy needed to achieve that velocity. This was brought home to residents of the United States when, in the 1970s, exporting countries imposed an embargo on oil deliveries to the United States. Long lines quickly formed at gas stations. To reduce import dependency by conserving motor fuels, a new national law quickly imposed a speed limit of 55 mph. Why 55? Because this speed offers the best blend of speed and fuel efficiency. The faster the vehicle, the bigger the necessary engine, and the greater the amount of energy needed to overcome aerodynamic drag (Figure 3.20).

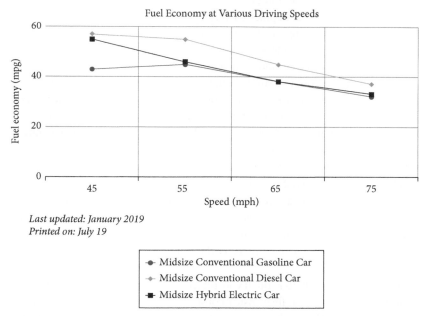

Last updated: January 2019
Printed on: July 19

Figure 3.20 This chart shows how fuel economy varies with driving speed for various vehicle categories, as modeled by Argonne National Laboratory's Powertrain System Analysis Toolkit (PSAT). All conventional gasoline vehicles (and the midsize diesel car) achieve their best fuel economy at 55 mph. The selected hybrid electric vehicles (HEVs) and the large diesel SUV achieve better fuel economy at 45 mph than at 55 mph.

Source: http://www.afdc.energy.gov/data/#tab/all/data_set/10312.

As you can see, the relationship is not linear. Above 55 mph, mileage drops off steadily. At 70 mph, for example, cars are 17 percent less efficient than they are at 55 mph. At 80 mph, they are 28 percent less efficient. For heavy-duty trucks, declining mileage is even more pronounced.

Another example, this time with aircraft, further helps illustrate the relationship of energy cost to higher speed. A typical subsonic 747-400 comes in at 91 passenger-miles per U.S. gallon, but the now-abandoned supersonic Concorde—which traveled at over twice the speed of a 747—operated at 14 passenger-miles per U.S. gallon.[41] This higher fuel consumption helps explain why the Concorde was never a moneymaking venture.

Keeping speed in check saves energy, but there are many other adjustments to help as well. Two that continue to attract a lot of attention are improving overall vehicle efficiency and decreasing the time engines idle. Looking first into the matter of efficiency, ask yourself what percentage of the energy held in your tank moves your car. A correct answer would be around 20 percent. In other words, between 70 and 80 percent is wasted.

You might think we should be able to do better, and you would be correct, at least in theory. An excellent place to start is by looking under the hood. When we do that, we find a hot engine. This should not be surprising because more than two-thirds of the energy released when we burn the fuel is lost to the inefficiencies inherent to internal combustion engines.

If that is news to you, you surely realize that car engines are hot. They pour off so much heat that cars must have radiators to keep them from overheating. What does all that excess heat imply? If that much heat is coming off the engine, it is not being used to turn the wheels. Simply stated: a hot engine is an inefficient engine. If we can improve engine efficiency, we will burn less fuel (and need less cooling as well). Besides, we would also reap financial, environmental, and geopolitical benefits from reducing the need to import oil.

The question is, how can we do even better? One way is to make our cars "slippery"—that is, improve their design so they pass through the air with the least possible resistance. Car manufacturers have been working on this approach. If the engine does not have to work as hard to move air out of the way, overall efficiency will improve.

But we can do even better still. If you ask energy expert Amory Lovins, he would recommend using carbon fiber instead of steel. It is as strong as steel but at a fraction of the weight. The resulting cars are just as safe, but they need smaller engines. Lovins also suggests using the braking system of the vehicle to recharge a battery. Such a system is now standard on hybrid and electric cars. For his "hypercar," Lovins has also suggested many other steps, including minimizing rolling resistance and reducing engine idling. What you get is an efficient, lightweight, safe car with better performance and better mileage. Look it up at www.RMI.org.

What about idling? Think about it in personal terms: While you are driving your car, you have undoubtedly experienced the frustration of being stuck in traffic, burning fuel, but going nowhere fast. Such car idling is mostly the result of having too many vehicles on the road or an accident. There is not much you can do about the cause in that instance, but in others, you have choices. If you are waiting for someone or have pulled off the road to take a photo and left the car running, you are burning fuel, it costs you money, and it is creating pollution. One study estimates that if all Canadian motorists reduced their idling by just 3 minutes a day, it would be equivalent to taking 320,000 vehicles off the road for an entire year.[42] The United States has about ten times the population of Canada, so just imagine the benefit of taking that step here.

Truck idling is even a more severe drain of resources. As you can guess, heavy big-rig trucks burn more fuel than automobiles, including during idle moments. According to Argonne National Laboratory, a Class 8 tractor uses 0.8 gallons of fuel per hour when idling.[43] Over time, this seemingly small fuel use adds up because trucks are often parked during long trips. If a driver parks 14 hours each day with the truck left idling, for example, the vehicle would use 11.2 gallons of diesel with no miles traveled. If the driver's vehicle averages 5 miles per gallon (a low estimate with today's modern engines), the driver will lose 56 miles' worth of fuel at the cost of $44.80 when fuel prices are at $4.00/gallon. If the driver is on the road just 200 days a year, that is a loss of almost $9,000 for the owner.[44]

There is little doubt: it takes a lot of energy to move things around, it produces a lot of pollution and noise, and it is a drain on the pocketbook. How much money are we talking about? That depends on several factors, among them the cost of fuel. It also depends upon what you include in your calculations. Should you, for example, include the cost of pollution; or building and maintaining highways, airports, railroads, and parking structures; or the costs of dumping or recycling vehicles at the end of their useful lives?

It is a complicated calculation, so we will limit ourselves here just to personal expenditures for energy. The average cost for transportation is about $2,200 per person per year, or about two-thirds of the average national expenses for energy. For those of you who think you pay a lot to your local utility companies for your electricity and natural gas, consider this: transportation expenditures are twice as much.[45] The $2,200 figure is, of course, a national average. Each state will deviate from this average, some more than others. For example, about $5,800 per person per year is spent on transportation in Wyoming. In Alaska, it is over $8,000 a year.

Keep in mind that these are just the costs for the energy expenditures to operate your vehicle. You really must consider other expenses as well, including depreciation, interest on your loan, taxes and fees, insurance premiums, maintenance, and repairs. If you were to total all those expenses, you would find that a private car costs much more than you have imagined. The five-year total for a 2013 Subaru Outback four-cylinder car, for example, is almost $45,000, or about $9,000 per year. For a Mercedes-Benz E63 AMG 4MATIC S-Model Sedan (5.5L V8 twin-turbo AWD seven-speed automatic), the cost is over $145,000 for five years or $29,000 per year. You may be shocked when you find how much you pay—willingly—for the convenience, enjoyment, security, and comfort of owning a car. You can use the following website to estimate true costs for your circumstance: http://www.edmunds.com/tco.html. After you do that, maybe you will make a different choice when you need to move around. Perhaps a bicycle?

Summing Up

My intent in this chapter has been to bring to your attention the "everyday" threads of energy, those that, without much notice, connect you to the needs and activities that make up your everyday life. We have considered food, reliability and safety, comfort and shelter, and transportation. Now that you have read this review, I hope that you have acquired some sense for the subtle but essential role energy plays in everyday, ordinary activities. I also hope you understand some of the ways you can live well with less. Most of all, I hope you know more deeply than ever why it is cheaper, safer, more secure, more comfortable, less threatening, and more responsible to do so.

4

Art

Reflections of Meaning

What Is the Point of Art?

What is the point of art? Why do people elect to devote their time and energy to creating things and indulging in behaviors that have little direct survival value?[1] As art is self-expression, another purpose of art is to communicate with others. At its best, art can evoke many emotions, including adoration, amusement, anxiety, awe, calmness, craving, disgust, envy, excitement, fear, horror, interest, joy, nostalgia, romance, sadness, satisfaction, sympathy, and triumph.[2]

The breadth of media is equally wide. While the earliest discovered art was as graffiti and simple paintings using natural dyes, today's art takes form in wood-carving, ceramics, textiles, ironworking, sculpture, writing, musical composition, photography, motion pictures, and computer graphics, just to name the most obvious.

The subject of the earliest art—however you might define the term—is speculative, but the oldest yet discovered depicts vital activities such as hunting. Over the years, the breadth of subject matter has widened to encompass two broad themes: inanimate subjects such as landscapes and activities of economic significance, such as farming. One of the most common "threads," if you will, that has consistently run in and out of our lives since the Industrial Revolution has been energy. Today, energy in all its forms and phases has found expression in all media, traditional to electronic. One of the most evocative and impactful form is the commercial motion picture, if for no other reason than that movies are the most immersive, realistic, and broadly viewed art form available. So let's start there.

Motion Pictures

Humans have always been a storytelling species, and watching stories unfold on a big screen in a darkened theater plunges us into a form of intense realism unachievable by other media. Indeed, that is a large part of their appeal. Entering movie houses, we willingly suspend disbelief while images flicker to life before us. Movies have the potential to move a wider audience emotionally than do

The Thread of Energy. Martin J. Pasqualetti, Oxford University Press. © Oxford University Press 2021. DOI: 10.1093/oso/9780199394807.003.0004

static exhibits of sculpture, painting, drawing, or photography. Movies are likely the most potent medium to convey the influence of energy on our lives, our environment, and our future.

A recent example of an "energy movie" is the Academy Award–winning film *There Will Be Blood,* a film adaptation of Upton Sinclair's social and political satire *Oil.* The film depicts cutthroat competition, backstabbing corruption, and a wide assortment of illegal activities that did not stop short of mayhem and murder. If you watch this sprawling tale of American capitalism, you will see Daniel Day-Lewis portray the opportunistic oil developer Daniel Plainview. Plainview is a figure based on the real-life oil tycoon Edward Doheny, one of those who started the Southern California oil boom in 1892 (Figure 4.1).[3]

Given that the Southern California of today is nothing like Southern California 125 years ago, the producers of *There Will Be Blood* had to find a stand-in location to do their shooting. They chose Marfa, Texas, which, by coincidence, was also the backdrop for another classic oil movie, *Giant,* staring Hollywood superstars Elizabeth Taylor, James Dean, and Rock Hudson. In *Giant,* "black gold" seemed to be everywhere drillers spudded a well.

Figure 4.1 Historical photo of early years of rampant and corrupted oil development in the Los Angeles Basin, a time depicted in the movie There Will Be Blood. *Source:* The Huntington Library.

There were many common themes. Both movies depicted greed and avarice by showing the leading characters drenched in oil, a visual signal of the mind-numbing wealth that was the stuff of legend for those in the "oil patch." Both movies repeated a common rags-to-riches legend of wildcatters finding oil, wealth, and the power that came with it. And both movies reinforced the public impression that energy development was, and remains, a surefire pathway to success, security, and power. The oil business prompted both movies and added to the enduring legends of the relationship between wealth and power. Most recently, this relationship converged in early 2017 when Rex Tillerson was confirmed as US secretary of state. Prior to that, Tillerson was the CEO of ExxonMobil, one of the largest oil companies in the world.

Looking deeper into the underpinning of both movies reveals the historical context that keeps them alive in the public imagination. With the release of *Giant* in 1955, the 59,000 active oil wells in Texas were producing about 1 million barrels per day (bpd). Soon afterward, oil drilling accelerated, but the days of the huge discoveries had primarily passed, despite the dreams of riches that still excited the latecomers. When *There Will Be Blood* hit the screens fifty years later, 151,000 wells were producing only 344,000 bpd. People at that time were predicting the "end of oil."[4] By the beginning of 2020, however, improved methods of hydraulic fracking helped swell production to about 5 million bpd just in the Permian Basin of West Texas.[5]

Giant and more recently *There Will Be Blood* projected several messages about energy. Among these has been that the era of oil riches still exists, at least for the hardiest and most optimistic among us, even after 150 years of development. Another message is that the control of energy resources can bring enviable wealth, personal prestige, respect, notoriety, and political influence. For these reasons, oil resources are considered to be worth fighting over, regardless of whether the protagonists live in a wildcatter's trailer or a presidential palace.

And things have not changed much. The attractions of oil persist, just as demand continues growing because it remains the fuel needed by cars, trucks, trains, boats, and airplanes. For this reason, even though oil is increasingly expensive to bring to market, it still propels its own compelling narrative. Such momentum is likely to continue rolling into the future—despite the setbacks caused by the Covid-19 pandemic of the first quarter of the twenty-first century—motivating creative new techniques of exploration and production, all as a means to tempt developers to keep pushing into deeper waters, denser jungles, and permanently frozen tundra.

Giant and *There Will Be Blood* are two of the most prominent energy movies. Still, there have been many others, and all of them reinforced the bonds between energy and society as well as provided the lubricant for the political machine. A sample of these films includes *Oklahoma Crude* (about oil); *The Molly*

Maguires, Matewan, How Green Was My Valley, Coal Miner's Daughter (all of which address inequities forced upon coal miners); and *Silkwood* (about the dangerous industrial sloppiness within the nuclear fuel chain).

No energy movie likely had a greater degree of serendipitous—and ultimately unexpected—social and political impact than *The China Syndrome*, the story of a meltdown at a fictional nuclear reactor in California. The large impact of the film rested in part on the timing of its release. A showcase vehicle for such Hollywood stars as Jane Fonda, Jack Lemmon, and Michael Douglas, *The China Syndrome* premiered just twelve days before an actual meltdown on March 28, 1979, at Unit #2 of Three Mile Island (TMI), a 906 MWe (megawatts, electric) nuclear power plant operating between the Pennsylvania capital, Harrisburg, and the Philadelphia metropolitan area (Figure 4.2). The accident at TMI triggered huge public anxiety, several bureaucratic missteps, assurances of technical control despite a mandatory public evacuation order, and an abundance of negative news coverage about nuclear power. It brought to a halt an aggressive twenty-year program of nuclear power plant construction in the United States, creating a hiatus that continues today.

The TMI accident, amplified by *The China Syndrome*, prompted a thorough evaluation of established nuclear power plant procedures and protocols. The examination revealed many shortcomings of operator training and emergency

Figure 4.2 A sign dedicated in 1999 in Middletown, Pennsylvania, near the Three Mile Island nuclear plant describing the accident and the evacuation of the area. *Source:* Courtesy of Michael Anderson.

response plans, and it accentuated the critical importance of assessing more thoroughly the role of public perceptions of risk. The need to develop, receive approval for, and implement revamped policies and procedures had significant and costly consequences. For example, at the Shoreham nuclear power plant on Long Island, New York, no emergency plan could meet the revised evacuation requirements. As a result, the plant never operated commercially, and the public had to absorb the entire cost of its construction.

Many scholars have noted the long-term impacts on public policy produced by *The China Syndrome*. Writing in 1981, Gary Weimberg suggested that the film had entered the realm of "explicit politics": "Regardless of the intentions of the filmmakers, Three Mile Island has forced audiences to become aware of the inevitable connection between film and politics."[6] John Wills of the University of Essex coined the phrase "celluloid chain reaction" to express the links between the movie and the real event. He pointed out that the power of the film to galvanize incipient public distrust of nuclear power, with all the subsequent political consequences, came from the thrust of the storyline:

> *The China Syndrome* diverged from traditional nuclear movies by exposing the political machinations of the American nuclear industry. Rather than provide audiences with colourful examples of manifest atomic destructiveness, of radioactive mutants and frightening war prophecies . . . the picture offered an informed account of troubles within the commercial atomic industry. To create an authentic and convincing storyline, the producers . . . relied on documented instances of corruption and misfortune at US facilities for their inspiration. . . . *The China Syndrome* served as a celluloid witness to salient fears in American society.[7]

In 2013, Tony Shaw reinforced the connection between film and energy policy in an article for *Cinema Journal*. He referenced a 1997 PBS documentary, *Nuclear Reaction*, that explored why Americans so feared nuclear power. Using a clip from *The China Syndrome,* he illustrated what he called Hollywood's "undue hostility to nuclear power" and "fearmongering."[8] He also referred to a *New York Times* article about climate change, titled "The Jane Fonda Effect" and written by Stephen J. Dubner and Steven D. Levitt, authors of the book *Freakonomics*, that he said "playfully linked the widespread panic about nuclear energy induced by *The China Syndrome* . . . to the United States' continued reliance on inefficient and dirty power plants burning coal and fossil fuels."[9] Shaw then coined the phrase "energy-media complex" to magnify awareness of the thread that connects films with energy policy.[10]

The epithet "dirty" is often used as a modifying adjective before the word "coal." The travails of coal miners everywhere have fortified the implications of this narrative in several movies, most notable among them *Matewan* and *The Molly Maguires*.[11] Both tell powerful stories about miners killed because they attempted to organize against coal companies in bids for higher wages and safer working conditions. Nowadays, the plight of such miners continues to be the subject of artistic expression (Figure 4.3).

Mine safety in the United States has improved substantially from the shameful conditions that predominated for more than a century. Given recent trends, such improvements may no longer be as significant; mining jobs are disappearing as coal use wanes, at least in the United States and Western Europe. This reduction

Figure 4.3 An artist portrait of this coal miner conveys the sad circumstances in which miners worked.
Source: © Valerie Ganz. Used with permission.

is reflected in employment statistics: renewable energy markets employed more than 43 percent of the electric power generation sector's workforce in 2016, while fossil fuels combined accounted for just 22 percent.[12] With the declining profits from coal mining, new movies that center around coal would seem to be less likely in the future.

Paintings, Sculpture, Photography

Despite the relatively small audiences that most artists attract, many have found energy a compelling subject. For example, the French artist Claude Monet not only painted the smog of nineteenth-century London but also paid witness to arduous coal-loading operations in Newcastle in England's North Country in that era (Figure 4.4).

While Monet was familiar with England's industrial scenes and the environmental costs of coal mining, he probably did not descend into the dangerous confines of the mines themselves. However, we do have the work of many other artists who did work in the mines, called the "Pitmen Painters."[13] Known more formally as the Ashington Group, the Pitmen Painters began their work in the 1930s in evening classes in a village in the north of England to increase their appreciation of art. This activity included trying a hand at painting scenes that they knew best—coal mining.

The wretched conditions experienced by coal miners served as an alluring theme for painters—particularly around the turn of the twentieth century—but the subject matter continued to inspire painters over the years. For example, in April 2015, the BBC aired *Visions of the Valleys,* in which Kim Howells looked at how artists have responded in the past to the industrial landscape of coal mining in the Welsh valleys and to the lives of those who lived in the mining communities.[14]

The story Howells presented began with J. M. W. Turner in 1790, offering him as an example of artists who were attracted to paint natural wilderness "in their quest for the sublime." "Soon, though, the imagination of artists was fueled by a different kind of sublimity that stirred awe, namely the dramatic landscapes of the Industrial Revolution."[15] Before the advent of storytelling with moving pictures, labor actions attracted painters and their brushes to record the scenes for the rest of us to ponder.

Even more recently, the plight of coal miners continued to inspire artists. For example, the British coal miners' strike of 1984–1985 was depicted on canvas by Valerie Ganz. After that tumultuous strike, coal mining shrank in the United Kingdom almost to extinction. "Now [visual depictions of coal mining] appear to us as images of a vanished way of life." There are "no working mines left in the

Figure 4.4 "The Coal Workers" by Claude Monet, 1875. This painting is the only painting Monet executed showing people in an industrial context. The people on the planks are reduced to anonymity and become simply a decorative element within the composition. Monet shared the preoccupations of some of his contemporaries, such as the novelist Émile Zola, who tried to describe a full spectrum of life at the time in books such as *Germinal*. The Seine River is not the lighthearted setting for regattas, but a river plied by heavy barges. The banks are lined not with trees, but smoking chimneys. Sunday strollers have given way to workers unloading coal from the barges to supply the nearby factory. Text from https://www.claude-monet.com/the-coal-workers.jsp
Source: Illustration courtesy of Chris Robart. Used with permission.

valleys, where there were once hundreds."[16] It is a tale repeated in many areas of the United States as well, particularly in Appalachia.

Painting has taken energy as a principal theme, but there are other art forms to consider as well, such as sculpture. Each sculpture tells its story not in two dimensions but three, allowing visitors to experience them from several directions, and perhaps even climb on them. One example of such a sculpture is "Oil Drop," located in the Siberian city of Kogalym, 1,400 miles northeast of Moscow and home to 60,000 residents. The sculpture celebrates oil and its importance to the local economy in a place (at latitude 62°N) where winter temperatures are commonly -40°F or colder, and where permafrost and seasonal standing water make roads and homes difficult to build and challenging to maintain. Kogalym is, however, the headquarters of the energy corporation Lukoil, and oil development is profitable enough to constitute one of the few economic activities that can entice the creation and continued support of large settlements in some of the most remote locations on earth.

Kogalym, sometimes called "a town at the end of the earth," is a testament to the continued demand and profitability of oil and its enduring importance to national economies.[17] It also foretells that no area—no matter how inhospitable, challenging, or ecologically fragile—is secure in its untrammeled state. The city's name means "wretched hole" in the language of the Khanty, the indigenous people of the region. If even such a harsh and remote site could be transformed into a vibrant city, no place is immune to the changes energy can impose on the land. The artist used "Oil Drop" as a message to everyone who might question why the city exists there in the first place.

Gerd Ludwig's photo of "Oil Drop" reminds us that photography can be used to illustrate remote energy activities (Figure 4.5).[18] However, other

Figure 4.5 A monument in the shape of an oil drop, "Oil Drop" bears the logo of the Lukoil oil company. Sculpted in bronze and given to the city of Kogalym by Lukoil, Russia's largest private oil company, in 2001.
Source: https://media-cdn.tripadvisor.com/media/photo-p/0c/fc/70/8a/caption.jpg

Figure 4.6 Bernard Lang created this collage of coal mine images for a Paris art gallery exhibition. The original photos, and more, can be found on Bernard Lang's website, https://www.behance.net/bernhardlang. See plates for color version.

photographers, notably the German photographer Bernhard Lang, have married energy and photography more abstractly. Lang recently managed to transform the massive disruptions caused by the Garzweiler open-cast coal mines west of Cologne, Germany, into an art exhibit displayed at Galerie Liusa Wang in Paris (Figure 4.6). In a similar vein, the Fine Art America website presents the work of many other photographers who have focused on a variety of energy activities.[19]

Artists of all stripes continually look for new subject matter, and in recent years wind energy has been well positioned to accommodate them. So evocative are wind turbines that motorists often risk their safety by stopping along busy highways to photograph them. Movie producers use them to enhance the visual

Figure 4.7 "Wind Is the Key," a painting by Jennifer Kim, Rosemont Middle School, La Crescenta, California

impacts in their films (see, for example, *Rain Man*). Wind turbines have appeared in children's art, city promotions, and advertisements (Figure 4.7). They have even provided a backdrop for wedding photos. The popularity of wind turbines as a subject of artistic expression is explained in part by their eye-catching movement, but also by their novelty, their sleek modernity, and their transformative impact on the landscape (Figures 4.8–4.11).

Art interprets reality. The abundant images of wind turbines and solar power indicate a change in energy technology just as it signals a shift in public attitudes. Paintings and photographs of energy activities in the past depicted lives of drudgery, the subjugation of the working class, degraded environments, compromised health, and lives of everlasting difficulty. Wind turbines and solar installations mirror none of these costs. Instead, they project an image of a world that is more responsible and optimistic. Renewable energy has taken on the mantle of landscape art, environmental stability, and a more hopeful future. In so doing, these images are chronicling the shift between the environmentally *burdensome* resources of the past and the more environmentally *benign* resources in our future.

Figure 4.8 Wind used in the promotion of Palm Springs, California, a community that initially rejected the introduction of wind turbines.

Figure 4.9 Wind turbines on a traditional small family farm near Dixon, Illinois, the boyhood home of President Ronald Reagan.

Figure 4.10 Field of wind turbines transforms the landscape at Palm Springs, California, which is backed on the west by the 11,000-foot Mt. San Jacinto.

Figure 4.11 Wind energy replacing fossil fuel at Wolf Ridge, Texas.

Architecture

The personal and social costs of energy are echoed in the creative designs of buildings, especially considering that we are living within the envelope of energy-directed architecture. Such designs are a relatively new form of energy art, one that illustrates how public awareness of energy's many costs can powerfully influence where we live, work, and play. Whereas in the United States the architecture of the past paid little heed to energy needs, the architecture of the present is often predicated on increasing energy efficiency. It does this by implementing design principles and strategies that are tuned to the aspirations and directives of energy efficiency. Such attention is manifest in myriad ways, including materials, insulation, building orientation, landscaping, and fenestration. Some of the approaches are grouped under the label "passive design" to distinguish them from the type of design that obliges "active" mechanics to maintain desired levels of light, temperature, function, and comfort.

Ironically, many design principles that are commonly employed today are just modern versions of old practices that were abandoned because mechanical and electrical equipment could accomplish the same goals regardless of the intention. Many of these principles are a practical application of simple physics, such as using thermal mass to serve as insulation, utilizing overhanging roofs for shading, or creating columns of rising air resulting in natural drafts, as with "wind towers" (Figure 4.12).[20]

Today, with modern materials and engineering, we can apply old principles to new buildings with pleasing results, as in this schematic illustration of natural shading and thermal mass (Figure 4.13). Many other simple adjustments promote air movement and the use of ambient lighting. My point here is to emphasize that the art inherent in architecture can complement the energy needs of users. In this case, form follows function. In creative hands, energy-efficient houses can also be artistically pleasing (Figure 4.14).

Perhaps the most recent and most creative melding of artistic design with energy efficiency is found in the work commissioned and supported by the Land Art Generator Initiative. Their tag line is "Renewable Energy Can Be Beautiful," and they hold biennial competitions. The most recent ones were in Santa Monica, California; Dubai, UAE; and Melbourne, Australia (Figures 4.15 and 4.16).

The Written Word

Artists normally project their purpose through one medium at a time, and people react differently to each. You might prefer painting, for example, or sculpture. You might be moved more by one form than another, even if the subject matter is the same. If, however, you encounter the same subject matter repeatedly in

Figure 4.12 An ab anbar (water reservoir) with double domes and windcatchers (openings near the top of the towers) in the central desert city of Yazd, Iran. It is a passive form of space cooling.
Source: https://en.wikipedia.org/wiki/Windcatcher.

many different media, your normal reaction is likely to be that the topic must have special—usually societal—significance. As we have already illustrated, energy themes have often prompted artistic reflection across a broad spectrum, whether in painting, photography, sculpture, or film. All these have something in common. Do you know what it is? We absorb all these art forms visually and in the moment. They exist in space. If you try to name an art form that does not, writing should pop to mind.

In contrast to music and the visual arts, reading tends to be a more reasoned activity, less visceral, usually solitary, and entirely cerebral. This predominantly individual experience is processed with less speed but more care. The written word is individually neutral. Each of us takes meaning from it in our way. Collectively, works can convey nuance, opinion, preconceptions, cultural tunings, splashes of color, and theatrics. We write to record, to inspire, to motivate.

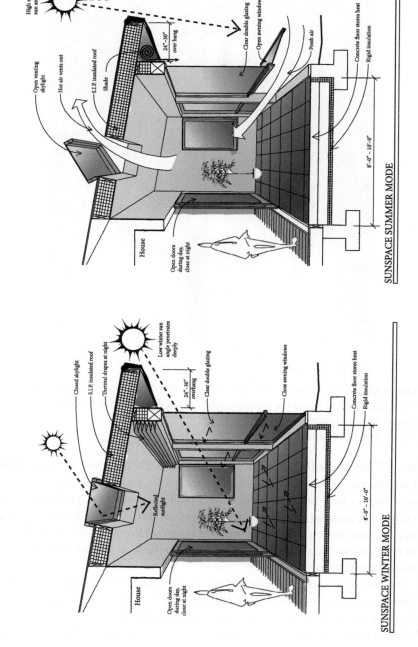

Figure 4.13 Illustrations of some passive solar designs.
Source: http://davidwrightarchitect.com/news/.

Figure 4.14 Energy-efficient house with appealing architecture.
Source: https://adorable-home.com/architecture/energy-efficient-house-21588.

Figure 4.15 Sun Towers, a submission to the 2016 Land Art Generator Initiative design competition for Santa Monica. John Perry, Matteo Melioli, Ramone Dixon, Terie Harrison, Kristina Butkute (BLDA Architects), Tom Kordel, Sherleen Pang, Kostas Mastronikolaou (XCO2), Steven Scott Studio. Team location: London, UK. Energy technology: photovoltaic panels, wave energy converter, tidal underwater turbine. Water technology: solar distillation, reverse osmosis desalination. Annual capacity: 4,000 MWh, 110 million liters of drinking water.
Source: Full details at https://landartgenerator.org/LAGI-2016/mmrdkb16/. Used with permission of LAGI.

Figure 4.16 The Solar Seesaw, 2019. Team: Luca Fraccalvieri, Ahmad Nouraldeen. Team Location: Santeramo in Colle, Italy. Energy Technologies: flexible thin-film CZTS solar photovoltaic, kinetic energy harvesting pavers. Annual Capacity: 2,800 MWh. A submission to the Land Art Generator Initiative 2019 competition for Abu Dhabi, UAE.
Source: "https://urldefense.com/v3/__https://landartgenerator.org/LAGI-2019/the-solar-seesaw/__;!!IKRxdwAv5BmarQ!JYTi-whg6_8oTtCCaTc5C5YIvYJMNDFIq FTEBFzMeDq2telz9kL8nzUlOIZ2BOHa5w$" https://landartgenerator.org/LAGI-2019/the-solar-seesaw/. Used with permission of LAGI.

One of the most valuable characteristics of words is that they have spatial and temporal longevity. They can be replicated, transmitted, and nowadays clicked on instantly around the world. Words can last for centuries and have potentially the largest audience of all of the arts. Within the world of creative writing, energy is a common topic. For over a century—in novels, poetry, plays, and nonfiction—coal was one of the most common themes.

Coal miners have worked for centuries in dangerous conditions. Mostly they worked with mute resignation to their plight and the social station that accompanies their profession. The pay they received for the work they did would frequently fall short of providing fully for their families. Always worried, spouses and children would wait at the end of every shift for the daily return of their blackened and exhausted loved ones from underground.

The feelings of angst that accompany those in and around coal mining have been prevalent for so long in countries near and far that it should not surprise you to find that it has been the central theme of several classic novels. Among these are *How Green Was My Valley* (by Richard Llewellyn, about coal in Wales), *Germinal* (by Emile Zola, about coal mining's impacts on the miners of Montsou in northern France), *The Road to Wigan Pier* (by George Orwell, on coal mining in Wales and northwestern England), and the novel simply titled *Coal* (by Upton Sinclair on coal mining in the United States).

While all of the books just mentioned reached print before 1940, the dangers and lifestyles that motivated their writing persisted for many decades, without improved conditions. In 1870, 1,000 miners died in Britain. By 1910, the death toll was still 1,000 per year, as Paul Poplawski reminds us in his discussion of the coal-mining references within the work of the literary giant D. H. Lawrence, a writer who himself came of age in the Black Counties of England.[21]

So compelling is the "us vs. them" theme of miners and management that books are still forthcoming. A recent one is *The King of Coal*, by Jeremy Figgins, whose protagonist strives to become his country's richest and most powerful person on the back of a worldwide and unstoppable appetite for coal.[22] Another is *The Coal King's, Slaves,* which tells the story of a father and his three sons who face "blackness, filth, hardships, and extreme danger in the anthracite coal mines of eastern Pennsylvania while the woman of their home struggles to keep her family alive."[23] In *The Hollow Ground,* Natalie S. Harnett writes: "We walk on fire or air, so Daddy liked to say. Basement floors too hot to touch. Steaming green lawns in the dead of winter. Sinkholes, quick and sudden, plunging open at your feet."[24]

These books, like all the other works of fiction that circulate on an energy theme, have achieved momentum by reflecting the central place of energy in the workings of society, especially in relation to strife, danger, subjugation, poverty, injustice, and conflict. And all of them—while fictional—have their basis in fact; indeed, that these books reflect actual conditions is what makes them even more gripping. They all tell stories of powerless people working long, dangerous hours with little hope of fairness from a system that uses them as machines having no feelings or limits of endurance. Instead, "their lives are regulated by economic determinants that grow into deeply engineered social practices that corral generations of such miners by the fatalistic attitudes that have been fashioned for them."[25]

Many additional book-length, non-fiction treatments of the plight of coal miners also exist. In this vein, no book has been as successful in conveying the continuing plight of coal miners in Appalachia as *Night Comes to the Cumberlands* by Harry Caudill. Caudill, like Anthony Bimba in *The Molly Maguires,* lamented that miners lived like indentured servants, owing almost everything to the company for which they worked, and owning almost nothing themselves. The coal companies owned the housing, equipment, grocery stores, health services, and more. It was not uncommon for a miner to work deep underground in unhealthy and cramped spaces to make just enough money to cover the expenses that were being charged him by the company. Bimba gave the following example from 1896:[26]

Coal mined, 49 tons at 71½ cents [per ton]	$35.03
Supplies	$8.25
Blacksmith	.30
Fixing two drills	.30
Rent	6.00
Groceries, etc.	20.18
Total	$35.03
Net Balance	**$00.00**

Coal is not the only energy resource that is at the foundation of novels about hardship and suffering. Although there are relatively few about oil and gas, nuclear power has again picked up the pace. Nuclear power started to kindle its literature in earnest with the response to the April 26, 1986, accident at the Chernobyl #4 reactor in present-day Ukraine. The accident resulted in the permanent evacuation of the city of Pripyat. Airborne contaminants from the accident wafted into Turkey, Europe, and North Africa, continued eastward across the Soviet Union, and eventually made their way around the entire Northern Hemisphere. Reindeer in Lapland and sheep in Great Britain were slaughtered as a precaution against consumption of meat contaminated with harmful amounts of radioactive isotopes. Cow milk produced within this area was discarded, often dumped in the streets. The resulting death toll from Chernobyl continues to rise; estimates range between 4,000, by the International Atomic Energy Agency and World Health Organization, and almost 1 million, as reported in a book by three Russian scientists.[27]

Chernobyl is the topic of hundreds of books, including one by Svetlana Alexievich, who won the Nobel Prize for Literature for her body of work, including *Voices from Chernobyl: The Oral History of a Nuclear Disaster*.[28] Other notable contributions include Julia Voznesenskaya's *The Star Chernobyl* and Irene Zabytko's *The Sky Unwashed*.[29] The accident also provided the subject matter for a powerful stage play called *Sarcophagus* by Vladimir Gubaryev, after the term for the protective barrier that was quickly constructed around the damaged reactor.[30] The setting for the play is a hospital where those critically injured in the immediate aftermath of the event were treated. As the play progresses, fewer and fewer actors occupy the stage as the characters they portray surrender to their injuries. Thirty years after the accident, the original sarcophagus of the title was entombed by a 31,000-ton permanent shell, rolled into place on October 4, 2017 (Figure 4.17).

A nuclear power plant accident of comparable severity occurred at Fukushima, Japan, in March 2011. It followed a 9.0-magnitude earthquake and a tsunami that reached a height of as much as 133 feet in Miyako. The Japanese National Police Agency reported 15,861 deaths, 6,107 injured individuals, and more than 3,000

Figure 4.17 Permanent shell covering Unit #4 at Chernobyl. The largest movable structure ever built, it is designed to provide a safe environment for more than a century, allowing specialists to clean up inside.
Source: https://en.wikipedia.org/wiki/Chernobyl_New_Safe_Confinement.

people unaccounted for because of these events, almost all of which were from the tsunami. About 129,225 buildings fell, with nearly double that number sustaining significant damage and nearly 700,000 buildings damaged in some way. Roads, railways, and dams collapsed. In the northeastern portion of the country 4.4 million homes had no power, and 1.5 million had no water.[31]

The meltdown of the reactors occurred after the tsunami topped the protective wall surrounding the power plant, causing three reactors to lose cooling water and explode, spewing radioactive material into the atmosphere. The toxic dust settled in an area encompassing part of the Pacific Ocean and several quickly abandoned communities that are still unpopulated. The Tokyo Electric Power Company (TEPCO) estimates that about 900,000 terabecquerels of radioactivity (9×10^{12} becquerels, or 24,324,324 curies; becquerels and curies are both measures of the activity of radioactive material) entered the air between March 12 and March 31. This level of radioactivity compares to 5,200 petabecquerels (5.2×10^{15} becquerels, or 140,540,540 curies) released at Chernobyl.[32] Owing to the continuous release of radioactivity at Fukushima, TEPCO has recently conceded that the total amount of radioactivity released from the site might eventually exceed that from Chernobyl.[33] For comparison, the accident at Three Mile Island, which effectively halted nuclear power plant construction in the United States, released 15–20 curies.

So devastating was the Fukushima disaster that the literary community started picking up their collective pens. They produced novels such as *Horses, Horses, in the End the Light Remains Pure: A Tale That Begins with Fukushima,* which won UNESCO's Noma Literary Prize and was the first major novel in response to the nuclear meltdown that so rocked northeast Honshu, Japan. There are many other books in the same vein, including *Fukushima's Stolen Lives: A Dairy Farmer's Story; Reverberations from Fukushima: 50 Japanese Poets Speak Out;* and *Facing the Wave.*[34]

While fiction and non-fiction help bind society and energy, poetry adds a closely related strand. As a literary genre, poetry tends to be minutely crafted, shorter, and sharper. As a literary art form, it is a medium preferred for the communication of human emotions, and as such, it provides a window on the human costs of meeting our energy needs. Given coal's long history of use, in poetry too it is the most common resource depicted in accounts of energy, even appearing within one of the poems by the literary giant W. H. Auden. In this excerpt from "The Watershed," the plight of coal miners holds central focus:[35]

> . . . two there were
> Cleaned out a damaged shaft by hand, clutching
> The winch a gale would tear them from; one died
> During a storm, the fells impassable,
> Not at his village, but in wooden shape
> Through long abandoned levels nosed his way
> And in his final valley went to ground.

In discussing this poem, Marsha Bryant reminds us yet again that coal mining is a risky profession for the few who work to meet the energy needs of the many:

> All three of the workers that Auden describes died in the mines. The "wooden shape" in "long abandoned levels" not only means a coffin . . . but also the support beams that once held the tunnel ceilings in place. Auden's death scenes . . . point to the dangers the miners faced at work.[36]

Those in the 1930s and 1940s reading Auden in England's non-industrial south learned of activities and lifestyles in the Black Counties that would otherwise seldom penetrate the genteel fascia of their lives. The north-south divide that reigned in Auden's day—and before, of course—carries into the present day, even as coal mining in England has ceased almost entirely. This divide was demonstrated most recently by the public vote in Britain in 2017 that favored leaving the European Union. The tide turned toward exiting the EU largely because of voters in the working-class north who harbored lingering resentments against the elite who used the energy the miners produced without dirtying their hands.

The human costs of coal mining are evident wherever and whenever the resource has been transferred from mines to millions. It has been no different in the United States. Even a mite of research would uncover stories of people risking life and limb, earning little money, suffering debilitating diseases, living in ramshackle houses, and enduring constant worry that their livelihood could vanish at any time, upended by technological improvements, stricter environmental regulations, or the whims of politicians and corporate management (Figure 4.18).

These conditions are depicted in movies, novels, and reportage, as mentioned earlier, but they are also evident in the poetry coming out of Appalachia. The 157 poems in a collection titled *Coal: A Poetry Anthology* are presented in groups whose titles describe the stresses and sacrifices of coal miners: "Miners and Work," "Disasters and Mining," "Families and Community," "Life After the Mines," "Environmental Degradation," and "Resistance."[37] In a review of this collection, Gurney Norman remarked: "As soon as I read the first poems, I felt the power of the words coming from the heart of the nation."[38] His was a feeling of empathy for the miners who had been dedicating their lives to keeping the country running.

Among the many writers in the anthology who chronicled living conditions, the brief poem by Jason Frye exposes the raw tragedy of the lives lost along Buffalo Creek, West Virginia, in February 1972 when a wall of water and mine waste swept away, in a few minutes, more than 130 people in the dark of night (Figure 4.19).[39]

Figure 4.18 Modest homes of coal miners in a hollow near Neon, Kentucky.

Figure 4.19 Memorial to those who perished along Buffalo Creek, West Virginia, after an upstream coal mine impoundment failed and sent a torrent of water and debris downstream, wiping out most of the town.

Buffalo Creek

The river belches out its dead, and they flood
the banks with dragging feet asking *What happened to us?*
What happened? It was morning and time
for breakfast. They gather in puddles outside your window
and listen closely for the *hmm-click* of the burner
as you cook your morning eggs. They gather in the yard
behind the church and in the Foodland parking lot. They gather
in the fog and rockdust and orange pall of tipple lights.
They gather on the banks from Kistler to Chapmanville
to Salt Rock and the Ohio and ask *What happened?*
It was morning and time for breakfast.

The attention poets gave the physical, social, and natural costs of coal—which are easily recognized and largely confined—eventually widened to include nuclear power, whose costs are more difficult to detect, spatially more extensive, and not always apparent until years into the future. Recently, fifty poems conveying the human costs of nuclear power were published as *Reverberations from Fukushima*. Those who read these poems cannot help consider the lie of the "peaceful atom," a Faustian bargain ignorantly accepted when we repurposed the technology of warfare to become the technology of electricity. However, the poets this time were expressing not fears of atomic bombs but fears of electrical generation:

The Pollution of Our Ancestral Land

We picked edible wild plants in spring
and mushrooms in autumn.
But now it has approached
our mountains.

We can neither see it, nor hear it,
And it does not make any sound.

It comes suddenly on the wind,
falling on the earth,
infiltrating the soil in a flash,
lurking beneath the feet of animals,
the foretaste of an opaque death.

The mountains of our ancestral land are mounting.
Their angry voices ripple through the trees.
Leaves fall on the earth and sink in the water
accompanied by the melody of a koto.

Our ancestral land will become a wasteland.
These mountains will refuse even woodsmen,
now that no one is taking care of them.

Without saying anything,
these mountains cut a scar into the page of our future.
This demands a reply
from us haughty humans:

What in the world
have you done?[40]

Music

Music has been a consistent and potent truth-telling medium through which to express the emotions evoked by energy development. Owing to its widespread and lengthy use, coal again received the majority of the attention from songwriters. One of the most celebrated of these composers was Woody Guthrie. While he was most famous for his song "This Land Is Your Land," he was also well-known for protest songs that championed the cause of the dispossessed, disregarded, and dismissed. Such was the theme of Guthrie's sorrowful 1949 song "Ludlow Massacre," which memorialized the loss of at least twenty-six lives in Ludlow, Colorado, at the end of a fourteen-month coal miners' strike in 1914.[41] The Colorado National Guard and company guards strafed and burned the tent colony of 1,200 striking coal miners and their families (Figure 4.20).

Figure 4.20 "Ludlow, Colorado, 1914," by John Sloan. Sloan memorialized the bloodbath by depicting a miner, gun in hand, firing back at the Guardsmen who had murdered his family. Cover for *Masses*, June 1914. Dartmouth College collection, a gift of John and Helen Farr Sloan, courtesy of the Trustees of Dartmouth College.
Source: http://art-for-a-change.com/blog/2008/02/apostles-of-ugliness-100-years-later.html.

The United Mine Workers of America (UMWA) organized the Ludlow strike against three companies: the Colorado Fuel & Iron Company, owned by the Rockefeller family; the Rocky Mountain Fuel Company; and the Victor-American Fuel Company. The publicity about the carnage at Ludlow played a significant role in strengthening the hand of the UMWA at coal mining sites throughout the United States.

Ludlow Massacre

It was early springtime when the strike was on,
They drove us miners out of doors,
Out from the houses that the Company owned,
We moved into tents up at old Ludlow.

I was worried bad about my children,
Soldiers guarding the railroad bridge,
Every once in a while a bullet would fly,
Kick up gravel under my feet.

We were so afraid you would kill our children,
We dug us a cave that was seven foot deep,
Carried our young ones and pregnant women
Down inside the cave to sleep.

That very night your soldiers waited,
Until all us miners were asleep,
You snuck around our little tent town,
Soaked our tents with your kerosene.

You struck a match and in the blaze that started,
You pulled the triggers of your gatling guns,
I made a run for the children but the fire wall stopped me.
Thirteen children died from your guns.

I carried my blanket to a wire fence corner,
Watched the fire till the blaze died down,
I helped some people drag their belongings,
While your bullets killed us all around.

I never will forget the look on the faces
Of the men and women that awful day,
When we stood around to preach their funerals,
And lay the corpses of the dead away.

We told the Colorado Governor to call the President,
Tell him to call off his National Guard,
But the National Guard belonged to the Governor,
So he didn't try so very hard.

Our women from Trinidad they hauled some potatoes,
Up to Walsenburg in a little cart,
They sold their potatoes and brought some guns back,
And they put a gun in every hand.

The state soldiers jumped us in a wire fence corners,
They did not know we had these guns,

And the Red-neck Miners mowed down these troopers,
You should have seen those poor boys run.

We took some cement and walled that cave up,
Where you killed these thirteen children inside,
I said, "God bless the Mine Workers' Union,"
And then I hung my head and cried.

Woody Guthrie, copyright © BMG Rights Management US, LLC

Another composition, this one penned by John Prine in 1971, wails against the conditions braved by those who provide coal to power plants. Many recording artists have covered his song, which must be considered evidence of how well it captures the mood of the region and the human costs of coal. The power plant in the song carried the name Paradise, and it is located in western Kentucky. It was the largest coal-fired power in the world at one time. Commissioned in 1963, its unslacking thirst for a river of coal transformed the nearby communities, and not just in terms of its contribution to economic development. In 2009, it emitted 10 million pounds of toxic chemicals, part of the 58 million pounds of hazardous chemicals that impact the health of all people living in western Kentucky and southwestern Indiana. This and other effects prompted Prine to write the song.[42] In so doing, he brought the dirtier images of coal to the attention of millions of listeners, a prime example of the power of the arts to make the environmental costs of energy vivid. "Peabody" in the lyrics refers to the Peabody coal company (Figure 4.21).

Figure 4.21 Tennessee Valley Authority's Paradise power plant near Central City, Kentucky, spewed forth nearly 10 million pounds of toxic chemicals in 2009. By 2017, it had been replaced by a combined-cycle natural gas power plant. *Source:* Photo by John Blair. Used with permission.

Paradise

When I was a child my family would travel
Down to Western Kentucky where my parents were born
And there's a backwards old town that's often remembered
So many times that my memories are worn

[Chorus]
And daddy won't you take me back to Muhlenberg County
Down by the Green River where Paradise lay
Well, I'm sorry my son, but you're too late in asking
Mister Peabody's coal train has hauled it away

Well, sometimes we'd travel right down the Green River
To the abandoned old prison down by Airdrie Hill
Where the air smelled like snakes and we'd shoot with our pistols
But empty pop bottles was all we would kill
[Chorus]

Then the coal company came with the world's largest shovel
And they tortured the timber and stripped all the land
Well, they dug for their coal till the land was forsaken
Then they wrote it all down as the progress of man
[Chorus]

When I die let my ashes float down the Green River
Let my soul roll on up to the Rochester dam
I'll be halfway to Heaven with Paradise waitin'
Just five miles away from wherever I am
[Chorus]

John Prine, copyright © Warner/Chappell Music, Inc.

While "Ludlow Massacre" and "Paradise" tend to attract audiences within folk and country communities of music, it was another song about coal mining called that was to become a cross-cutting national hit. "Sixteen Tons," written in 1947 by Merle Travis, has been recorded by artists as diverse as LeAnn Rimes, ZZ Top, and even the Red Army Choir, but the version that made it famous was performed by "Tennessee" Ernie Ford. Ford provided the cadence and the rhythm that recalls the sound of a hammer striking rock.[43]

16 Tons

Some people say a man is made out of mud
A poor man's made out of muscle and blood

Muscle and blood, skin and bones . . .
A mind that's weak and a back that's strong

[Chorus]
You load sixteen tons, and what do you get?
another day older and deeper in debt
St. Peter, don't you call me, 'cause I can't go
I owe my soul to the company store

I was born one mornin' and the sun didn't shine
I picked up my shovel and I walked to the mine
I loaded sixteen tons of number nine coal and
the straw boss said, "well bless my soul!"
[Chorus]

I was born one mornin' it was drizzlin' rain
fightin' and trouble are my middle name
I was raised in a cane-brake by an old mama lion
can't no high-toned woman make me walk no line
[Chorus]

If you see me comin', better step aside
A lot of men didn't, a lot of men died
One fist of iron, the other of steel
If the right one don't get you, then the left one will
[Chorus]

Merle Travis

While these and many other songs are fifty years old or more, the impacts of
coal mining that stimulated them have disappeared neither from our memory
nor from our national heritage. Indeed, they are still absorbed and vividly
remembered, especially in the coal country of Appalachia. As if in recognition
of the everlasting human and environmental cost of coal mining, Julia Wolfe recently was awarded the 2015 Pulitzer Prize in Music for her oratorio for choir
and chamber ensemble, *Anthracite Fields*. A commission of the Mendelssohn
Club with contributions from New Music USA, the work was first performed
in Philadelphia on April 26, 2014. The oratorio commemorates the history of
the northeastern Pennsylvania coal fields in what the Pulitzer Prize citation
described as "a powerful oratorio . . . evoking Pennsylvania coal-mining life
around the turn of the 20th Century."[44] Music critic Mark Swed of the *Los Angeles
Times* lauded the composition as "an unforgettably haunting, harrowing evo-
cation of the plight of Pennsylvania's coal miners, incorporating many musical
styles and effectively shadowy visuals."[45]

Compositions about other energy resources continue. One of these is about oil and gas extraction, called "The Fracking Song."[46] Others have focused on nuclear power. Noriko Manabe's recently published book *The Revolution Will Not Be Televised: Protest Music After Fukushima* describes several dozen compositions and performances that, for the most part, mock nuclear power and those who manage and promote it.[47]

Fukushima, as bad as it was, was not the first accident to be referred to in musical form; Chernobyl preceded it by twenty-five years. The "cultural impact of the Chernobyl disaster" includes films, books, and just about every other art form.[48] There are numerous songs, including ones by Paul Simon and David Bowie, but perhaps the best-known song, called "Ghost Town," was hauntingly performed by Huns and Dr. Beeker in 2005.[49] The title refers to the town of Pripyat, evacuated three days after the accident and—despite what the hastily gathered residents were assured—was never to be reoccupied.

Ghost Town (Chernobyl)

There was no way they could have known that morning
That they awoke upon a fateful day
The killer wind came down without a warning
And no one had the chance to get away
The firemen were brave, they fought with honor
But the blaze was more than it appeared to be
And one by one they fell beside their comrades
Victims of a foe they could not see

[Chorus]
Mama, where are you? Papa, where did you go?
And where are all the children who used to play here?
Only Heaven knows

What they saw defied all explanation
Someone said the trees were glowing red
They say the light came from the radiation
But maybe it's the spirits of the dead . . .
[Chorus]

Gone the homes, the gardens, and the playgrounds
Gone the souls who made their livings here
They say this place will always be a ghost town
It will be for at least six hundred years
[Chorus]

Huns and Dr. Beeker

Final Thoughts

The arts enhance our lives and stimulate our imagination. We use them to communicate what we consider important, to navigate our emotions, and to bring attention to events to which we we wish to award longevity. They provide outlets for our thoughts, our reactions, our misgivings, our worries, our passions, and our need to commiserate with others. Energy themes have long been represented in the arts in every possible form, especially whenever we encounter events and ongoing practices that are particularly alarming and threatening, as we have with the Chernobyl accident. They are in black and white, in color, in music and theater, in prose and verse, in two dimensions and three, in still life and moving images, in ephemera and perpetuity. Energy is everywhere in the arts because energy is everywhere in our lives. The arts are the bonds between energy and us.

5

Environment

Natural Costs of Energy

Energy and the Environment

Even if your days are not devoted to following such things, you can surely name some environmental impacts of energy demand. It is getting simple to do this because they have become so numerous and well publicized. For your convenience, here's a starter list:

- Stronger storms
- Melting glaciers
- Extirpated species
- Inundated coasts
- Damaged reefs
- Radioactive fallout
- Flooded canyons
- Smoggy skies
- Altered ecologies
- Forced migrations
- Melting permafrost
- Relocated villages
- Dying forests
- Acidic rivers
- Oil-stained beaches
- Desertification
- Mine-scarred hillsides
- Mine fires
- Shrinking wilderness

As you see, the list is long, and its implications are becoming increasingly menacing to all of us. It was not always so. There was a time when the impacts of energy development and use did not bother us much—or, more likely, we thought we could not do much about them anyway. Perhaps, for the most part, people were less aware of the accumulating price that was being exacted from nature.

The Thread of Energy. Martin J. Pasqualetti, Oxford University Press. © Oxford University Press 2021.
DOI: 10.1093/oso/9780199394807.003.0005

There were still places on the planet where you could escape the environmental degradation and revel in pristine nature. Even in the last quarter of the nineteenth century, John Muir could walk from San Francisco to the Sierra Nevada and wonder at the native wildflowers and untouched mountain landscapes. Experiencing anything close to that experience today is difficult to imagine now that an additional 6 billion of us are using the same water, land, and air that has always been here.

Yet it was not that people were ignorant of the damage they were doing to the environment. In 450 BCE Herodotus described dismal deforestation in sub-Saharan Africa.[1] At the time, what he observed was considered an unpleasant but unavoidable consequence of human need. What finally changed in recent years is that there is now a broader and deeper understanding that environmental damage is becoming both consequential and more visible, that scientific study strongly validates our perception of the damage that is being done, and that the phenomenon is more widely understood by private citizens and their representatives. It has become increasingly apparent that delinquent and intentional ignorance of consequences is no longer a sensible, reasonable, or justifiable stance. Although there is a growing intolerance for this damage, it has taken quite a while for us to admit we have a problem. Only a generation or two past, the majority of us living on the "third rock from the sun" were unwilling (or unable) to give these impacts much serious attention. Even for those who noticed what was happening, critical levels of understanding, development of appropriate tools, and sufficient political will to fix things were lacking. We had a lot of other things on our mind.

Such wide-scale disinterest was understandable. It is safe to say that for just about everyone, not only was there little recognition that anything *should* be done, there was little awareness that anything useful *could* be done. Moreover, more important day-to-day things to worry about prevailed, such as personal security and where the next meal was coming from. Today, there is a different reality. The accumulating environmental penalties of our energy demand have overtaken us. Impacts we could once ignore are now right in front of us. We cannot escape the fact that the accumulating momentum of energy demand is regularly bumping against the unforgiving limits of environmental recuperation of air, water, and land.

The Air We Breathe

Health Effects

As soon as people started clustering in cities, the wastes they produce started concentrating there as well. It is one of the drawbacks of urban lifestyles. Why is that? No process is 100 percent efficient. Waste is inevitable. There is no avoiding it. Animals produce waste. Plants produce waste. Humans produce waste. Cars, trucks, trains, ships, planes, and factories emit poisonous gases and particulates.

If you find American cities today chaotic and filthy places, they would seem sparkling clean compared to certain cities elsewhere or cities in centuries past.[2] Travel back in time to the preindustrial cities of Ur, Jericho, Mohenjo Daro, or even to the industrial eighteenth-century cities of Britain, France, or China, and you would encounter in all of them open sewers, trash-lined streets, and billowing black smoke from cooking stoves and warming fires. In most major cities of the world today, conditions are generally much better, although there are plenty of exceptions.

Many of the environmental changes from our use of energy were apparent early on. They included visibility changes, labored breathing, and various odors. These changes began thousands of years ago, and they accelerated as soon as humans learned to control fire. For many reasons, however, it is unlikely that they were considered insurmountable or intolerable problems. After all, population densities were low, and people lived in relatively small and mobile groups.

Once a more settled life evolved, however, population clusters grew in size, wood became the common fuel, and soot and odors lingered like an overstaying house guest. Eventually, as large urban concentrations developed, concentrations of pollution soon followed. The Roman philosopher Seneca in the first century CE, for example, described what he called "heavy air" in Rome. It was a reference to the air pollution that plagued that city.

The advent of coal burning only worsened the problem of urban air pollution. Consider medieval England. One of the first recorded instances we have of air pollution caused by coal comes from King Edward's mother, Queen Eleanor. The queen was made so sick by the coal fumes wafting up from the town below that she had to flee Nottingham Castle.[3] Reacting to the darkened and fouled British skies of 1306, the English king, Edward I, banned the burning of "sea coal" in artisan furnaces. He even imposed a penalty of death for burning coal, although the practice continued out of necessity nonetheless. Fast-forward 250 years and we find Queen Elizabeth I recognizing the need to take action to relieve some of the smoggy conditions in London. She banned coal burning while Parliament was in session.

Despite royal decrees, ill health rose. The culprit was coal, and everyone knew it. Nonetheless, there was little that could be done about it more than temporarily. Gradually, London became darker and darker. Visitors like Claude Monet noticed, as illustrated in his "Parliament Series."[4]

Between the 1840s and the mid-twentieth century, conditions in London worsened from bad to intolerable, as everyone in the rapidly growing city burned coal to stay warm. Noxious smoke from millions of chimney pots was polluting

London's air, dirtying residents' clothes, smothering vegetation, and staining the walls of all the buildings. So notorious did London's air quality become that references to it often seeped into the literature of the time, including that of Charles Dickens, and it was said to have emboldened criminal activity. For example, women found that venturing along London's streets and into its parks could be risky. Consider, for example, how London's sulfurous "fog" is believed to have cloaked the movements of the notorious Jack the Ripper.[5]

This fog was also becoming increasingly deadly, through a variety of pathways. Every week that a smog episode lingered, hundreds of people died of lung infections. More than 11,000 Londoners died of bronchitis in 1886 alone.[6] About 4,000 died in 1952 after a three-day smog episode, with an additional 12,000 deaths resulting from the long-term effects. It was this event that finally sparked the beginning of public environmental responsibility and research into the intertwined conditions that helped produce the incidents in the first place, including thermal inversions and the local topography (Figure 5.1). After that, coal burning for heating was banned in England.

With the curtailment of the most apparent major source of air pollution, it became reasonable to clean the centuries of grime from the stone facings of Parliament and other London buildings (Figure 5.2). Similar cleanups were undertaken in other European cities such as Rome, where the Colosseum has been meticulously scrubbed clean with every possible implement and tool, including toothbrushes.[7]

Despite such actions, coal use continued to expand. By the early decades of the twentieth century, it was widely used in power plants to generate electricity. Partly this choice was a function of inadequate availability of alternative fuels, and partly it resulted from the strong role coal played in continued economic development.[8] Although the United Kingdom is closing most of its coal-fired power plants, the human cost is mounting elsewhere. For example, it is estimated that

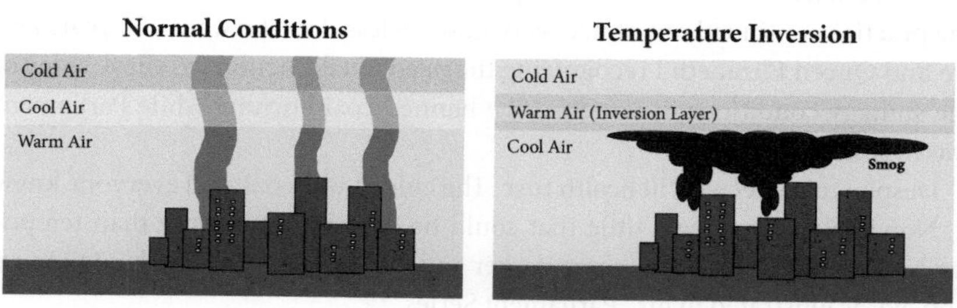

Figure 5.1 When an inversion occurs, pollutants are trapped and concentrated, producing unhealthy conditions.
Source: http://geographygems.blogspot.com/2011/09/smog.html.

(a)   (b)

Figure 5.2 Before and after cleaning the pollution stains from coal burning off Elizabeth Tower, London, affectionally known as Big Ben.

coal pollution killed 366,000 people in China in 2013.[9] In India, "the breakneck pace of industrialization is causing a public health crisis with 80,000–120,000 premature deaths and 20 million new asthma cases a year due to air pollution from coal power plants."[10]

Visibility

While many of the ingredients in coal make burning it harmful to humans and other living things, it is often a reduction in air clarity that first attracts public attention.[11] The most remarkable impact of Beijing's air pollution problem, for example, is how it restricts visibility. The city has become notorious for its poor air quality, and people around the world are shocked by photographic evidence of the problem.

Such alarming loss of visibility is not confined to China. Consider the following pair of photographs taken on different days from the same spot in San

Francisco (Figures 5.3a 5.3b). If you saw just the one on the top, you might not even be aware of the existence of San Francisco Bay! Days like that, especially in such picture-postcard places, provoke citizens and their political representatives to action.

For over a century, visitors to western U.S. states such as Arizona, Utah, and Wyoming have marveled at the clarity of the skies. It is the same area of the country that hosts one of the world's great accumulation of scenic treasures, including Grand Canyon, Zion, Arches, and Yellowstone National Parks. For this reason, it is not surprising that it was here that strong voices of concern and protest demanded action.

Finally awakened, the U.S. Congress gave air quality there (and elsewhere) some official protection with the passage of the Clean Air Act in 1970.[12] The provisions of the act have provided the authority to require cleaner emissions from regional coal-burning power plants. Such view-erasing emissions led to mandatory improvements in emission control at the Four Corners power plant in New Mexico and the Navajo Generating Station (NGS) in Arizona (Figures 5.4 and 5.5).

Whereas deteriorating visibility can attract public derision and regulatory action—especially in sensitive habitats—early remedial measures often focused on local improvements, with little attention to impacts further away. This approach was apparent in the early days of pollution control, when utility companies responded to public and official complaints by installing taller chimneys at their power plants. The leaders of these companies reasoned that the faster-moving air currents found aloft would help disperse air pollutants (Figure 5.6). It was a successful strategy, to a point.

Evidence that taller chimneys would not be the ultimate solution came to public attention in the last third of the twentieth century. Although the stronger winds indeed dispersed emissions from the taller chimneys, this approach created unexpected new problems downwind when the emissions affected forests and lakes in the Adirondack Mountains of upper New York State. In the 1970s, older campers who remembered that the croaking of bullfrogs used to keep them awake at night in past years noticed that there no longer seemed to be any noise of bullfrogs at all. Everything was quiet. What had happened? Where were the frogs? Testing eventually discovered that the trees, soil, and water around their traditional camping spots had become acidified.

That frogs could not survive such acidification was just the beginning of our recognition of the problem. Subsequent studies revealed that very little else was living in the water either, other than acid-tolerant moss. With acidity identified as the cause of the ecological disruption, a related question quickly emerged: there were no significant industrial polluters nearby, so where was the pollution coming from? Eventually, scientists discovered that the source of the

(a)

(b)

Figures 5.3a and 5.3b Looking north of Twin Peaks in central San Francisco, the same view on different days. Comparing the two illustrates that air pollution has an aesthetic as well as a health cost.

Figure 5.4 The coal-burning Navajo Generating Station in northern Arizona. This photo was taken just prior to the demolition of the original (narrower) chimneys. The original chimneys could not accommodate the greater air volumes produced from the new sulfur-removal equipment that was installed to improve air quality over nearby Grand Canyon. It was closed in 2019. As of July 2021, most of the power plant had been demolished.

Figure 5.5 The Vermillion Cliffs and Marble Canyon, an area whose visibility improved after installation of new emission control devices on the Navajo Generating Station. It will improve more now that the NGS has closed.

Figure 5.6 Taller chimneys replaced the shorter chimneys (to the right) at the Kingston coal-burning power plant in Tennessee in an effort to disperse emissions into faster-moving air and improve local air quality.

problem was the large cluster of coal-burning power plants hundreds of miles away, mainly in the Ohio River valley. They learned that emissions of sulfur dioxide and oxides of nitrogen, if they stayed aloft even a few days, could chemically change to produce sulfuric and nitric acid (Figure 5.7). By then, the wind had carried the culprits to the Adirondacks.

The publicity around the deteriorating ecological quality in the Adirondacks gave a needed lift of recognition to the fact that energy demand and environmental quality are intimately entangled. Most specifically, the circumstances in the Adirondacks underscore that solving one problem can create another problem, and the newly created problem might be more severe than the original one. In this particular case, hundreds of miles from the source of the emissions, in areas otherwise wild and trackless, trees were being stressed and mortally damaged, heavy metals were being leached from the soils in surface and groundwater, lakes were being rendered lifeless, and an entire ecological web was on the verge of collapse. Thus began a new era of concern, especially about burning coal. The popular media called the problem "acid rain."

Once coal burning in the Ohio River valley was convincingly tied to acidification in the Adirondacks, comparable links were documented elsewhere. In the rural town of Pitlochry, Scotland, pH levels in the precipitation dropped to that of vinegar. Lakes in southern Norway and Sweden started dying. Acid-tolerant

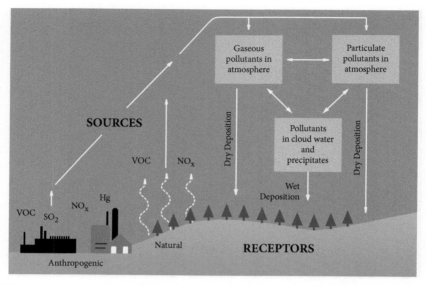

Figure 5.7 A typical pathway for the creation of acid rain: (1) Emissions of SO_2 and NOx are released into the air. (2) The pollutants are transformed into acid particles that may be transported long distances. (3) At some point downwind, these acid particles fall back to earth as wet and dry deposition (dust, rain, snow, etc.). (4) They may cause harmful effects on soil, forests, streams and lakes.
Source: https://www.epa.gov/acidrain/what-acid-rain.

grasses were displacing forests between Dresden and Prague. Worried residents near the cherished Black Forest of Germany began tacking skull-and-crossbones signs bearing the words *hilfen mir* (help me) to their beloved trees. Pastoral, often highly valued places all over Europe were being victimized by continued reliance on coal-fired electrical generation located in distant countries. Soon, politicians and private citizens were complaining that they were suffering the effects of pollution they had had no role in creating. Acid rain had become an international issue. So alarming were the stories that the term "acid rain" soon appeared as part of pop culture; a 2020 search of a lyrics website finds the phrase appearing in 306 lyrics from forty-nine artists.[13] Acid rain confirmed that one form of air pollution could morph into a form that is even more damaging.

Sulfur oxide emissions are only one of the problems of coal-burning power plants. Another has to do with the need to control the toxic fly ash that lands on soils and trees in the immediate neighborhood. However, as soon as devices were installed at the power plants to capture the fly ash before it reached the chimneys, a new question emerged: how to safely discard the collected ash. Once again, solving one problem only produced another problem. Typically,

Figure 5.8 Pollution control devices that capture fly ash at coal-burning power plants must be discarded somewhere nearby. This is a dedicated impoundment for fly ash disposal at the Ghent coal power plant, along the Ohio River, Kentucky.

utility companies truck the ash as a solid or pipe it as a slurry to a dump site nearby (Figure 5.8). While this seems like a reasoned approach, it too is not fail-safe. On December 22, 2008, an ash disposal dike broke at the Tennessee Valley Authority's Kingston Fossil Plant in Harriman, Tennessee. The resulting spill of 5.4 million cubic yards of coal ash sludge covered nearly 300 acres, killed wildlife, polluted an extensive area, and damaged forty nearby homes, some irretrievably. The cost of cleanup from this event has exceeded $1 billion.

For as long as utility companies have been capturing fly ash, they have been trying to identify a solution other than just merely piling it up somewhere. It may be that they have now found a solution. In some locations, especially ones that were not too remote, it proved profitable for coal plants to sell the fly ash for use in manufacturing concrete blocks. It has been a particularly attractive solution, as such blocks are not only lighter in weight than conventional blocks but stronger. As the adage goes, "If life gives you lemons, make lemonade."

Responses

As we have become increasingly aware of the impact of our energy demand on air quality, we have also been developing an array of reactions. We are, for example, raising our efforts to increase energy efficiency, to educate the public about the need to provide funds to control emissions, and to pass legislation to codify these realities and concerns.[14]

We have installed pollution control equipment at coal plants to trap fly ash and sulfur dioxide. These devices include electrostatic precipitators and baghouses for fly ash, and flue-gas desulfurization apparatus for sulfur dioxide control. These devices and strategies are welcome, but the coal industry continues to decline steadily in many countries, such as the United States and a number of European countries.

Control of the pollutants nitrogen dioxide, particulates, and sulfur dioxide has reduced emissions from power plants, improving air quality. Nonetheless, other troublesome emissions continue, such as methane, nitrous oxide, and carbon dioxide (Figure 5.9). These potent greenhouse gases contribute to global warming. They have several sources, and we are not yet very successful in controlling them. As for carbon dioxide, one of the big problems we face is that the volume of the emissions is enormous and expensive to control. For example, until it was turned off in November 2019, the Navajo Generating Station in northern Arizona could emit up to 19 million metric tons of CO_2 per year.

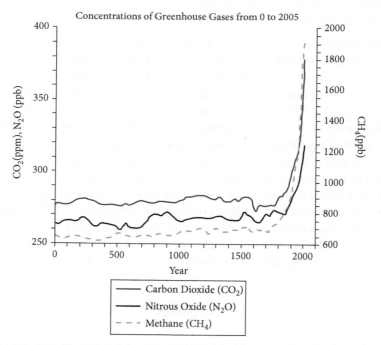

Figure 5.9 Greenhouse gases have been rising precipitously since about 1750. *Source:* https://www.industrytap.com/the-big-picture-breakdown-of-greenhouse-gases/6201.

Several paths exist to reducing the detrimental impacts of energy development on air quality. They fall into three general categories: increased energy efficiency, fuel switching, and further pollution controls. All these will require attention to economics, technology, policy, and public acceptability. These are discussed in some detail in Chapter 6, on climate change. Still, if none of these measures are considered sufficient by the residents being affected, there is another remedy that one would hope is not often replicated. It is an unusual, intriguing, surprising, and drastic approach that demonstrates how dire the impacts of energy development can sometimes get: to abandon the town.

Surprising as it may seem, in the case of Carlisle, Ohio, the least-cost option for the utility company responsible for the pollution was to buy and vacate the entire town. Carlisle sits on the Ohio River close to the Gavin power plant (built in the early 1970s) and the older Kyger Creek plant (operating since 1954). The agreed $20 million price tag for the town was to absolve "the utility from any future liability for damage to the locals' health or their property."[15] Eventually, almost all Carlisle residents moved away.[16]

While Carlisle was abandoned because of air pollution from power plants, cities elsewhere have also been abandoned due to energy activities. Three have received the most attention. The first one is Centralia, Pennsylvania, which had to be evacuated because of a subterranean mine fire that could not be extinguished despite several decades of expensive effort. A second one is Pripyat, Ukraine, which was abandoned after the Chernobyl accident of 1986 and which is slowly being reclaimed by nature. A third instance is the large area downwind of the Fukushima nuclear accident of 2011.

Hundreds of other energy sites around the world have been abandoned to the elements, either because of perceived hazards or because the energy resources have become uneconomic. Fifteen of these are in West Virginia.[17] Most remain in unremediated form, an accumulating reminder of the environmental legacy for the energy we use.

The Water We Drink

Energy is just as entangled with water as it is with air, although the points of intersection are more tangible. Unlike air, water is a substance we can feel, see, and cup in our hands. Over the years of human use, we have learned to utilize the kinetic energy of falling water in many ways, such as for grinding, sawing, lifting, and transportation. More recently, we learned to turn those relationships inside out. That is, we now use extra energy to pump water and to cool machinery. In the past several years, we have started grouping these examples of co-dependency under the rubric "the water/energy nexus." It takes water to make energy, and it takes energy to make water. A few examples will help define this term more fully.

Water Pumping

Perhaps the simplest example of the water/energy nexus is to consider rural water wells. Nowadays, most water wells require pumping. If the water table is shallow, pumping can be accomplished by human muscle. For deeper wells, a wind-driven water pump is sufficient, as was first demonstrated over 1200 years ago in Persia. Later this technology was picked up by people in China, the Middle East, and Europe, culminating when the Dutch dewatered the wetlands of the Netherlands. Such wind-powered devices could lift water several hundred feet in some places.

So successful and sensible were such water pumpers that a great windmill boom in the United States lasted for over fifty years. Between 1880 and 1935, more than 6 million windmills were sold by about twenty manufacturers, for use primarily in the Great Plains. These water pumpers are good early examples of the water/energy nexus; that is, differences in air temperatures and pressures created the wind, then people used the power of the wind to lift water to the surface where it was stored in tanks. In that way, the water amounted to a form of stored solar energy. It was a perfect renewable energy system!

The deeper the water table, the greater the amount of energy needed to bring it to the surface. That growing energy requirement translates into higher monetary costs, which rise rapidly because water is heavy. As water tables dropped, more powerful pumps were needed. Fortunately, electricity became available to rural Great Plains farmers in the 1930s. Rather quickly, the wind-driven water pumpers gave way to pumps operating on electricity or diesel, with 1 gallon of diesel equivalent to 10 kWh of electricity.[18] As a rule of thumb, the electrical energy cost of lifting 1 acre-foot of water 1 foot is 1.02 kWh.[19]

Lifting from greater depths requires even more energy, as does lifting greater volumes of water. Along the Central Arizona Project are fourteen pumping plants called "lift stations." They are used to raise water to a level that allows it to flow by gravity until the next lift station. The first lift on the CAP is the largest. It takes six 60,000-horsepower pumps to lift the water nearly 825 feet out of Lake Havasu into the CAP canal. The average lift of other pumping plants is 150 feet.

For several decades, the Navajo Generating Station provided the electricity needed to pump the CAP water. Located near Page, Arizona, the power plant ceased operation in November 2019. But until then, about 500 MW of its generating capacity served the CAP, which delivers an average of 1.5 million acre-feet of Colorado River water annually, much of it to Tucson, 336 miles away and more than 1900 feet higher than Lake Havasu. Pumping that water uses 2.8 million megawatt-hours of electricity every year, making it the most significant single user of power in Arizona. If it were not being used to pump water, the required power would meet the needs of at least 250,000 suburban homes.[20] To give this number a more solid feel, 500 MW requires burning about 5,000 tons of coal per day. So when everything was running at full capacity, those living in Phoenix and Tucson were responsible for burning 10 million pounds of coal each day to provide themselves with water. That means that every year that the Navajo Generating Station was supplying water to central and southern Arizona, it was degrading the air quality in northern Arizona, hundreds of miles distant.

Condenser Cooling

The ties between the Navajo Generating Station and the CAP are but one illustration of the complex relationships between water and energy. For example, generating electricity at thermal power plants such as NGS requires continuous supplies of water to condense the steam as it exits the turbine. The amount that is needed is inversely related to the efficiency of the power plant: when the efficiency goes up, the amount of water required goes down. A coal-fired power plant such as NGS evaporates 500 gallons in its cooling towers for each megawatt-hour generated (Table 5.1). That means that at full throttle, the NGS evaporated into the atmosphere 27 million gallons of water each day, or the yearly water consumption of about 150 homes in the Phoenix area. Of course, pumping the water out of Lake Powell itself requires electricity, and a large pump meets that need. So, as you realize, this system has a rather circular nature: the power plant needs water to generate electricity efficiently, and some of the electricity is used to pump the water to the cooling towers that, in turn, are necessary to allow the power plant to run efficiently so that it can supply the electricity to pump the CAP water to central and southern Arizona. This circular arrangement is one example of the water/energy nexus.

Once-through cooling is an approach that withdraws water from the ocean, a river, or a lake. It evaporates less water but requires higher volumes (i.e., 15,000 gallons/MWh for once-through cooling vs. 500 gallons/MWh for cooling towers). Instead of evaporating the water, the water is returned to its source at a higher temperature than when it was withdrawn. As oceans are the largest heat sink, power plants tend to concentrate along such shorelines.

Several ecological changes may result when cooling water is used in a power plant. In the case of once-through cooling, the form and severity of the impacts themselves depend on factors such as the temperature difference between the discharged water and the host water, the volume of the water body receiving the

Table 5.1 Water Consumption Factors for Non-Renewable Technologies (gal/MWh)

Fuel Type	Cooling	Technology	Median	Min	Max
Nuclear	Tower	Generic	672	581	845
	Once-Through	Generic	269	100	400
	Pond	Generic	610	560	720
Natural Gas	Tower	Combined Cycle	198	130	300
		Steam	826	662	1170
		Combined Cycle with CCS*	378	378	378
	Once-Through	Combined Cycle	100	20	100
		Steam	240	95	291
	Pond	Combined Cycle	240	240	240
	Dry	Combined Cycle	2	0	4
	Inlet	Steam	340	80	600
Coal	Tower	Generic	687	480	1100
		Subcritical	471	394	664
		Supercritical	493	458	594
		IGCC	372	318	439
		Subcritical with CCS	942	942	942
		Supercritical with CCS	846	846	846
		IGCC with CCS	540	522	558
	Once-Through	Generic	250	100	317
		Subcritical	113	71	138
		Supercritical	103	64	124
	Pond	Generic	545	300	700
		Subcritical	779	737	804
		Supercritical	42	4	64

Source: Jordan Macknick et al., "Review of Operational Water Consumption and Withdrawal Factors for Electricity Generating Technologies," NREL/TP-6A20-50900, 2011, National Renewable Energy Laboratory, Golden, CO. * Carbon Capture and Storage

discharged water, the resulting range of temperatures produced, the species in the affected water, and a number of other considerations. Considering these many factors individually and collectively, the ecological changes that result from cooling-water needs may range from trivial to deadly.

Each case warrants individual evaluation, but the impact on aquatic creatures is settled science. For fish, such thermal discharges can interfere with migratory patterns, reproductive behavior, vulnerability to disease, swimming speed, and metabolic rates. Any of these impacts, individually or in various combinations, can result in large-scale fish kills. Where cooling towers are in use, the effects on fish can be less if no water is returned to the water source, which was the case at the Navajo Generating Station in Arizona. Operators of NGS had rights to pump up to 34,100 acre-feet per year from Lake Powell.

Acid Mine Drainage

Water pumping and power plant cooling are two of the many examples of how energy and water entwine. Many more ties exist elsewhere in the fuel chain. One example comes from coal mining. Centuries of coal mining have contaminated thousands of miles of surface water in Appalachia and other places, which will have devasting effects on rivers, streams, and aquatic life for hundreds or even thousands of years. This contamination is called acid mine drainage (AMD), and it comes from abandoned coal mines as well as active mining (Figure 5.10a and 5.10b).

What is the source of the acid? It comes from mining ores often rich in sulfide minerals. When the mining process exposes the sulfides to water and air together, they react to form sulfuric acid. So prevalent has been coal use in the mid-Atlantic area that sulfuric acid is currently the primary pollutant of surface water in the region.

Fracking

Oil and natural gas production has boomed in recent years in the United States and several other countries. The primary cause of this boom is the application of refined methods of hydraulic fracturing. Commonly called "fracking," the procedure forces water, small-grain sand, and some proprietary chemicals underground under high pressure. Innumerable tiny cracks created in the host material increase the oil flow. So successful is this technique in the United States that natural gas is at record low prices, liquid natural gas is now exported, and oil is loaded onto tankers for shipment outside the country in an unexpected reversal after decades of net imports.

(a)

(b)

Figures 5.10a and 5.10b Colored water indicating acidified streams resulting from coal mining near Wilkes-Barre, Pennsylvania. See plates for color versions.

Many people consider fracking a genuine environmental threat on several fronts. One concern is the potential for groundwater contamination. In June 2013, scientists published a study in the *Proceedings of the National Academy of Sciences* showing that drinking-water wells within a kilometer of high-volume fracking operations in Pennsylvania had methane concentrations an average of six times higher than in wells farther away. Many examples exist of methane-contaminated water catching fire. In response, public opposition to fracking is mounting.[21]

Fracking causes problems not just with water quality but also with water quantity. For example, the fracking boom in Barnhart, Texas, caused such an increased demand for water for drilling that water for human needs fell short. Looking at the volumes involved in fracking, one can see why. By one estimate, it can take 10,000 cubic meters (about 2,643,000 gallons) of water to drill one gas well.[22] As one newspaper account put it: "The town's wells appear to have run dry because of water being extracted for fracking use. Residents report that water troubles began when fracking began in the area. Farmers have lost livestock and crops, and the town was water-rationed."[23] One owner's livestock mysteriously began having motor-skill breakdowns followed by sudden death. "A veterinarian said the deaths may have been caused by arsenic, high levels of which were found in water on the farmer's property."[24]

In recent years, fracking has also been linked to increased seismic activity, especially in an area of the country not generally associated with earthquakes, Oklahoma. Before 2009, there were, on average, two earthquakes a year in Oklahoma that were magnitude 3.0 or higher. In 2014, there were 907. The cause appears to be the sharp increase in the reinjection of fracking waste-water deep underground, which many believe lubricates geologic faults, causing earthquakes (and which some suspect results in groundwater contamination as well). At the beginning of September 2016, an earthquake with a magnitude of 5.6 was recorded near Ponca City, Oklahoma.[25] So alarmed has the public become that Oklahoma has ordered thirty-seven wells shut down.

Oil Pollution

Although it is well understood that water quality can be compromised by continual thermal discharges, acid drainage, and fracking, occasional accidental oil spills receive the most media attention, especially when they are large and occur near sensitive coastal areas.[26] The most notorious spill in U.S. history and the world's largest accidental spill occurred in 2010 when a blowout 5,000 feet beneath the offshore platform called Deepwater Horizon spewed almost 5 million barrels (about 200 million gallons) of oil into the Gulf of Mexico. Giant slicks closed recreation beaches in five states, killed several people on the rig plus thousands of animals, and forced the shuttering of hundreds of businesses.

The total environmental and economic consequences of this spill have not yet been fully calculated. Still, in 2011 the Center for Biological Diversity estimated that the spill "likely harmed or killed approximately 82,000 birds of 102 species, approximately 6,165 sea turtles, and up to 25,900 marine mammals, including bottlenose dolphins, spinner dolphins, melon-headed whales and sperm whales."[27] The knock-on effect on the local economy has been substantial: "It severely affected the many small businesses that rely on the Gulf, including those dealing with tourists, recreational boating, and those that collect fish, crabs, shrimp, and many other creatures."[28] Thousands of claims for relief, mostly from aggrieved coastal inhabitants, are still pending. The cost of fines, damages, and repairs had already summed to over $62 billion by mid-2016.

The environmental threats of oil spills are not limited to blowouts. Accidental spills occur from all the many modes of oil transport, including trucks, trains, and pipelines on land, and ocean tankers on the sea. Deciding which form of transportation to use is driven by factors such as cost, politics, environmental risk, and the availability of transport equipment. In the United States, 70 percent of crude oil and petroleum products are moved by pipeline, and 23 percent by tankers and barge; trucking accounts for 4 percent of shipments, and rail for just 3 percent. While accidents from sea transport do not result in the greatest volume of spilled oil, tanker accidents have produced some of the best-remembered spills in history. Among these spills is that of the *Exxon Valdez* in 1989, which spilled almost 11 million gallons into the sensitive fishing waters of Alaska's Prince William Sound.[29] It was the largest oil spill in the United States up until that of the Deepwater Horizon.

Although tanker accidents attract substantial media attention because of the visible and long-term consequences they produce, they have become much less common over the years (Figure 5.11). Such welcome reduction resulted from the codification of operational improvements as well as the additional margin of safety that has come from the construction and operation of double-hulled ships, all because of the MARPOL agreement. This agreement, developed by the International Maritime Organization, had the objective of minimizing pollution of the oceans and seas, including from dumping, oil, and air pollution. It entered into force on October 2, 1983. As of January 2018, 156 states are parties to the convention, accounting for 99.42 percent of the world's shipping tonnage.[30]

Given that tanker spills are down and offshore oil rig blowouts are rare, as are railroad derailments, which mode would you recommend as the safest regardless of price? Your response should be something like this: truck is worse than train, which is worse than pipeline, which is worse than tanker.[31]

Accidental spills such as those just described are locally damaging and attract headlines. However, the largest oil spill in history was no accident. It occurred during the closing days of the Gulf War in January 1991 when retreating Iraqi forces purposely opened valves at the Sea Island oil terminal and dumped oil from several tankers into the Persian Gulf. It produced an oil slick 101 miles by

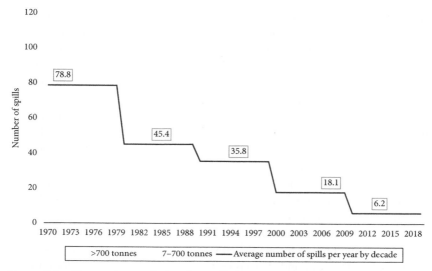

Figure 5.11 The incidence of oil tanker spills has declined substantially in recent years in response to public alarm and industry liability, including improved operating procedures and the use of double-hulled vessels.
Source: http://www.itopf.com/knowledge-resources/data-statistics/statistics/.

42 miles in size, comprising 300 million gallons—enough oil to fuel an average automobile for over 200,000 years.[32] Millions more barrels spilled on land.

Hydroelectric Dams

Hydropower is the leading renewable energy source of electricity in many countries, even as wind and solar power development continues surging. The leading countries for hydropower facilities include China, Canada, Brazil, and the United States, with a global total of 1.21 terawatts of installed capacity in 2018. One reason for their popularity is their multi-level benefits: they allow rapid dispatch while diminishing flooding threats, they convert the kinetic energy of the falling water into electricity with high efficiency, they provide potential recreational opportunities as their reservoirs buffer against dry years, and they are a form of low-carbon power generation. For three decades starting in the 1930s, large dam construction in the United States was popular and public opposition rare. What was not to like?

Investigating that question from a present-day perch reveals numerous threads between hydroelectric dams and the environmental impacts they inevitably create. Along with their advantages, they provide an example of how erstwhile noble goals can also carry unintended consequences. In addition to sometimes creating international watershed conflicts—such as is occurring right now along the Nile—dams plug up rivers, block natural flows, drown canyons, submerge towns, force relocation, flood archeological remnants, trap sediments, accelerate downstream erosion, increase evaporation, alter plant and animal habitat, and block migrating fish.[33]

Through the late 1950s, the United States basked in the admiration that came from our technical prowess in building monumental structures such as Hoover Dam, Grand Coulee Dam, Fort Peck Dam, and many, many others. However, resentment started gaining traction in the 1960s. In large part, such a shift in attitude resulted from the construction of Glen Canyon Dam in northern Arizona. The dam, over 700 feet tall, impounded the Colorado River and created Lake Powell, extending over 180 miles upstream.

The reservoir was named, with some irony, after John Wesley Powell, who led a small band of men through Glen Canyon a few years after the Civil War. Compared to the wild and dangerous river conditions they had found elsewhere on the Colorado River, Glen Canyon offered the party relative calm and unequaled beauty. It held springs, natural amphitheaters, reminders of the vanished Anasazi, alcoves with hanging ferns (the "glens" of Powell's description), and unmatched serenity. It was "the place no one knew," as Elliot Porter called it.[34] Until it was too late.

The deeply held objections that are now tightly laced to the existence and operation of Glen Canyon Dam were personified—even before the dam existed—by David Brower. At the time, Brower headed the Sierra Club, the most influential environmental advocacy organization of the day. His principal objection to the dam was that Lake Powell inundated what many considered the single most naturally scenic and archaeologically valuable canyon in the country. In a personal conversation I had with him in San Francisco in the 1980s, he expressed his sadness that he had not successfully opposed the dam when he had the chance as executive director of the Sierra Club. It was a disappointment that haunted Brower for the last fifty years of his life.[35]

In subsequent years, many others lent their voice in opposition to the continued operation of Glen Canyon Dam. They hoped to drain Lake Powell itself and return the canyon to its preexisting condition. The most vocal proponent of this movement was the writer Edward Abbey. He had become a famous "voice in the wilderness" after publishing *The Monkey Wrench Gang*, a novel about the shenanigans of a small band of anarchists who worked fearlessly to release the river to run free again. The Glen Canyon Institute continues to champion the same idea, using a more science-based approach.[36]

The lessons from the Glen Canyon Dam may be applied to large dams outside the United States in almost every way, except that the creation of Lake Powell forced virtually no one to move. In contrast, large-scale forced relocation has been the tremendous human cost of large hydroelectric dams. In the most notable recent example, the construction of China's Three Gorges Dam on the Yangtze River pushed over 1.5 million people from villages their ancestors had occupied for centuries. Today, many of those displaced people report that their living circumstances are worse than when they were living in the settlements now under the lake.[37] For these people, damming the river had irretrievable human and environmental costs, some unpredicted (Figure 5.12).

The Three Gorges Dam continues to generate as much controversy as it does electricity, but it is not the only such project that is prompting hand-wringing. Other projects of similar grand scale include those on the Romaine River in Canada, the Blue Nile River in Ethiopia, the Narmada River in India, the Euphrates River in Turkey, and the Zambezi River in Zambia. When completed, such projects will have displaced millions of people, submerged many

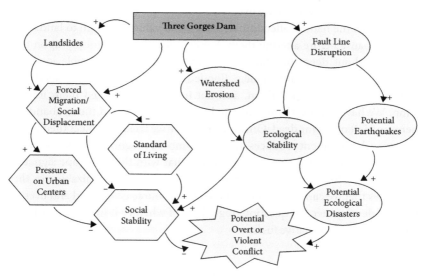

Figure 5.12 A depiction of the impacts of the Three Gorges Dam, China. The circles are ecological issues, and hexagonal shapes are sociological issues.
Source: Blake Campbell-Hyde, ICE Case 254: "Breaking Ground: Environmental and Social Issues of the Three Gorges Dam in China," http://mandalaprojects.com/ice/ice-cases/china-dam-impact.htm.

communities, and forfeited several designated World Heritage Sites.[38] Whether the benefits of irrigation, flood control, and cheap electricity justify the environmental and social costs remains an open question. It is the type of fundamental question that appears once we tie energy development to human costs, many of which are permanent.

The Land We Inhabit

Just as the development of energy can affect the air we breathe and the water we drink, it can also test the integrity of the land we inhabit. The core feature of the land we occupy, farm, and admire is its tangibility. It is what we touch, what we covet, what we aspire to possess, and what we sometimes fight to defend. It is where we build our factories, our museums, our schools, our businesses, our ballfields, our homes, our cities, and our cemeteries. It is the backdrop for our activities, the primary source of our food, and the object of our admiration. We consider land solid, permanent, and durable. We consider its ownership a symbol of wealth. We have a reverence for its beauty and the opportunities it affords us. The land is where we live.

Despite all the natural affection, necessity, and value we assign to landscapes, we rarely pause before altering them. For centuries we have foisted upon them changes without deep thought or noticeable concern for possible long-term consequences. We have mined them, flooded them, planted them, scraped them, contoured them, reshaped them, and not infrequently blasted them into oblivion. We have probed them, exploited them, contaminated them, abandoned them, and left many for dead. Only recently have we tried to protect them.

Energy development produces some of the most-recognized landscapes on the planet. We have, over the years, become familiar with the landscape signatures of the various forms of extraction and processing. So recognizable are these signatures that we can usually pinpoint which resource is involved and even the phase of development in question, regardless of whether it involves fossil fuels, nuclear fuels, or alternative fuels.

Fossil Fuels

Just about everyone can recognize the landscape signature of an oil or gas field. In places like California, Texas, Oklahoma, Kansas, Colorado, North Dakota, and Louisiana, several landscape features stand out. These include tall drilling rigs, nodding pump jacks (Figure 5.13), clusters of storage tanks, a patchwork of drilling pads, and the convoluted plumbing that characterizes refineries (Figure 5.14), recognizable even from 35,000 feet (Figure 5.15).

Figure 5.13 An intensely transformed landscape produced by a concentration of pump jacks in Oildale, California, immediately north of Bakersfield at the southern extreme of the San Joaquin Valley.

Figure 5.14 OMV-Schwachman Oil Refinery along the Danube River, Vienna, Austria, is one of the largest inland refineries in Europe, with a capacity of 175,000 barrels per day. It has a distinct landscape imprint.

Coal mines produce landscape features that are readily recognizable as well, whether open pit, strip (including contour), or mountaintop removal (Figures 5.16, 5.17, 5.18). Even coal mines that have been reclaimed leave their unique imprint on the land (Figure 5.19). For those living among the "coal measures," such energy landscapes are all too familiar.

Figure 5.15 Oil wells in the Permian Basin of West Texas. The familiar and recognizable patchwork leaves a distinct imprint on the land.

Figure 5.16 Wadesville pit, an open-pit anthracite mine, south of Wilkes-Barre, Pennsylvania. Anthracite has high energy density and low waste byproducts, making it a valuable resource for industrial processes and home heating.

Figure 5.17 Cordero Rojo coal strip mine, 25 miles south of Gillette, Wyoming. One of the most productive coal mines in the United States, currently owned but not controlled by the Navajo Nation. It produced about 34 million tons in 2008, dropping to 11 million tons in 2019, according to the US Energy Information Administration.

Figure 5.18 Last remaining house next to mountaintop removal coal mining, Mud, West Virginia.
Source: Photo © Melissa Farlow. Used with permission.

Figure 5.19 Bales of straw trucked in from the Great Plains to be used as part of the ongoing reclamation of mined lands on the Navajo Nation, Black Mesa, Arizona.

Although energy landscapes from traditional fossil fuel development abound, emerging forms of energy landscapes are making their own mark. Each new type brands the land in its own way. Without question, the most extensive new energy landscapes are produced from separating the oil from the oil sands of Alberta, Canada (Figure 5.20). The scale of these landscapes and the related demands on water and air quality have been attracting strong adverse public reactions, not just in Canada but in other places where such resources are known to exist, such as east-central Utah (Figure 5.21).

Power Plant Profiles

Power plants are easy to spot, especially where they impose their industrial profile within an otherwise rural setting. Their appearance often signals the fuel they use and the environmental penalties these fuels carry (Figure 5.22). Imagining a coal-fired power plant, for example, would conjure up images of mines and miners, company towns, unit trains, acid mine drainage, tall chimneys, a limestone supply chain, a fleet of ash-disposal trucks, at least one disposal site, noticeable downstream changes, and reduced visibility.

Figure 5.20 A sample of the one of the most extensive industrial landscapes in the world, produced by oil sands recovery north of Ft. McMurray, Alberta, Canada.

Figure 5.21 A sign erected by protesters camped close to oil sands development in east-central Utah, the only state in the United States with active oil-sand development activities. Their organization's website is http://www.tarsandsresist.org/.

Figure 5.22 Hunter Power Plant is a 1,320 MW coal-fired power station operated by PacifiCorp near Castle Dale, Utah. The Clean Air Task Force estimates that, annually, fine particulate matter (PM2.5) is responsible for many deaths annually, as well as multiple illnesses. Using the most recent emissions data (2016), MSB Energy Associates on behalf of Clean Air Task Force used the Powerplant Impact Estimator (PIE) tool developed by Abt Associates to estimate the death and disease due to coal plant fine particulate matter (PM2.5) in that year. They estimate over 13,000 deaths and tens of thousands of illnesses attributable to fine particle pollution from U.S. coal plant emissions.

Nuclear power plants present a largely different landscape signature (Figure 5.23). The most obvious feature of many such power plants is their dome-shaped containment building. Inside the massive concrete walls of the containment structure lies a devilishly complex system of equipment that is the "business end" of a long chain of steps that includes mining, fuel processing, fabrication plants, and waste disposal sites. Hidden from view are the inherent risks of handling and transporting the fuel, the vexing worry that enriched materials might fall into the wrong hands, and the possibility of an accident that could put the public at risk, even those at substantial remove.

In contrast to the inherent drawbacks of coal and nuclear plants, those fueled by natural gas are comparatively benign, though they too have a unique visual profile (Figure 5.24). Admittedly, they produce emissions of various sorts, and yes, they must rely on an underground natural gas delivery system. Still, their

Figure 5.23 The dome-shaped containment structures are an unmistakable signature of nuclear power plants. These are the three units at the 4,000 MW Palo Verde Generating Station, 50 miles west of downtown Phoenix, Arizona.

Figure 5.24 Arlington Valley is a 577 MW natural-gas-fueled, combined-cycle plant located on a 3,000-acre tract approximately 50 miles west of Phoenix, Arizona, near the Palo Verde hub. Its profile differs from coal and nuclear power plants by having much lower chimneys, no conveyor system, no need for massive containment structures, and a compact configuration.

contribution to greenhouse gases is much smaller than a coal plant's. Plus, they do not require significant arrangements for waste disposal. Because of these characteristics—as well as the public's relatively positive responses to them—siting them near population centers usually attracts little disfavor. If you live in a major city, natural gas power plants are probably nearby.

Derivative Landscapes

As suggested earlier, the development of energy resources produces more than direct landscape impacts. It produces indirect consequences as well. I call these *derivative landscapes,* and you will find that the changes they demonstrate can be profound, even if their origin is not always obvious. If you were to encounter them, would you recognize their origins? Most people would not. It might take a bit more discernment and keener observational skills, but you will get better with practice. Here are just a few examples:

- Forests decimated by acid rain (Figure 5.25)
- Heaps of non-combustible waste materials from coal mining, such as the "culm banks" in Wilkes Barre, Pennsylvania (Figure 5.26), and the small mountains that accumulated over decades north of Essen, Germany (Figures 5.27 and 5.28)
- Damaged artworks (Figure 5.29)
- Pockmarked landscape from subsidence caused by underground mining
- Abandoned communities made uninhabitable by energy activities and danger (Figure 5.30)

Renewable Landscapes

The rising contribution from renewable energy resources is creating a new generation of energy landscapes, ones that are impossible to overlook. A good example is those formed from the development of wind power. Because using wind power to generate electricity is a site-specific activity, the installed turbines must coexist with whatever land use surrounds them. The preexisting land use can be anything from an abandoned oil field to a transportation corridor, from a place where people crowd together to a place revered for the quiet and solitude it offers. Because of wind's low energy density, "wind farms" tend to occupy large areas, thereby transforming landscapes wherever adequate resources are available.

The rapid expansion of wind power is making such landscapes increasingly commonplace. As such, they are teaching us a lesson. Their visibility serves as

Figure 5.25   Forest killed by acid rain in the Jizera Mountains, Czech Republic. Such damaged forests might give way to other more acid-tolerant trees or grass, thereby changing the landscape appearance in such a fundamental manner that its transformed origin might not be recognized.

Source: Lovecz | Wikimedia.org. Public domain. https://www.usgs.gov/media/images/forest-affected-acid-rain. .

a constant reminder that electricity, even if generated carbon-free, does not emerge magically from wall sockets. Instead, their high profile goads us to accept that there is always a price attached to the conveniences that electricity provides. This price has been somewhat of a revelation to those living in cities who are often blissfully unaware where the electricity they use originates or what its production costs the environment.

Although solar installations also have a low energy density, they have a different appearance in several respects. Instead of rising vertically, as do wind turbines, solar modules spread horizontally. So smooth and uniform are they that they can appear like a lake or even the shadow of a cloud when seen from a distance. Of more environmental significance, large solar installations require removing vegetation, reshaping the land, disturbing animal habitat, altering natural drainage, and possibly infringing on or displacing age-old archaeological sites (Figure 5.31). Even in the twenty-first century, where travelers often view desolate stretches of the southwestern deserts as ripe for uncontested solar installation, noteworthy public resistance surfaces. As we begin looking with increasing hope toward solar

Figure 5.26 Piles of accumulated non-combustible materials—locally called "culm banks"—discarded from decades of coal mining activity in Wilkes-Barre, Pennsylvania.

Figure 5.27 The hills on the horizon north of Essen, Germany, were created in large part by the accumulation of the waste materials from many decades of regional coal mining. They are similar in origin to the culm banks of Pennsylvania, but more massive.

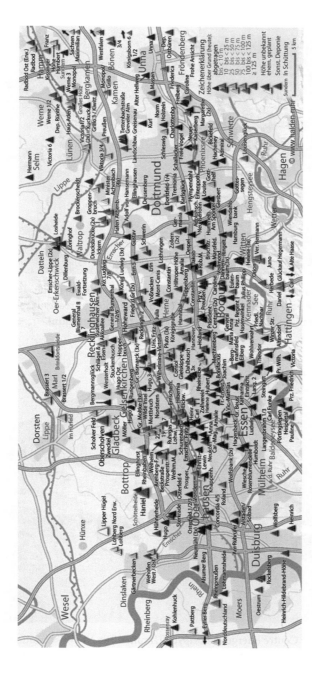

Figure 5.28 A map of the large number of waste dumps north of Essen, Germany between the Ruhr River and the Lippe River. Primarily created by coal mining, they dominate the horizon in every direction. Several disposal sites have been converted into parks and concert venues See plates for color version.

Source: Pictures, maps, text and graphics: S. Hellmann • www.ruhrgebiet-industriekultur.de & www.halden.ruhr.

(a)   (b)

Figure 5.29 German statue from 1704. Two hundred years later, it had the appearance at left. As a result of acid rain, by 1968 it was almost unrecognizable.
Source: http://acidraingermany.weebly.com/acid-rain-in-germany.html.

Figure 5.30 Culture Centre Energetika Pripyat, Ukraine. The city was abandoned in 1986 after being contaminated by material from the explosion and subsequent fire at the Chernobyl nuclear power plant.
Source: Tiia Monto, Wikimedia Commons, CC 3.0, https://commons.wikimedia.org/wiki/File:Pripyat_-_Palace_of_culture.jpg.

Figure 5.31 Solar energy has been touted as an environmentally clean alternative to fossil fuels. The imposing 380 MWe Ivanpah central receiver solar generators in California alongside a desolate stretch of Interstate 15, 45 miles southwest of Las Vegas, Nevada.
Source: Photo by R. Sullivan, Argonne National Laboratory.

energy as a benign alternative to conventional fuels, such reactions prompt us to realize that even places that might appear utterly worthless to the casual observer are often valuable to someone. This is a core lesson we need to absorb as we begin transitioning toward a sustainable energy future.

What Have We Learned?

Now that we have reviewed some of the fundamental ties between energy and environment, what have we learned? One predominant lesson is that "there is no free lunch." Our energy supplies all come from somewhere. There are always environmental costs to developing them. Sometimes these costs are apparent and close, as when pollution belches from power plants. Other times they are subtler and farther away, as when forests hundreds of miles distant suffer damage and possibly die. We have learned that the threads of energy are entangled with every sector of the environment. We have learned that each energy resource holds its own particular assemblage of environmental challenges, risks, dangers, and advantages. We have learned to recognize that environmental costs, while not always apparent, are accumulating everywhere. We have learned that such costs come in a wide variety of forms, that they rarely recognize political boundaries, and that sometimes they are more worrisome collectively than they are individually. We have learned that it has been customary to rely heavily on nature's recuperative powers to shield ourselves from accepting the responsibility for the harm we cause and what we should do to fix it. We have learned that our demand for energy is now producing ecological disruptions and imbalances at all scales, from next door to worldwide. And we have learned that we can and must adjust our traditional understanding of environmental assets from a pattern of linear exploitation to a goal of circular harmony. There is also another, more optimistic viewpoint to consider. It is this: As we turn the corner toward a more sustainable energy future, the impacts of each calorie or kilowatt-hour we use will decrease. Such a cognitive reset will help us survive and prosper in the future.

This reset leads us to an existential question: can we persist without adjusting our present path? Despite the increasing frequency and volume of calls to slow and even reverse present trends, disagreements abound about what should be done. Some believe that the worries over environmental damage are overblown, and that continuing our pattern of energy usage unchanged is possible if we just apply our innate ingenuity to the problems we are encountering. Others argue that moving up the ramp to ever-greater prosperity will lead to irreversible ruin if we do not rededicate ourselves to the principles of sustainability.[39] I take the latter view. The goal of this chapter has been to provide a starting point, a baseline of understanding about the environmental costs of energy to our air, our water, and our land.

6

Climate

Growing Threats of a Warming Planet

Worrying About Climate Change

Are you worried by climate change? If you are not, you must be ignoring newsfeeds, social media, rumors, conversations with friends and neighbors, tens of thousands of news reports, myriads of articles, hundreds of books, the conclusions of international panels of some of the planet's top scientists, local discussions, television programs, award-winning movies, and the ceaseless work of hundreds if not thousands of non-governmental organizations. The sheer mass of attention that is directed to climate change is astounding. Not only are such warnings about the perils of climate change everywhere, but they are personal, life-altering, long-lasting, calamitous, and soon to be irreversible. They threaten changes to the planet and everything on it. Everyone with a pulse should be worried, if not actively trying to convince the global community to fix the problem before it is too late. So far, however, actions on a scale large enough to make a significant difference continue to be out of reach. And the earth continues warming.

Wherever you live, you will be affected. For those living in the southwestern deserts of the United States, it is going to get particularly ugly. Phoenix, for example, will not only be hot in the summers, as expected by those who choose to live there; it is going to get hotter than even such "desert rats" can tolerate. If you are among the billions of people who live in low-lying areas of the world—Miami Beach, Tampa, New Orleans, New York City, Atlantic City, Boston, Kolkata, Dhaka, Jakarta, or Venice, for example—you are probably already experiencing impacts from rising sea level, along with occasional massive flooding. If you are one of billions who count on the snowpack and glaciers in the Himalayas to release freshwater into streams that are the lifeline of your countries, get ready for an increasing number of other climate anomalies (Figure 6.1).

To avoid the worst climate impacts, global emissions of greenhouse gases (GHGs) not only will need to drop by half in the next ten years, but will then have to reach net zero around 2050. The World Resources Institute calls net zero emissions "climate neutrality." Achieving it will require two categories of adjustment: deep cuts in emissions and possibly the removal of atmospheric overloads directly by direct air capture.

The Thread of Energy. Martin J. Pasqualetti, Oxford University Press. © Oxford University Press 2021.
DOI: 10.1093/oso/9780199394807.003.0006

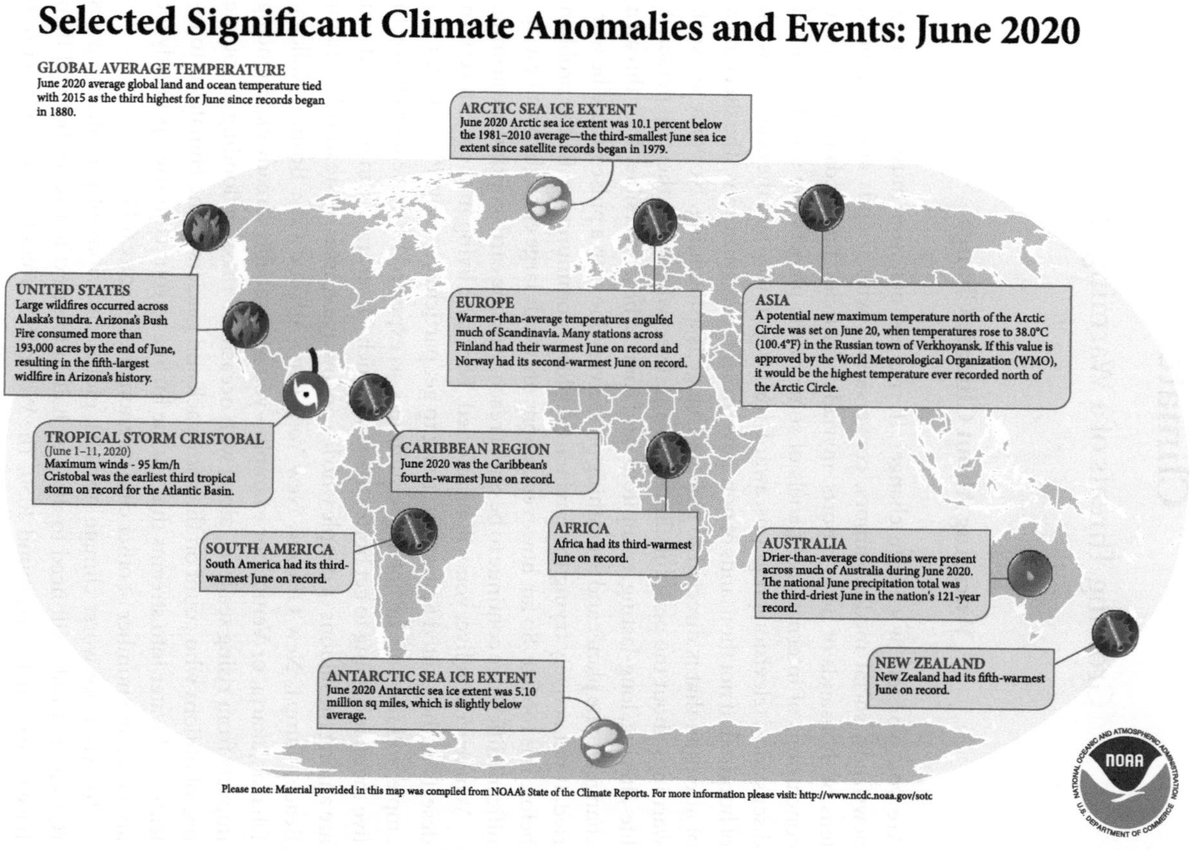

Selected Significant Climate Anomalies and Events: June 2020

GLOBAL AVERAGE TEMPERATURE
June 2020 average global land and ocean temperature tied with 2015 as the third highest for June since records began in 1880.

UNITED STATES
Large wildfires occurred across Alaska's tundra. Arizona's Bush Fire consumed more than 193,000 acres by the end of June, resulting in the fifth-largest wildfire in Arizona's history.

TROPICAL STORM CRISTOBAL
(June 1-11, 2020)
Maximum winds - 95 km/h
Cristobal was the earliest third tropical storm on record for the Atlantic Basin.

SOUTH AMERICA
South America had its third-warmest June on record.

ARCTIC SEA ICE EXTENT
June 2020 Arctic sea ice extent was 10.1 percent below the 1981-2010 average—the third-smallest June sea ice extent since satellite records began in 1979.

EUROPE
Warmer-than-average temperatures engulfed much of Scandinavia. Many stations across Finland had their warmest June on record and Norway had its second-warmest June on record.

CARIBBEAN REGION
June 2020 was the Caribbean's fourth-warmest June on record.

AFRICA
Africa had its third-warmest June on record.

ANTARCTIC SEA ICE EXTENT
June 2020 Antarctic sea ice extent was 5.10 million sq miles, which is slightly below average.

ASIA
A potential new maximum temperature north of the Arctic Circle was set on June 20, when temperatures rose to 38.0ºC (100.4ºF) in the Russian town of Verkhoyansk. If this value is approved by the World Meteorological Organization (WMO), it would be the highest temperature ever recorded north of the Arctic Circle.

AUSTRALIA
Drier-than-average conditions were present across much of Australia during June 2020. The national June precipitation total was the third-driest June in the nation's 121-year record.

NEW ZEALAND
New Zealand had its fifth-warmest June on record.

Please note: Material provided in this map was compiled from NOAA's State of the Climate Reports. For more information please visit: http://www.ncdc.noaa.gov/sotc

Figure 6.1 Selected significant climate anomalies and events in 2019.
Source: https://www.ncdc.noaa.gov/sotc/global/201913.

If all this makes you nervous and at least a bit depressed, it is understandable. I would not be surprised if you have already given at least passing consideration to how climate change will affect you and your family moving into the future. This chapter lays out some of the fundamental principles about climate change, including the evidence about what is causing it, how it will affect us, and what we are (and are not) doing about it and why.

The Worry

We are worried about climate change, and for good reason. The amount of carbon dioxide in the atmosphere has risen by 31 percent since 1958, and by about 48 percent since the start of the First Industrial Revolution. As a result, the estimated global mean temperature for 2018 is 0.99 ± 0.13°C above the Second Industrial Revolution baseline (1850–1900). These estimates comprise five independently maintained global temperature datasets, and the range represents their spread. They all tell the same story. There is no doubt: the global temperature is rising (Figure 6.2).

And, in the opinion of well over 98 percent of the world's finest climate scientists, it is caused by us.[1] The few remaining deniers continue to assert we have nothing to worry about. We would not be pressing the issue much if we

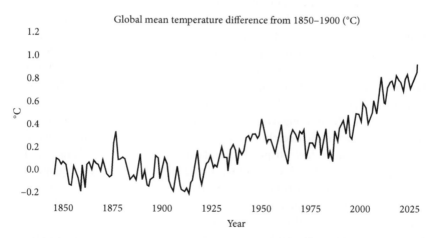

Global mean temperature difference from 1850–1900 (°C)

Figure 6.2 Observed sea level since the start of the satellite altimeter record in 1993, plus independent estimates of the different contributions to sea level rise: thermal expansion and added water, mostly due to glacier melt. Added together, these separate estimates match the observed sea level very well.
Source: NOAA, Climate.gov graphic, adapted from Figure 3.15a in *State of the Climate in 2018.*

lumped these few people into the same category as those who believed that the Covid-19 pandemic was a "hoax." As with the virus, it is beyond dispute that we will learn—perhaps later than we should—that climate change is an even graver threat than we thought when experts first sounded the alarm.

Evidence for climate change is massive and growing. It is so abundant that it is a bit tough to decide where to start. Perhaps the best first step is to just look around. Unless you are living under a rock, you cannot miss reports about melting Arctic and Greenland ice sheets, Antarctic heatwaves, retreating mountain glaciers, submerging areas of Bangladesh, submerging islands of the Maldives, greater hurricane strength in the Atlantic and Pacific Oceans, bleaching coral reefs, plagues of locusts in Africa, expanding deserts in Africa, shrinking polar bear habitat in northern Canada, and the poleward expansion of disease-carrying mosquitoes. And, it must be stressed, these examples are but a few of the most apparent, most obvious, and most alarming effects that are catching our attention. Matters are likely to get worse.

If you prefer data-driven rather than anecdotal evidence, look at the conclusions of the 2018 report from the Intergovernmental Panel on Climate Change (IPCC).[2] This authoritative publication included over 6,000 scientific references and the input of ninety-one authors from forty countries. Its key finding? That holding the world's average temperature increase to only 1.5°C (2.7°F) above pre-industrial levels is still possible but would require deep emissions reductions and "rapid, far-reaching and unprecedented changes in all aspects of society."[3] The report also states that limiting global warming to 1.5°C compared with 2°C (the latter of which should be a more achievable though still extremely difficult goal to reach) would reduce challenging impacts on ecosystems and on human health and well-being. The IPCC experts who wrote this report also show that for global warming to be limited to 1.5°C, "global net human-caused emissions of carbon dioxide (CO_2) would need to fall by about 45 percent from 2010 levels by 2030, reaching 'net zero' around 2050."[4] The emissions target for 2030, along with the associated changes and challenges involved in achieving such rapid decarbonization, was a key focus all around the world.[5]

Many indicators have tripped the alarms about climate change and its consequences. The World Meteorological Organization and other organizations sponsor the Global Climate Observing System (GCOS), which tracks seven main indicators: surface temperature, ocean heat content, atmospheric CO_2, ocean acidification, sea level, glacier mass balance, and Arctic and Antarctic sea ice extent. Additional indicators allow a more detailed picture of the changes in the respective domain. These indicators include—but are not limited to—precipitation, GHGs other than CO_2, snow cover, extent of deserts, ice sheets, extreme events, and climate impacts.[6] All of these indicators point to a changing global climate.

Each of these indicators is evidence of human-caused global warming. Among the most dramatic and well-publicized indicators of climate change is the sea-level rise from melting ice. According to the U.S. National Oceanic and Atmospheric Administration (NOAA), since 1880 the global sea level has risen 8–9 inches (210–240 mm).[7] In 2018, the global sea level was 3.2 inches (81 mm) above the 1993 average—the highest annual average in the satellite record (1993–present).

What gives us even more worry: the rate of rise is accelerating. The rate more than doubled from 0.06 inches (1.4 mm) per year throughout most of the twentieth century to 0.14 inches (3.6 mm) per year from 2006 to 2015. In many locations along the U.S. coastline, high-tide flooding is now 300–900 percent more frequent than it was fifty years ago. Even if the world follows a low-GHG pathway, the global sea level will likely rise at least 12 inches (0.3 m) above the 2000 level by 2100. If we follow a pathway with even higher emissions, a worst-case rise of as much as 8.2 feet (2.5 m) above the 2000 level by 2100 cannot be ruled out. This rise could cause the forced migration of at least 300 million, mostly in Asia.[8]

As for the Arctic Ocean, the area covered by sea ice continues to shrink, a change that is visually obvious when seen from above (Figure 6.3).

Today we understand the various mechanisms that threaten the protective layer that shields us from harmful radiation and moderate temperature swings that would otherwise occur diurnally and seasonally. After many years of study, thousands of scientists are now convinced that the threats are resulting from our excessive emission of heat-trapping GHGs into the atmosphere. That is, while most of these gases exist in the atmosphere naturally in some measure, we have been disrupting their equilibrium by emitting enormous volumes artificially. This interruption in the natural order began in earnest in the mid-eighteenth century, when the shift to coal ushered in industrialization on a grand scale, and it has become more troublesome with each passing year. As a result, we have put in jeopardy the effectiveness of the thin sheltering atmospheric skin that wraps around our planetary home and keeps us safe.

The Earth's atmosphere consists of a mixture of gases. They all serve a function, and we need them all. Most of this mixture is nitrogen (about 78 percent). Almost all the remaining amount is oxygen (about 21 percent). Together they account for 99 percent of our atmosphere. The problem we are having is with the other 1 percent. This is where we find water vapor, CO_2, nitrous oxide (N_2O), ozone, helium, neon, argon, methane (CH_4), and several other molecules, including a group known as the halocarbons (e.g., chlorofluorocarbons, hydrochlorofluorocarbons, hydrofluorocarbons, sulfur hexafluoride, perfluorocarbons, and nitrogen trifluoride).

2019 SUMMER MINIMUM

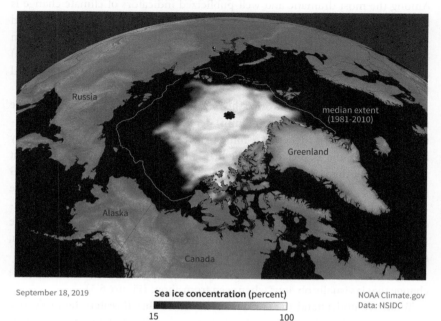

September 18, 2019   **Sea ice concentration** (percent)   NOAA Climate.gov
                                                                         Data: NSIDC
                            15   100

Figure 6.3 2019 summer minimum of Arctic sea ice extent (area of ocean with at least 15 percent sea ice).
Source: NOAA, https://www.climate.gov/news-features/understanding-climate/climate-change-minimum-arctic-sea-ice-extent.

Ironically, we created new halocarbons to help wean ourselves off earlier ones. The chlorofluorocarbons (CFCs) were invented in 1928 and developed in the 1930s as non-flammable and non-toxic refrigerants and propellants. CFCs, we discovered in the 1970s, were harmful to the stratospheric ozone layer that filters out harmful ultraviolet light before it reaches the Earth's surface. But it turns out that the newer halocarbons are extremely potent GHGs.

Were there such a thing as an atmospheric police department patrolling the skies, their highest priority would be to take down a gang of excess molecules for the crime of global warming. Even though these molecules are largely invisible and odorless, a little detective work would identify the ringleader as CO_2. Today we estimate that the artificial emission of this gas totals over 36 gigatons globally per year, the majority of which comes from fossil fuel and biomass energy combustion. That is 180 times greater than the weight of all the discarded material Australia sends to landfills each year. If we go back to the beginning of the Industrial Revolutions, we estimate that humans emitted into the atmosphere a total of 1.5 trillion metric tons of CO_2 between 1751 and 2017.[9] That is a weight equivalent to about six times all the estimated recoverable coal reserves within the borders of the United States in 2019.[10] And we are not stopping: we continue to add four times more CO_2 to the atmosphere each year than the other GHGs combined. No wonder we have a problem!

Many people have noticed that the rise in CO_2 seems to coincide with an increasing frequency of harsh weather events around the world, including in the United States (Figure 6.4). For example, just in the past few years, hurricanes of enormous power have devastated parts of the southeastern United States, Puerto Rico, and the Bahamas. Are rising sea temperatures responsible? It is impossible to ascribe causation with certainty in the short run. Still, we do know that hurricanes strengthen in warmer waters, and we do know that such warming is occurring in the Atlantic Ocean. Even if warming oceans were somehow not directly responsible for the increasing incidence of harsh weather, rising sea levels will produce more powerful storm surges when a hurricane of any strength strikes us. Recent modeling supports these suppositions.[11]

Whereas the rising level of CO_2 must shoulder much of the blame for global warming, we understand that for most of our time on the planet, CO_2 has protected us from life-threatening swings in temperature. In its natural concentration, it is a critical constituent of our atmosphere rather than a pollutant; we have benefitted from it because it plays a vital role in keeping temperatures within a range that allows plants and animals to thrive. Without CO_2, the earth

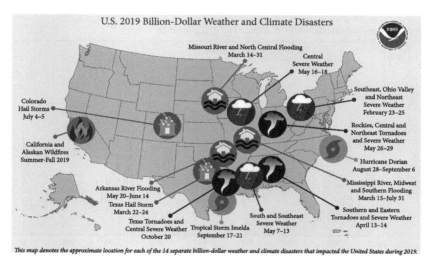

Figure 6.4 Billion-dollar weather and climate disasters in 2019 suggests a disrupted climate balance, caused by global warming.
Source: https://www.ncdc.noaa.gov/billions/.

would not be very accommodating or livable, at least for us. One way to think about rising CO_2 is that there can be too much of a good thing. Our immediate problem is that the atmospheric concentration of CO_2 is increasing too quickly for us to manage all the innumerable readjustments we need to make to keep up with the pace of change, along with adverse effects on animal and plant life.

What is the origin of such an unwelcome upward spike in CO_2? The answer is the ballooning use of carbon-rich fuels that we burn to provide most of the energy we need. The blame starts with coal, beginning with the Industrial Revolution. The use of coal climbed rose for over 150 years and then started climbing ever more steeply around the turn of the twentieth century. On a global basis, it continued going up. The more we used coal, oil, and gas, the higher the concentration of CO_2 we reached.

Some people might think that the rising level of GHGs and global temperatures are all part of natural cycles and not caused by human activities.[12] We do know that there have been fluctuations in global temperatures in the past; without such fluctuations, we would not have had glacial advances and retreats. However, while significant fluctuations *have* occurred, we now know that such fluctuations occurred not over a few dozen years but over very long periods (Figure 6.5). Moreover, it has been a long time since CO_2 levels were at the height we see today. The last time we reached the current CO_2 concentration of 415 ppm was 3 million years ago.

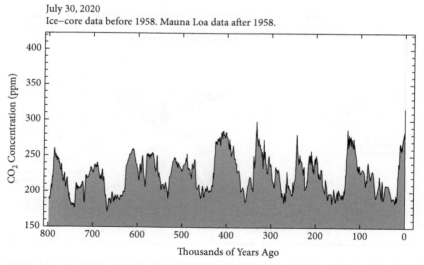

July 30, 2020
Ice−core data before 1958. Mauna Loa data after 1958.

Figure 6.5 Even if we measure CO_2 back 800,000 years, it has rarely even reached 300 ppm, let alone the 415 ppm that it is now.
Source: https://scripps.ucsd.edu/programs/keelingcurve/.

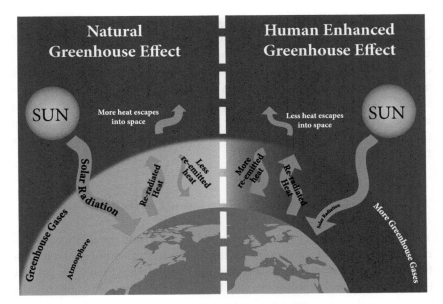

Figure 6.6 Generalized depiction of the greenhouse effect.
Source: NPS.gov, https://www.nps.gov/goga/learn/nature/climate-change-causes.htm.
See plates for color version.

You might wonder why CO_2 and other contributing emissions have been la-beled "greenhouse gases" (Figure 6.6). The reasoning behind this nickname is that it is an understandable, if somewhat inaccurate, analogy. As the name implies, CO_2 acts similarly to the glass on a greenhouse in several steps. As the first step, shortwave radiation from the sun penetrates the atmosphere. Then, once it is ab-sorbed, the energy is reradiated back as longwave radiation. CO_2 and the other GHGs act like a blanket holding in the heat before it escapes back into space. This mechanism maintains the temperature on the planet within a tolerable and usu-ally comfortable range for humans and other animals; life as we now find it on the planet has evolved to survive within this range. We should be eternally grateful that the system works this way; without GHGs, the daily swings of temperature would resemble those found on the moon, rising and falling hundreds of degrees.

Absent human interference over long periods, the many natural systems on the planet established their balance. Until the last few decades, we rarely paid much attention to these balances or what we were doing to disturb them. Just as a lobster in a pot of warming water might not notice what is happening until it is too late, we have also been slow to notice our warming planet. Being smarter than a lobster, however, we have noticed that the rising global temperature might threaten us before it was too late. What we do and when we do it in response re-main open questions.

One of the most visible signals of change has been the documented evidence of melting glaciers and thinning ice sheets. The average thickness of thirty well-studied glaciers has decreased more than 60 feet since 1980, according to NOAA.[13] This thinning, in turn, results in rising sea levels that are already occurring in coastal areas, each with dreadful implications. In addition to millions of plants and animals living along the oceanic littoral, hundreds of millions of people will have to flee the rising waters. Where will they go? Can they muster up the financial wherewithal that will be necessary to move away from the danger?

Out of Balance and Resisting Correction

Today, the world's energy system has strayed from its natural stability. More energy is coming in than is going out, and the whole planet is warming. And it is our fault—or, to use less intentional language, our responsibility. Even if it is possible to reestablish balance, it will not be easy. Despite all the actions we might consider taking to fix things, we must be able to apply our remedies on a planetary scale, overcoming the massive inertia that has accumulated over the past several decades.

People in the wealthy countries will resist giving up the way of life they enjoy. With equal resolve, people in the poorer countries will resist losing hope in their dreams of the life that their future might bring. Neither group wants to lose what sustains them. One group wishes to maintain the security of what it has. The other group wishes to maintain the image of what it wants.

As you know, everything in the biosphere connects to everything else in ways direct and indirect. Knowing this, you should not be surprised that climate changes have been producing insidious and worrisome knock-on effects. While we understand and can track some of these effects, some undoubtedly have so far escaped our notice. As the IPCC phrases it: "Taken as a whole, the range of published evidence indicates that the net damage costs of climate change are likely to be significant and to increase over time."[14]

Atmospheric concentrations of CO_2 are increasing as you read this, and they are triggering responses in the biosphere that are dangerous and potentially deadly for many organisms, including us (Figure 6.7). Coastal flooding, diminished river flow, expanding deserts, lost arable land, widening extremes of weather, fluctuating plant and animal ranges, stronger storms, and many more impacts are likely.[15]

And the threats are accumulating. Not only are we dumping excessive amounts of CO_2 and other GHGs into the atmosphere, but we have been doing very little to curtail the practice. We have also not mastered the thorny problem of how to reduce the high concentrations already in place. Why do we have to do

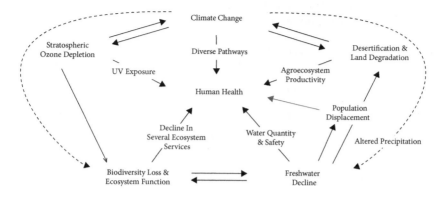

Figure 6.7 The multifaceted relationships and consequences of climate change. *Source:* https://sites.psu.edu/etclementrcl/2015/04/10/what-exactly-is-climate-change/.

that? Because once GHGs make their way into the atmosphere, they tend to re-main there for many years. Such residence time means that we must not only *cur-tail* emissions but also *reduce* the concentration of these gases, principally CO_2, through direct removal.

Direct removal of CO_2 is essential to slowing global warming because emis-sion control alone is not going to solve the problem. Even if we could magi-cally stop all emissions of GHGs from human-based sources tomorrow, global temperatures would continue to rise because GHG concentrations are already too high. We are paying the price for decades of inaction. If we do not control emissions and remove GHGs that are already clogging the atmosphere, the en-ergy imbalance will worsen and so too will the consequences we have to accept. Such a scenario should be seriously worrisome for all of us. It portends an exis-tential threat to humanity.

The impacts of various types of GHGs differ in many ways. Two differences dominate: (1) the ability to absorb energy (their radiative efficiency) and (2) how long they stay in the atmosphere (their residence time). The first difference be-tween gases is in their global warming potential (GWP). The GWP is a measure of how much energy one additional ton of a gas will absorb over a given time rel-ative to one additional ton of CO_2. This number is calculated by the IPCC, based on the intensity of infrared absorption by each GHG and how long emissions re-main in the atmosphere. One calculates GWPs using a set time horizon, usually over a 100-year timeframe. GWPs are updated periodically as improvements are made to the underlying science. The higher the GWP, the more that a given GHG warms the earth compared to the warming that CO_2 produces (Table 6.1).

Table 6.1 100-Year GWPs from the IPCC's Second Assessment Report (SAR) and Fourth Assessment Report (AR4)

Gas Name	Formula	Lifetime (years)	SAR GWP	AR4 GWP	Percent Change
Carbon Dioxide	CO_2		1	1	
Methane	CH_4	12	21	25	+19%
Nitrous Oxide	N_2O	114	310	298	−3.9%
Sulfur Hexafluoride	SF_6	3,200	23,900	22,800	−4.6%
Nitrogen Trifluoride	NF_3	740	n/a	17,200	
Hexafluoroethane (PFC-116)	C_2F_6	10,000	9,200	12,200	+33%
Octafluoropropane (PFC-218)	C_3F_8	2,600	7,000	8,830	+26%
Octafluorocyclobutane (PFC-318)	C_4F_8	3,200	8,700	10,300	+18%
Tetrafluoromethane (PFC-14)	CF_4	50,000	6,500	7,390	+14%
Hydrofluorocarbon 125	HFC-125	29	2,800	3,500	+25%
Hydrofluorocarbon 134a	HFC-134a	14	1,300	1,430	+10%
Hydrofluorocarbon 143a	HFC-143a	52	3,800	4,470	+18%
Hydrofluorocarbon 152a	HFC-152a	1	140	124	−11%
Hydrofluorocarbon 227ea	HFC-227ea	34	2,900	3,220	+11%
Hydrofluorocarbon 23	HFC-23	270	11,700	14,800	+26%
Hydrofluorocarbon 236fa	HFC-246fa	240	6,300	9,810	+56%
Hydrofluorocarbon 245fa	HFC-245fa	8	n/a	1,030	
Hydrofluorocarbon 32	HFC-32	5	650	675	+3.8%
Hydrofluorocarbon 365mfc	HFC-265mfc	9	n/a	794	
Hydrofluorocarbon 43-10mee	HFC-43-10mee	16	1,300	1,640	+26%

Source: California Air Resources Board, "GHG Global Warming Potentials," https://ww2.arb.ca.gov/ghg-gwps.

The second difference between one GHG and another is in their atmospheric residence time—that is, how long they remain in the atmosphere. This residence time can vary from days to millennia. For example, the residence time for water vapor, H_2O (which is a natural GHG), can be short. In addition, its concentration in the atmosphere varies depending on latitude and other factors. Its global concentrations are nevertheless going up, and this is worrisome. While not an emission, water vapor concentrations are affected by increased emissions of other GHGs because warmer air holds more water vapor than cooler air does. As the planet warms from the increasing concentration of the other gases, the ability of water vapor to hold more heat rises in response. We can think of the greenhouse properties of water vapor as part of a feedback loop, rather than a direct cause of climate change.

Each of the other gases has its own residency properties. For example, 65–80 percent of the CO_2 in the atmosphere dissolves into the ocean over a period of between 20 and 200 years. The rest dissipates much more slowly, meaning it can affect the climate for thousands of years. This effect means that we are not going to be able to count on its natural removal to solve our problem, at least in a timeframe that is relevant to people now living or soon to be born. For its part, CH_4's residence time is much shorter, about twelve years. Although methane is more potent CO_2 on a per-unit basis (with a GWP twenty-one to twenty-five times greater), its residence time is much shorter (about twelve years). Another potent GHG, N_2O, is an even bigger problem on a per-unit basis. It is removed from the atmosphere more slowly than methane, persisting for 114 years, but it has a GWP about 300 times greater than CO_2. Worse still, multiple artificial halocarbon compounds, such as sulfur hexafluoride, can linger in the atmosphere for many thousands of years, even though the volume of their emissions is much smaller.[16]

Sources and Variability of GHG Emissions

Despite a great deal of rhetoric about the need for an energy transition, we still depend on fossil fuels to satisfy most of our energy demand. This dependency is responsible for the largest portion of CO_2 emissions we are producing, amounting most recently to almost 35 gigatons emitted globally. Most of the excess emissions come from burning fossil fuels, either in power plants or in transport vehicles. While CO_2 and other GHGs are a problem everywhere on the planet, emission rates vary from one place to another. For example, on a global basis, the percentage of CO_2 emissions from transportation is 14 percent, while in the United States it is 34 percent. Industry on a global basis is responsible for 21 percent, while it is 15 percent in the United States (Figures 6.8 and 6.9). Such comparisons highlight why, despite the global scale of the problem of

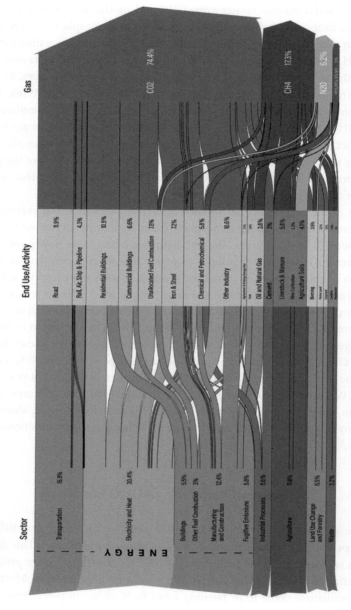

Figure 6.8 Global GHG emissions by sector.
Source: World Resources Institute.

The following text appears within the figure:

World Greenhouse Gas Emissions in 2016
Total: 49.4 GtCO₂e

Sector

ENERGY

- Transportation 15.9%
- Electricity and Heat 30.4%
- Buildings 5.5%
- Other Fuel Combustion 3%
- Manufacturing and Construction 12.4%
- Fugitive Emissions 5.8%
- Industrial Processes 5.6%
- Agriculture 11.8%
- Land Use Change and Forestry 6.5%
- Waste 3.2%

End Use/Activity

- Road 11.9%
- Rail, Air, Ship & Pipeline 4.3%
- Residential Buildings 10.9%
- Commercial Buildings 6.6%
- Unallocated Fuel Combustion 7.8%
- Iron & Steel 7.2%
- Chemical and Petrochemical 5.8%
- Other Industry 10.6%
- Oil and Natural Gas 3.8%
- Cement 3%
- Livestock & Manure 5.8%
- Rice Cultivation 1.3%
- Agriculture Soils 4.1%
- Burning 3.5%

Gas

- CO2 74.4%
- CH4 17.3%
- N2O 6.2%

Source: Greenhouse gas emissions on Climate Watch. Available at: https://www.climatewatchdata.org

WORLD RESOURCES INSTITUTE

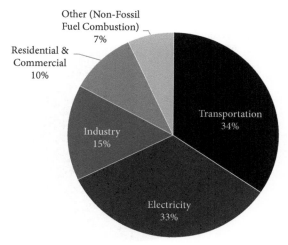

Figure 6.9 U.S. CO_2 emissions by source, 2018.
Source: U.S. EPA. All emission estimates from the Inventory of U.S. Greenhouse Gas Emissions and Sinks: 1990–2018, https://www.epa.gov/ghgemissions/overview-greenhouse-gases.

global warming, the emphasis of whatever mitigation and emissions reduction response plans we attempt might differ from one country to another. Some countries will emphasize emission controls, some will concentrate on converting to electric cars, and some will target agricultural practices. In the end, of course, we need all those responses because the problem is global.

The exact emphasis of our responses will depend on the types of GHGs being emitted. As I already mentioned, the largest GHG emissions are of CO_2. The U.S. Environmental Protection Agency (EPA) reports that, taken together, transportation, industry, and electricity generation account for 82 percent of all the GHGs emitted in the United States (Figure 6.10). But we have to worry about other gases as well, such as—in the United States, for example—CH_4 (which accounts for 10 percent of the GHGs emitted), N_2O (6 percent), and fluorinated gases (3 percent).

Each of these gases has its dominant sources, and each poses a particular threat to climate stability. We are emitting CH_4, for example, at a rate that is much greater than is returning to the earth's sinks. With 558 million tons emitted annually between 2003 and 2012 and 548 million tons returning to the earth's sink during the same period, the burden of methane grows at a rate of 10 million tons per year, and it has a potency many times that of CO_2.

Before we can begin to lower the emission of CH_4, of course, we need to understand its origins. Most of it is natural; it is the excess that is the problem. In the United States, about one-third of the excess CH_4 comes from oil and gas

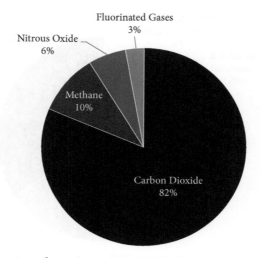

Figure 6.10 Overview of greenhouse gases, 2018.

Source: U.S. EPA. All emission estimates from the Inventory of U.S. Greenhouse Gas Emissions and Sinks: 1990–2018, https://www.epa.gov/ghgemissions/overview-greenhouse-gases.

production. This emission comes from unintended well and pipeline leakages, sloppiness, and other shortcomings. One of the most egregious and most obvious sources is the practice of "flaring" natural gas. When crude oil is being extracted, the natural gas that is associated with the oil is considered a low-value waste product. The solution? Simply burn off this natural resource right at the well (Figures 6.11 and 6.12). Not only is this a wasteful practice, but it is contributing to climate change.

Another artificial source of CH_4 emissions is animal meat raised for human consumption. Beef is the most carbon-emitting of the foods we eat.[17] So it should not be too surprising to learn that one-third of the excess CH_4 results from the global demand for a meat diet. The culprits in this abundance are ruminants, such as cattle, deer, sheep, and goats. Not only do they produce massive amounts of manure—which give off CH_4—but they also produce CH_4 in their digestive tracts; most of that is belched out, although a small percentage emerges as flatulence (Figure 6.13). With the realization that the demand for meat is a major source of this potent GHG, there has been movement away from eating meat in some societies. This suggests that a diet-focused approach to reducing CH_4 might be easier than reducing other types of GHGs. "Reduction in GHG emissions from ruminant animals could be done earlier than from other sectors and requires less effort and financial investment, thus making the strategy more feasible."[18] If you want protein in your diet, just about anything has a smaller carbon footprint than beef, including pork, fish, cheese, nuts, tofu, beans, and peas.

Figure 6.11 Natural gas flare at an oil pump jack near Roosevelt, Utah.

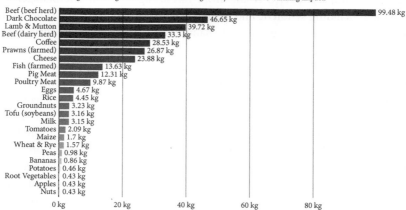

Figure 6.12 The carbon footprint of foods. *Source:* https://ourworldindata.org/carbon-footprint-food-methane.

Figure 6.13 Feedlots, such as this one near Dodge City, Kansas, are concentrated sources of methane emissions.

Another troublesome gas, N_2O, while amounting to only 6 percent of the total of emitted GHGs, is both potent and long-lasting. It originates mostly from human activities such as agriculture, and to a much lesser extent from fuel combustion, wastewater management, and industrial processes (Figure 6.14). Within the agricultural sector, N_2O results from various agricultural soil management activities, including synthetic and organic fertilizer application and other cropping practices, the management of manure, and the burning of agricultural residues.

Fluorinated gases, which are not natural, are a big problem. According to the U.S. EPA, there are four main categories—hydrofluorocarbons (HFCs), perfluorocarbons (PFCs), sulfur hexafluoride (SF_6, which is an excellent electrical insulator), and nitrogen trifluoride (NF_3).[19] HFCs are potent GHGs with generally very high GWPs (though their emissions levels are relatively low), and they are released into the atmosphere during manufacturing processes and through leaks, servicing, and disposal of equipment in which they are used. HFCs are used as aerosol propellants, foam blowing agents, solvents, and fire retardants. The major emission source of these compounds is their use as refrigerants in air-conditioning systems in both motor vehicles and buildings. The use of fluorinated gases is on the rise (Figure 6.15).

Ironically, these chemicals were developed as replacements for CFCs and hydrochlorofluorocarbons (HCFCs), both of which are linked to the depletion of

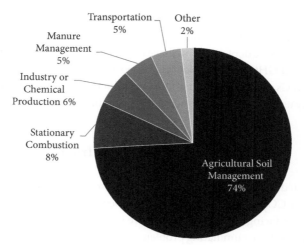

Figure 6.14 2018 U.S. nitrous oxide emissions, by source. *Source:* U.S. EPA, Inventory of U.S. Greenhouse Gas Emissions and Sinks: 1990–2018, https://www.epa.gov/ghgemissions/overview-greenhouse-gases.

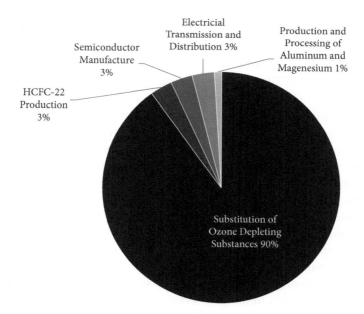

Figure 6.15 2017 U.S. fluorinated gas emissions, by source. *Source:* U.S. EPA, https://www.epa.gov/ghgemissions/overview-greenhouse-gases.

the stratospheric ozone layer. We are phasing out CFCs and HCFCs under an international treaty called the Montreal Protocol. The irony is that while addressing one problem—ozone depletion—we have introduced a new problem: fluorinated gases with great GHG potency. Newly developed hydrofluoroolefins (HFOs) are a subset of HFCs and have shorter atmospheric lifetimes and lower GWPs. HFOs are currently used as refrigerants, aerosol propellants, and foam blowing agents.

Where's the CO_2 Coming from?

As we've noted, the mixture of atmospheric gases has long included a small percentage of CO_2, and all life on earth has benefitted from its presence. The problem we are facing nowadays is that there is a surfeit of it, and we are responsible for most of the accumulating problems that are besetting us as a result. Two-thirds of total CO_2 emissions result from burning fossil fuels to provide us with heat, electricity, and fuel for our motor vehicles. Industry and buildings account for the remaining one-third. And, of course, we are also removing (and often burning) millions of trees that would otherwise pull CO_2 from the atmosphere, thereby making the problem even worse. We are attempting to offset this loss by planting new trees.

Let's consider the burning of fossil fuels. Since all fossil fuels are carbon-based, burning them produces CO_2. That much is obvious. What might be less apparent, however, is how much the volume of produced CO_2 varies from one form of fossil fuel to another. The three types are coal, natural gas, and oil. Coal burning produces more CO_2 than the others for an equal amount of energy released. It is responsible for 43 percent of CO_2 emissions, while 36 percent is produced by oil and 20 percent by natural gas. This amounts to approximately 2.5 metric tons (t) of carbon dioxide equivalent (CO_2e) for each ton of coal burned, if we consider the entire life cycle.[20] Natural gas produces less than half that amount. Solar photovoltaics would cut that amount by another order of magnitude, and wind power produces even less than that (Table 6.2).

Providing further insight into the variability of fossil fuel emissions, the International Energy Agency (IEA) states: "Shares in final consumption vary across countries: while transport is predominant in many American countries, in Asia, one-half of emissions derive from electric power generation and less than one-sixth from transport. The picture changes after reallocating emissions from power generation to the final sectors: industry accounts for slightly less than one-half of total emissions, buildings and transport for one-quarter each" (Figure 6.16).[21]

An energy-hungry country like China needs a great amount of energy to fuel its ambitious five-year plans, even as China hopes to reduce the contributing

Table 6.2 Lifecycle GHG Emissions by Electricity Source

Technology	Description	50th percentile, in g CO_2e/kWhe
Hydroelectric	Reservoir	4
Wind	Onshore	12
Nuclear	Various generation II reactor types	16
Biomass	Various	230
Solar Thermal	Parabolic trough	22
Geothermal	Hot dry rock	45
Solar Photovoltaic	Polycrystalline silicon	46
Natural Gas	Various combined cycle turbines without scrubbing	469
Coal	Various generator types without scrubbing	1,001

Source: W. Moomaw et al., "Annex II: Methodology," in IPCC, *Renewable Energy Sources and Climate Change Mitigation: Summary for Policymakers and Technical Summary*, 181–208 (New York: IPCC, 2011).

Global GHG Emissions by Sector (2014)

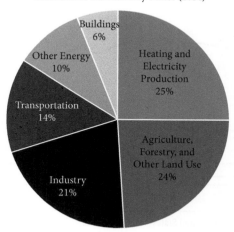

Figure 6.16 Global emissions by fuel and sector. *Source:* IEA, https://www.epa.gov/ghgemissions/overview-greenhouse-gases

share of coal to 58 percent by 2022 and 55 percent by 2030.[22] The Chinese leadership is aware that the consumption of coal contributes to global warming, so they are quickly building carbon-free alternatives in response. However, regardless of the rapid pace of construction of nuclear, solar, and wind power plants, meeting the burgeoning demand for electricity without coal is impossible. As a result, the Chinese (and we Americans) pay a heavy environmental price, and the coal miners and anyone living near the mines or power plants pay an even greater additional price.

Other fossil fuels are also important contributors to GHGs. Most motor vehicles burn fossil fuel, almost all in the form of refined petroleum products. That is the pattern everywhere in the world, despite ongoing programs to develop electric vehicles and those that can run on compressed natural gas (CNG). China has embraced the private automobile with open arms, figuring that doing so produces great numbers of jobs while also serving as status symbols of the nation's growing prosperity. Unfortunately, the rise in car ownership in China is responsible for even more emissions of CO_2 (Figure 6.17). There has been more than a tripling in these CO_2 emissions, from about 200 million metric tons (Mt) of CO_2 in 2000 to about 700 Mt CO_2 in 2017. Despite having four times the people, however, China produces less than half the CO_2 emissions of the road transport emission leader, the United States, which comes in at about 1,450 Mt CO_2.

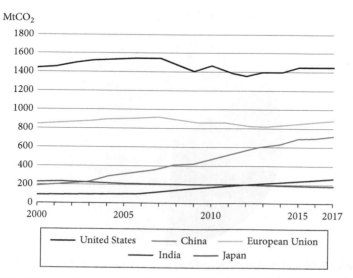

Figure 6.17 CO_2 emissions by selected countries. Note the rapid rise of emissions in China. *Source:* IEA, CO_2 Emissions from Fuel Combustion Highlights, 2019, https://www.epa.gov/ghgemissions/overview-greenhouse-gases. See plates for color version.

One of the causes of the substantial emission of CO_2 from transport vehicles is their lower conversion efficiency. Whereas electrical generating power plants contribute much more CO_2 in the aggregate sense, they are more efficient on a per-energy-unit basis than transportation in its most common forms. Specifically, motor vehicles in the United States are responsible for about 68 kilograms of CO_2 per million British thermal units (BTUs). In contrast, electrical generating stations are about a third better than motor vehicles, at about 48 kilograms of CO_2 per million BTUs (Figure 6.18).

Personal vehicles are a major contributor to GHGs and global warming. The Union of Concerned Scientists reports that cars and trucks account for nearly one-fifth of all U.S. emissions, emitting around 24 pounds of CO_2 and other global-warming gases for every gallon of gasoline.[23] About 5 pounds comes from the extraction, production, and delivery of the fuel, while the great bulk of heat-trapping emissions—more than 19 pounds per gallon—comes right out of a car's tailpipe. So if we estimate that a car travels 10,000 miles in a year and gets 25 miles per gallon, that would total 400 gallons multiplied by 24 pounds per gallon, or 9,600 pounds of CO_2 per year. That is almost 5 tons of a GHG per car! We need to do something about that, and we seem to be on our way with the uptick

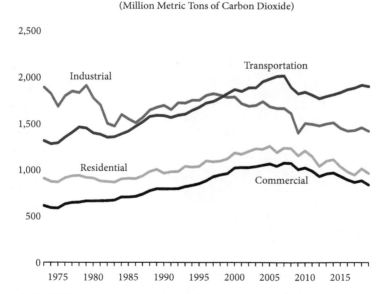

Carbon Dioxide Emissions Total by End-Use Sector 1973–2019
(Million Metric Tons of Carbon Dioxide)

Figure 6.18 While great attention is paid to emissions from power plants, in fact they emit about 20 kg less CO_2 per million Btu than the transportation sector. *Source:* U.S. EPA, https://www.eia.gov/todayinenergy/detail.php?id=31012.

of sales of electric vehicles. Two countries stand out: electric vehicles accounted for 56 percent of all cars sold in Norway in 2019, and China led the world in electric vehicle sales in 2018 at more than 1 million.[24] In contrast, the United States added 361,310 electric vehicles to its fleet in 2019.[25]

When we examine the total emission for the United States, the U.S. Energy Information Administration (EIA) estimates that in 2018, U.S. motor gasoline and diesel fuel consumption in the U.S. transportation sector resulted in the emission of about 1,099 Mt CO_2 and 461 Mt CO_2, respectively, for a total of 1,559 Mt CO_2. This total was equal to 81 percent of total U.S. transportation sector CO_2 emissions and 30 percent of total U.S. energy-related CO_2 emissions in 2018.[26]

Until now, we have been discussing the many ways we are increasing CO_2 concentrations in the atmosphere directly. Now let's turn to how CO_2 concentrations can increase not by adding more but by subtracting less. The primary focus of this branch of the problem is deforestation (Figure 6.19). As you know, green plants absorb CO_2 and release oxygen. When we clear forests, that absorption capacity decreases. Making matters worse, frequently the forests are cleared by burning. This activity itself releases additional CO_2—15 percent of the total, by some estimates.

Figure 6.19 Deforestation reduces the capacity of forests to absorb CO_2. Deforestation for oil palm plantation in Indragiri Hulu, Riau Province, Indonesia. *Source:* https://upload.wikimedia.org/wikipedia/commons/f/ff/Riau_deforestation_2006.jpg. flickr:Riau flickr user:Wakx. Creative Commons. For time-lapse images of deforestation in South America, view this NASA link: https://earthobservatory.nasa.gov/images/145988/tracking-amazon-deforestation-from-above.

Variations by Region

Every country using coal for heating and electricity generation produces CO_2, but the total emissions of CO_2 vary from place to place. The top three regions of the world are China, the European Union (EU), and the United States. Their emissions are voluminous. Together they contribute fourteen times the sum of emissions of the bottom 100 emitting countries. China leads the world in total CO_2 emissions, ranking first at 10,065 Mt CO_2 per year.[27] The United States is in second place, but it is far behind at 5,416 Mt per year. These three regions are home to about 2.2 billion people, so we would expect their contributions to be high. Examining emissions on a per capita basis, however, reveals a different story. Out of about 220 countries, Qatar ranks first at 38 Mt CO_2 per person. The United States comes in twelfth, with 17 Mt CO_2 per person. China ranks thirty-ninth, at 7.0 Mt CO_2 per person. While China emits more than any other country, the United States is emitting twice as much on a per-capita basis.

China's total CO_2 emissions are even more egregiously high when you consider how much of the total global emissions the country is responsible for (Figure 6.20). The World Resources Institute estimates that China is responsible for about 27 percent of global carbon dioxide equivalent (CO_2e). Of this amount,

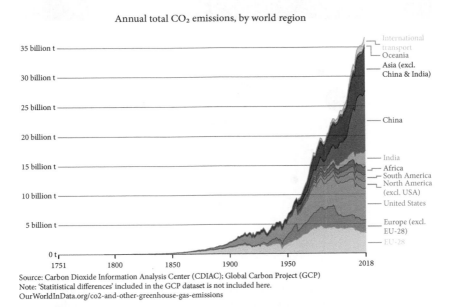

Annual total CO₂ emissions, by world region

Source: Carbon Dioxide Information Analysis Center (CDIAC); Global Carbon Project (GCP)
Note: 'Statitistical differences' included in the GCP dataset is not included here.
OurWorldInData.org/co2-and-other-greenhouse-gas-emissions

Figure 6.20 Annual total CO_2 emissions, by world region, through 2017.
Source: https://ourworldindata.org/co2-and-other-greenhouse-gas-emissions#how-have-global-co2-emissions-changed-over-time. See plates for color version.

about 22 percent comes from the energy sector and 5 percent from other sources. The United States does not get off easy, however. About 14 percent of the global CO_2e total is the responsibility of the United States, with 12.6 percent coming from energy and the remaining 1.4 percent coming from other sources such as agriculture. So if we total those numbers, more than half the global emissions of CO_2 come from energy activities in just two countries, China and the United States (Figure 6.21).[28]

We gain further insight into the matter of CO_2 emissions when we evaluate them based on national economic output, specifically emissions per gross domestic product (GDP) per capita (Figure 6.22). From this point of view, the United States does not look like such a bad actor, depending on one's point of view. While we are producing a lot of CO_2 per capita, we are also producing a lot of wealth as a "payoff" for such excess. Whether or not such levels of economic activity should give the United States license to emit such a high level of GHGs is another matter, and at least a questionable bargain.

Looking at the extremes, Saudi Arabia emits more CO_2 with about the same per capita GDP. In contrast, Qatar emits at about the same intensity but with a much higher per capita GDP. Hong Kong, Switzerland, and Singapore emit less

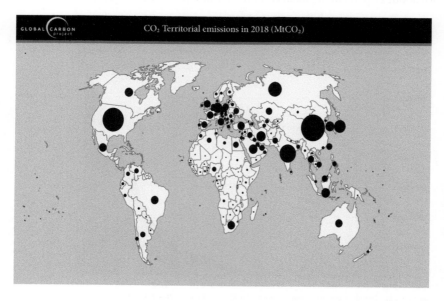

Figure 6.21 Global CO_2 fossil fuel emissions in 2019. You can see by the size of the circles the location and relative emission rates of CO_2 emissions. China, India, the United States, the EU, and Russia lead the world. For an interactive version of this map, where you can mix and match pollution levels, sources, countries and more, go to http://www.globalcarbonatlas.org/en/CO2-emissions.
Original source: Friedlingstein et al., 2020 : The Global Carbon Budget 2020, Earth System Science Data. Data for years 2018 and 2019 are preliminary.

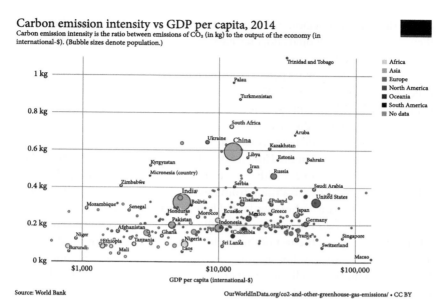

Figure 6.22 Carbon emission intensity vs. GDP per capita, 2014. *Source:* https://ourworldindata.org/co2-and-other-greenhouse-gas-emissions#how-do-we-measure-or-estimate-CO2-emissions.

with a higher per capita GDP. At the other extremes, countries such as Burundi, Malawi, Niger, and Ethiopia are low emitters but have a very low GDP per capita to match. Considering these numbers, you must decide which is more important, low pollution or high economic output. The challenge we face as a global population is to break down this correlation and get the countries with the wealthiest economies to emit less and still promote economic growth at home while helping the poorer countries to develop without adding environmental degradation. Doing both simultaneously is a difficult ask, considering that poorer countries often have higher priorities, such as satisfying their basic needs for healthy food, clean water, sanitary facilities, and energy. There are many organizations, such as the Bill and Melinda Gates Foundation, the World Bank, the International Monetary Fund, and the U.S. Agency for International Development, that are helping such countries achieve this goal. The development of renewable energy—particularly solar, wind, and geothermal—can play an important role in greatly lowering emissions per unit of economic output.[29]

As we have learned, the advanced economies of the world are responsible for over 75 percent of historical carbon emissions. This degree of emissions is their harsh burden and—some would say—an inevitable consequence of the lifestyle they enjoy. However, thanks to the in-depth work of organizations such as the

IPCC, they are also beginning to grasp the importance of lowering the weight they carry. And, one should add, they have seen some success, particularly in the United States and Europe.

Nonetheless, two worries persist. The first is our concern for what is sometimes called the "developing world." While we wish the people in these countries to live better, safer, healthier, and more secure lives, we fear that as they improve their socioeconomic status, they will follow the same disastrous path as the United States, Europe, and more recently China. And, as the developing countries often remind us, who are we in the United States to talk? That is, it is hypocritical for the countries that became wealthy in part because they were major polluters to ask the poorer countries to abstain from following along the same path—that is, to somehow achieve a higher standard of living while polluting less.

The trend lines do not look good. While CO_2 is trending downward in the advanced economies, the rest of the world is moving in the opposite direction, and at a faster pace. Insidiously, this pattern reveals an even more unfair pattern: some of the more polluting industries are increasingly outsourcing their activities to the poorer countries in order to take advantage of lower wages there.[30] By lowering the per-unit costs of the products that are consumed by the wealthy countries, they are shifting the pollution they used to produce in their backyard to the backyards of the poorer countries. As was concluded in one of the IPCC meetings, "Developed nations must take responsibility for their historical emissions and contribute the funds and transfer of technologies to developing countries needed to help avoid dangerous climate change."[31]

Our worries about global warming are compounding with each passing year. We are in a race to slow the pace of global warming, while at the same time testing the steps we can take to reduce levels of GHGs in the atmosphere. We understand the causes of climate change, including the sources of the GHGs that are creating the problem. We also know, at least in principle, what we need to do to try to fix the problem. For the electric power sector, several responses are under consideration, including emissions controls, fuel switching, direct removal, and reforestation. Each will play a role. We must embrace all possible methods to meet the goals set out by the IPCC.

Reducing Emissions via Carbon Capture

Reducing power plant emissions of greenhouse gases is possible—albeit not commercially feasible—by applying carbon capture and storage (CCS) technologies. The idea of CCS is increasingly popular with those countries and industries that would prefer to keep burning coal, as is the case at the San Juan Generating Station in the northwest corner of New Mexico (Figure 6.23).

Figure 6.23 San Juan Generating Station in northwest New Mexico. A fight to extend the life of the remaining 940 MWe of generating power—just one of many such deliberations in the U.S. and other countries—is playing out in a state regulatory proceeding where carbon capture may provide the key to the plant's continued operation. Unfavorable financials, however, may still doom the plant. Photo credit: Creative Commons on Wikipedia, https://en.wikipedia.org/wiki/San_Juan_Generating_Station#/media/File:SanJuanGeneratingStation_NM_2012.jpg.

Implementation of such techniques awaits the resolution of the remaining technical, economic, and environmental problems. Still, there is no shortage of ideas being promoted by companies eager to cash in on this lucrative opportunity. The CCS chain consists of three parts: capturing the CO_2, transporting the CO_2, and securely storing CO_2 emissions underground in depleted oil and gas fields or deep saline aquifer formations. The Carbon Capture and Storage Association promotes such an approach as a safe and secure option: "At every point in the CCS chain, from production to storage, industry has at its disposal several process technologies that are well understood and have excellent health and safety records. The commercial deployment of CCS will involve the widespread adoption of these CCS techniques, combined with robust monitoring techniques and Government regulation."[32]

Achieving widespread adoption is the problem, and there are many obstacles in the way. CCS costs will be folded into the price of electricity. Installing CCS equipment will significantly lower the efficiency of the power plants. Because it will help sustain the use of coal in power production, its implementation will raise concerns for the health and safety of the coal miners as it also results in additional landscape alterations. And, we don't know that sequestering carbon underground will be environmentally wise or stand up to public concerns and scrutiny. Most problematic of all, CCS cannot be deployed broadly enough and quickly enough to keep the rising global temperature within desired bounds. We need to do a lot more work on this technology.

Fuel Switching

Fuel switching—that is, shifting to an energy source that is more benign—could reduce carbon emissions, and it is an approach that is already under way in many countries worldwide. There are many options for using this strategy. Among them are natural gas, nuclear power, and renewable energy resources. As the United States continues shifting away from coal, the most popular alternative has been natural gas because it can provide baseload power with much lower GHG emissions (Table 6.3).

Despite many reservations, nuclear power is another low-carbon option that is often mentioned in connection with fuel switching. In 2020, 440 nuclear power reactors operating in thirty countries plus Taiwan had a combined capacity of about 400 GWe, providing more than 10 percent of the world's electricity. The global community has not yet reached consensus over the wisdom of expanding our use of fission. Although the use of nuclear power is winding down in some countries—including Germany, France, South Korea, and the United States—not all countries are following suit.

Table 6.3 Summary of Life Cycle GHG Emission Intensity

Technology	Mean	Low	High
	Metric tons of CO_2e/GWh		
Lignite*	1,054	790	1,372
Coal	888	756	1,310
Oil	733	547	935
Natural Gas	499	362	891
Solar Photovoltaic	85	13	731
Biomass	45	10	101
Nuclear	29	2	130
Hydroelectric	26	2	237
Wind	26	6	124

* The lowest rank of coal.

Source: World Nuclear Association, "Comparison of Lifecycle Greenhouse Gas Emissions of Various Electricity Generation Sources," July 2011, https://www.world-nuclear.org/uploadedFiles/org/WNA/Publications/Working_Group_Reports/comparison_of_lifecycle.pdf.

About fifty-five power reactors are currently under construction in fifteen countries, notably China (thirteen reactors for a total of about 11 GWe) and India (seven reactors for about 5.4 GWe). Over 100 power reactors with a total gross capacity of about 120,000 MWe are on order or planned, and over 300 more are proposed. Most are in Asia, a region with fast-growing economies and rapidly rising electricity demand.[33]

This scale of construction should not imply that the questions about nuclear power have disappeared. Many remain unresolved, including permanent disposal of spent fuel, nuclear weapons proliferation, matters of inequity, environmental justice, and high capital costs. Its lower carbon footprint, however, may lend the technology a new lease on life. The tortured question dragging against a nuclear renaissance can be framed like this: which concerns you more, global warming or the radioactivity from nuclear waste, weapons proliferation, terrorism, and questions of generational justice?

Renewable Energy and Energy Efficiency

Another type of fuel switching would require a massive scale-up in the development of wind, solar, geothermal, and biomass options. This approach is among the easiest, quickest, cheapest, and cleanest options to reduce GHGs. Moreover, all these options can be installed in a modular fashion, a bit at a time, allowing gradual scaling while avoiding the large initial capital commitments necessary for the development of fossil fuels.

The developing countries that exercise this option can leapfrog conventional energy development in much the same way that those without landline telephones have adapted so thoroughly to mobile phones. Low-polluting, cheaper, modular—what's not to like? Many of these countries have large populations with little or no access to electricity. Renewable energy can help close the gap between the haves and have-nots.

Deploying renewable energy technology, especially if we can perfect affordable energy storage capabilities, can go a long way toward reducing GHGs. We are making substantial progress on this front, but costs are still too high in most applications. However, many utility companies have found that existing electricity storage options are economical options in isolated, marginal, and remote areas where the cost of maintaining reliable reserve margins would be too expensive.

Another option, one that the U.S. Department of Energy lumps together with renewables, is energy efficiency, and it can have a significant impact on reducing GHGs. The attraction of this option is that it can power lights, appliances, air conditioners, and so forth without having to develop additional conventional energy resources. This option would, therefore, contribute to the GDP of countries without the penalty of emitting more GHGs. In other words, energy efficiency's attraction is that of satisfying energy demands without having to develop more energy supplies.[34]

Reforestation

A third and conceptually easier approach takes advantage of proven nature-based systems for sequestering carbon by instituting massive reforestation. Reforestation is the process of replanting historically wooded areas. (It is not the same as planting forests where none existed before. That is called afforestation.) To be successful, it requires knowledge of soil, climate, terrain, and other factors. Reforestation has multiple advantages—such as controlling erosion and reintroducing wildlife habitat—but for our purposes, it is an action that will increase the absorption of CO_2 from the atmosphere. The question is, does it work? The answer is yes. It does work, and such action does help reduce the CO_2 that contributes to global warming. The EPA estimates, for example, that America's trees, soils, and wetlands each year capture around 11 percent of the nation's emissions. Improving tree plantations, managing cropland nutrients, and restoring tidal wetlands also offer promising mitigation opportunities, but on a much smaller scale. Combining these approaches could sequester up to 21 percent of net annual emissions from the United States.[35]

The World Resources Institute has recently reported on some developments related to this nature-based approach to carbon removal:

> The people of Ethiopia set a world record . . . by planting over 350 million trees in 12 hours as part of Prime Minister Aiby Ahmed's Green Legacy campaign. In India's most populous state, Uttar Pradesh, a government-led campaign against climate change prompted citizens to plant 220 million trees in a single day in August. Combined, these efforts could remove 10 million tons of CO_2 per year from the atmosphere, based on IPCC data — equivalent to taking 2.2 million cars off the road.[36]

Globally, 0.9 billion hectares of canopy cover could store 205 billion metric tons of carbon in areas that would naturally support woodlands and forests.[37] Such a Herculean task, while feasible, seems unlikely in the short term or on the massive scale necessary, albeit any removal would be welcome. Perhaps adding a "price" to each ton of emitted carbon would more effectively spur action.

Direct Air Capture

Another approach to lowering GHGs is through direct air capture (DAC). We need to remove CO_2 from the atmosphere, not just avoid putting more in. Even if we maximize carbon capture and storage, energy efficiency, fuel switching, natural climate solutions, and every other method of emissions reduction, we still cannot remove enough CO_2 to avoid a rise in the global temperature of more than 1.5°C by 2050. For this reason, we must target direct air removal and embrace it wholeheartedly. Many technologies are being promoted that might accomplish this task, and we must perfect and deploy the best ones. To date, however, none of them has been implemented on more than an experimental scale.

One of many private companies working on this problem, Carbon Engineering (backed by Bill Gates), has a process that summarizes air capture this way: "Our Direct Air Capture technology does this by pulling in atmospheric air, then through a series of chemical reactions, extracts the CO_2 from it while returning the rest of the air to the environment. This is what plants and trees do every day as they photosynthesize, except Direct Air Capture technology does it much faster, with a smaller land footprint, and delivers the carbon dioxide in a pure, compressed form that can then be stored underground or reused." The estimated cost of capturing a ton of CO_2 was around $600 in 2011, way too high to be viable.

By about 2018, the cost had fallen to between $94 and $232, which is still too high because almost any source of renewable energy and energy efficiency can prevent a ton of carbon cheaper than that. But "beneath $100, DAC starts to look viable in a low-carbon world." Many universities are also working on DAC.[38]

Behavioral Changes

We can also help reduce carbon in the atmosphere by changing some of our behaviors. In some ways, these types of changes will be the most problematic to implement. Why? Because they fall within the worlds of politics, negotiations, financing, and compromise. The most logical place to start is to target the sources of GHG. That means convincing the three largest emitters (China, the European Union, and the United States) to take the necessary actions to reduce their emissions, particularly in the energy sector but elsewhere too. The question that remains is whether we can convince the Chinese, European, and U.S. populations to eschew fossil fuels. Alternatively, can we get them to accept nuclear fuels? Or can we plant enough trees, or get enough people to embrace the idea of electric cars?

One of the first steps in this litany of solutions is to convince leadership at all levels of government to commit to reducing their community's carbon footprint, and then to follow through to enact enabling legislation, subsidize necessary research, introduce incentives and penalties, and support and promote the changes that would be necessary. Sixteen countries have now adopted net-zero targets—Bhutan, Costa Rica, Denmark, Finland, France, Germany, Iceland, the Marshall Islands, New Zealand, Norway, Portugal, Sweden, Suriname, Switzerland, the United Kingdom, and Uruguay.[39] You will notice that the United States is not listed.

A related step is to identify the impacts of each person and the steps that each of us can take. A 2017 article summarized four recommended actions that the authors believe to be especially effective in reducing an individual's GHG emissions: having one fewer child, living car-free, avoiding airplane travel, and eating a plant-based diet. They also measured several other, lesser personal actions, such as hanging laundry in the sun instead of using a clothes dryer.[40]

Summing Up

In the past several decades, the global climate has been changing faster than ever. We now realize that these changes—especially global warming—result from the emission of massive amounts of global warming gases as a result of human activities. If the world community eschews drastic corrective measures, life is going to get downright nasty for just about everyone. Even for the very wealthy and privileged, there will be nowhere to hide from the tsunami of climate change. If we do not get global warming under control, no one will be spared. Can we save ourselves?

Now that you have a basic knowledge of the problem, its causes, and possible solutions, what are you going to do? The best starting point, as always seems the case, is for you to take personal action by understanding and reducing your own carbon footprint.[41]

7

Policy

Guiding the Direction of Energy

What Is Policy?

Have you ever tried to get a refund? Or, perhaps, be granted a personal exemption to a rule? If you have, you probably are accustomed to hearing something like "I am sorry, we can't do that. It's just our policy." Throughout our lives, this response is so typical that we are used to getting quite frustrated by the intransigence of such policies, because usually there is nothing we can do about it. It is just "policy"—always arguable, rarely with any legal standing, and often with no implications outside your personal life.

Now, consider how things change when we add the adjective "energy." Policies become more standardized and often carry provisos to stimulate some desired outcome. They may even be formalized with higher authority through legislation, international treaties, monetary incentives, taxes, occasionally even penalties. It is then that they become legally enforceable. When we consider policy, sooner or later we come to accept that *energy is a social issue with a technical component, rather than the other way around.*

Imagine a city council that declares: "It is the policy that our city becomes 100 percent reliant on renewable energy." That sounds like a laudable guiding principle, and the intended actions fundamentally voluntary. Compare that with "A building permit in this city requires you to certify to our satisfaction that the energy needs of your structure will be offset by the generation of an equal or greater amount from renewable resources over a calendar year." That statement not only is more robust but sounds somewhat punitive.

Energy policies are more numerous than ever. They affect every aspect of energy supply and consumption—from the efficiency of the appliances you use and the car you drive to matters of international trade and global geopolitics. Policy is fundamental to all discussions about energy, including growing demand, global competition for energy resources, mounting environmental impacts of their use, and how they are ultimately used. Matters of policy are rising in importance, and everyone is taking notice.

From an academic perspective, the growing public discourse on energy policy is reflected in the rising number of refereed scholarly journals now available on the subject (Table 7.1). Energy policy has become a hot topic, as people are

The Thread of Energy. Martin J. Pasqualetti, Oxford University Press. © Oxford University Press 2021.
DOI: 10.1093/oso/9780199394807.003.0007

Table 7.1 A Short List of Energy Policy Journals to Consult

Energy Policy (https://www.journals.elsevier.com/energy-policy)

Renewable Energy Law and Policy Review (https://relp.lexxion.eu/)

International Journal of Energy Technology and Policy (http://www.inderscience.com/jhome.php?jcode=ijetp)

The Journal of Energy Law and Policy (https://www.tandfonline.com/loi/rnrl20)

Economics of Energy and Environmental Policy (http://www.iaee.org/en/publications/eeeporder.aspx)

International Journal of Energy Economics and Policy (https://www.scimagojr.com/journalsearch.php?q=21100281302&tip=sid)

Energy Systems and Policy: An International Journal (https://www.researchgate.net/journal/0090-8347_Energy_systems_and_policy_an_international_interdisciplinary_journal)

---

becoming aware of its importance in decisions at all levels of jurisdiction and security. These journals are a rich source of information, and you might wish to thumb through them sometime. You will be surprised at how often what you read applies to you personally.

Energy policies vary in breadth and detail. They also vary in their impacts, a result of their position at the busy intersection of economic development, standards of living, financial requirements, environmental costs, and geopolitical impacts. Pushed to and fro by the forces of business, national interests, public whim, technological change, and world events, existing and proposed energy policy are often the focus of careful consideration and even heated debate.

When heated debate occurs, the most common focus is on how to identify and apportion winners and losers. Energy policy can strongly influence the portfolio of available energy resources, the environmental impacts of using those resources, which interest groups might benefit, and what the costs will be. So much rides on energy policy positions and adjustments that achieving consensus on the "best" balance of energy resources is a challenging assignment. One of the factors that encumber reaching any agreement is the large number of vested interests involved. Each one offers its perspective on policies proposed or implemented.

In this regard, energy developers, consumers, and the deliberative bodies that influence energy supply and demand can be immobilized by the avalanche of confusing and contradictory information. For some of these entities, money is the driving force motivating decisions. For other entities, the goal of environmental sustainability is the most critical influence. For still others, national security is the primary goal. Greed, altruism, survival, personal motivations, and even the health of the planet going forward are among the many factors that come into play in creating useful and helpful energy policy. For these reasons, enticing all parties to stay within a circle of compromise—a principal goal of negotiations—can be an elusive or even impossible ideal.

A second factor that slows the creation and acceptance of viable energy policy is characteristic indecision and incomplete transparency. For these reasons, negotiations often involve hedging by all sides. Imagine, for example, that your government representative adopts an anti-nuclear policy only to realize years later that the national economy is standing on the edge of collapse because alternative energy resource development was not able to meet expectations. Or imagine what might happen as we continue decommissioning coal-burning power plants. Perhaps only later will we discover that the expected abundance of cheap natural gas was incorrect, or that scientists found that the causal link between coal burning and global warming was not as critical as they once believed.

When we survey the multiple energy pathways that are available, plus the divergence of opinion about which path would be best, hesitancy reigns supreme. With such considerations in mind, what should we do? Which resources should we emphasize? How much money should we invest? What will pay off in the end? Which options carry more or less risk? Will one path be smoother than another? Is there enough political courage to wholeheartedly support one pathway over another? Which path do we think will provide for all future contingencies? Should we continue to emphasize intrusive wind farms and sweeping solar arrays? Should we accept that nuclear power is sufficiently safe to be used to lead us to a low-carbon future?

No country wants to come up short on meeting the energy needs of its citizens. Likewise, few would favor ripping up more land for the coal it hides, choking on the pollution produced when that coal is burned, heating the planet if we can avoid it, or leaving behind a legacy of hazardous waste.

Instead of taking the awkward position of choosing winners and losers, politicians often find it expedient to propose an "all of the above" approach. Such a choice means that funding sources, research activities, entrepreneurial spirit, and public backing become diluted. Is that wise? Arguably it is not, but it is undoubtedly the pathway we often accept, at least in democratic societies, where choosing sides is complicated, protracted, and politically risky. Meanwhile, China is the world leader in wind power and solar power, and it is currently building about a dozen nuclear power plants.

Within democratic societies, however, we know that political considerations rule the day and are always subject to change. Whatever direction one administration favors can be modified or even abandoned by a later administration. For example, in his 2011 State of the Union address, President Obama declared: "By 2035, 80 percent of America's electricity will come from clean energy sources." In contrast, his successor, Donald Trump, promised to protect coal and nuclear generation by offering government subsidies in the name of national security.[1] Obama, under the National Fuel Efficiency Policy, called for all cars to have an average fuel efficiency of 35.5 mpg by 2016 and for the fleet of American automobiles and trucks to reach 54.5 mpg by 2025, winning praise from environmental groups and a majority of the public.[2] President Trump called for substantially lower targets, much to the appreciation of automakers (and, presumably, oil companies).[3] The corporate average fuel economy (CAFE) standard is one example of how policies can change in significant ways, often from one administration to another.

Another example is the methane rule.[4] Under the Trump administration, the EPA loosened Obama-era national standards on the extraction of oil and natural gas. These had been implemented to limit methane from leaking into the atmosphere. The 2018 final rule revises the 2016 rule to reduce "unnecessary compliance burdens," such as waste minimization plans, well drilling and completion requirements, pneumatic controller and diaphragm pump requirements, storage vessel requirements, and leak detection and repair requirement. The 2018 final rule became effective on November 27, 2018.

Such shifts in policy reflect the varying ideas of interest groups, focused lobbying by industries, and the influence exerted on elected officials. Ultimately they waste money and momentum and often disrupt employment conditions. In the United States—but in other countries, too—no one seems willing to or capable of breaking this pattern of politically motivated shifts in policy. As a result, we suffer inefficiencies in meeting demand and prolong anxiety about the future, all the while snapping stock market values back and forth for thousands of companies that reel from the uncertainties.

Policy choices, especially those with substantial implications for future energy conditions, can be as perplexing as they are abundant, and of course they vary in terms of scale. For example, municipalities implement policies that affect efficiency at the scale of houses and transport through building permits and zoning, in addition to influence over utilities (which in turn influence electricity sources). State governments influence permitting of oil, gas, and renewables. Federal policies focus on interpretations of the Clean Air Act, Clean Water Act, and so on, and on developing new statutes. Whatever level we are dealing with, there will be a wide variety of opinions offered by the public, the press, and political rivals.

What actions should we take? Should we establish incentives and subsidies, bolster or curtail budgets, prompt more significant international trade and training? How do we measure the impact of energy policies at different scales: household, neighborhood, city, county, state, country? Will new policies overlap ones that are already in effect? Will they be redundant or even antagonistic to one another? How long should they last? Should the policies we are considering affect all our constituencies equally—governments, military, industry, commercial, residential—or should we apply a skewed measuring tool?

One of the most significant energy policy decisions we face is which technology options to favor and support. Should we cast aside fossil fuels because of their documented environmental and health risks? Or should we help guarantee their future by permitting more mines and pipelines? On the other hand, if we decide to promote renewable energy resources, can we manage the inconsistent flow of electrons that result from the natural fluctuations of solar and wind power? Looking into the future, how do you think your descendants would react if we got it all wrong? Should we worry about it? Does it even matter? What should we do? With these questions in mind, it is not surprising that we often seem helpless to overcome inertia and indecision.[5]

My hope for this chapter is that you grow to appreciate the importance of energy policy. Considerations of policy tend to meld a broad spectrum of factors and influences. They provide guidance and direction to developers, governments, international organizations, and private citizens. Without policies, energy development would be a disorganized and wasteful free-for-all.

With expanding energy access, the promulgation and influence of energy policies are growing as well. Such policies may be successful in creating jobs, influencing security, channeling trade, governing price, responding to environmental costs, steering development and availability, influencing issues of justice and access, and touching the personal lives of every person on the planet. Wound together into a single strand, policies make up one of the strongest and durable threads of energy.

The Energy Policy Trilemma

Energy policy is complex and variable, both from one place to another and from one social stratum to another. Likewise, it tends to consist of three things: security of supply, sustainability, and competitiveness (Figure 7.1). The challenge is how to keep everything in balance.

How can we achieve the best balance among all three influences? We must start by asking the most basic question: what are our objectives? In the broadest sense, the purpose of energy policy is to guide actions, encouraging or discouraging certain steps. Take the example of a policy that affects when and how students can drop a college class. The purpose of such a policy is to *encourage* you to graduate on time and to *discourage* you from demanding more and more time from those who must deal with your indecision if you don't.

Throughout your life, you will continuously be scrambling around within a web of policies. One of the first places you encounter this is in college. Say you want to graduate early, take an overload of classes, substitute experience for class time, or take a final exam earlier or later than officially scheduled. Most of the time, you will encounter resistance to granting exceptions from the policies already in place. You might have to satisfy a list of criteria, you might have to enroll for another semester, you might need a letter from a faculty sponsor, or you might not be permitted any deviation at all. If you flout the established policies,

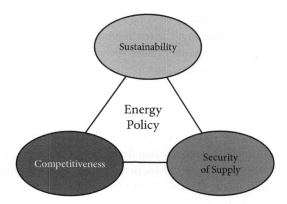

Figure 7.1 Three components of energy policy. *Source:* http://energy350.com/ energy-policy-a-bi-partisan-argument-for-energy-conservation/.

there are many points of leverage to compel you to comply, including denying a diploma. Such are the powers of policy.

What is the ultimate purpose of the incentives and disincentives that policies impose? Simply put, they are usually to maintain standards for the good of the whole and to avoid the chaos that would ensue if everyone could live by whatever they felt was in their interest (i.e., as if they were alone on the planet).

Imagine what would happen if there were no policies in place in your neighborhood and no policies to protect your well-being and property values. Your community, for example, may have policies against keeping noisy roosters (nuisance) or wild animals (dangerous) in your backyard.

City leaders—leaders you elected to represent you and to whom you have given specific authority—use policies to create incentives for certain types of behavior and actions. For example, you would not be allowed to keep a horse in your backyard in San Francisco, but if you moved to a house in Marin County, ten miles north, you would find that horses and many other animals are common. In these cases, policies are in place for some perceived public good. In the broadest sense, public good is the goal of all policy.

As you probably suspect, weaving the threads of energy policy into a beautiful tapestry of objectives and constraints can be a difficult challenge. It usually involves broad cooperation. For example, policies that support sustainability, security, and environmental stability tend to loathe isolationism. From house to house, city to city, state to state, and country to country, cooperation among all constituents—whether voluntary, recommended, encouraged, or coerced—is essential for the success of energy policy.

Because the stakes of inappropriate policies can be high at the international level, the importance of cooperation is most apparent there. Shifting policies can affect millions of jobs, billions of dollars, and global environmental health. For example, the United States has policies that favor importing oil from Canada. We have come to rely on Canadian oil, including how easy it is to transport it from Canada to the United States, its abundance and reasonable cost, and the fact that it is coming from a friendly and stable supplier.

Likewise, there are similar examples from the world of renewable energy. We import wind energy turbines from Denmark and solar modules from China. We favor these sources because they are less expensive than what we can obtain from U.S. manufacturers (unless protective tariffs are imposed). The result of such price-sensitive preferences is that domestic suppliers may go out of business. Reverse that policy to protect American industries, as President Trump favored, and domestic prices will rise. Nothing is perfect, and trying to follow the zigzags of energy policy—or predict their changing impacts—is difficult, time-consuming, and frustrating.

Energy policies are social instruments imposed on a technical enterprise. Doing so provides order and stability to all aspects of our lives, and it is part of the bargain we make to live in a civilized society. It is not surprising, then, to realize that social issues outnumber technical issues in most energy considerations. Policy is at the center of all factors affecting development, illustrating again *that energy is a social issue with a technical component, rather than the other way around*. We can see that when we look at some specific examples and several levels of government.

U.S. Energy Policies

If you were asked to identify or define our national energy policy, would you be stymied? Considering the implications of our energy decisions on energy security and the environment, you would be surprised to learn that a unified national policy does not exist. It is widely accepted that we must establish energy goals, pathways, purposes, and procedures, or we will inevitably promote a scattergun approach, hoping that we will somehow hit the bull's-eye of what we will need moving forward.

The fact that we do not have a national energy policy does not come from the absence of the need or a lack of effort. We have made several attempts to reach a consensus. As experts have pointed out, every U.S. president since Richard Nixon has pledged allegiance to the flag of "energy independence," even as the United States has remained dependent on imported oil. "A century and a half of an idea whose time has never come," said Charles Homans in an article in *Foreign Affairs* entitled "Energy Independence: A Short History."[6] As part of his argument, Homans usefully identified highlights in national energy policy development (Table 7.2).

Table 7.2 Highlights of National Energy Developments from 1859 to 2011[7]

1859	Edwin L. Drake drills the world's first oil well near Titusville, Pennsylvania.
1865	In *The Coal Question*, British economist William Stanley Jevons warns that Britain is in danger of running out of coal, threatening the country's strategic and economic preeminence—a matter, he writes, of "almost religious importance."
1912	Winston Churchill, then First Lord of the Admiralty, begins converting Britain's Royal Navy—the world's largest naval force—from coal to oil, exchanging energy independence for power and speed. Over the next several years, Britain wrestles the oil fields of the Persian Gulf away from a crumbling Ottoman Empire.
1960	Oil exporters Iran, Iraq, Kuwait, Saudi Arabia, and Venezuela formed the Organization of the Petroleum Exporting Countries (OPEC), which grows to fifteen countries by 2020.
1960	After taking power in a military coup, Muammar al-Qaddafi starts nationalizing Libya's oil fields—then the source of 30 percent of Europe's oil imports—thus beginning Western oil companies' expulsion from the Arab world.
1970	The United States becomes a net oil importer.
1971	Oil production in Texas, the motherland of U.S. crude, begins to decline. "Texas oil fields have been like a reliable old warrior that could rise to the task when needed," Texas Railroad Commission chair Byron Tunnell says. "That old warrior can't rise anymore."
1972	"The era of low-cost energy is almost dead," U.S. commerce secretary Peter G. Peterson declares. "Popeye is running out of spinach."
1973, April	In a *Foreign Affairs* article titled "The Oil Crisis: This Time the World Is Here," James E. Akins, who would become U.S. ambassador to Saudi Arabia that fall, warns that "the threat to use oil as a political weapon must be taken seriously."
1973, October	Egypt and Syria launch a surprise attack on Israel, beginning the Yom Kippur War. The United States responds with $2.2 billion in arms and aid to Israel. The following day, OPEC declares a halt to oil shipments to the United States, Western Europe, and Japan. By January 1974, oil prices had more than quadrupled.
1973, November 7	U.S. president Richard Nixon announces "Project Independence," a plan to wean the United States off foreign oil. The dream of American energy independence is born.
1975	The U.S. Congress passes the Energy Policy and Conservation Act, creating the Strategic Petroleum Reserve and imposing the first fuel-efficiency standards for vehicles.

Continued

Table 7.2 *Continued*

1977	U.S. President Jimmy Carter makes energy independence the central ambition of his presidency, later establishing the Department of Energy, investing billions of dollars in research and development, and installing solar panels at the White House.
1981	U.S. President Ronald Reagan announces his plan to lift price controls on oil and begin disassembling the renewable energy research programs begun under Carter.
1989	The *Exxon Valdez* oil tanker runs aground in Alaska's Prince William Sound. Americans' support for offshore drilling plummets and remains low well into the 1990s.
1990	After accusing Kuwait of stealing Iraqi oil with slant-drilling techniques, Saddam Hussein seizes Kuwaiti oil fields. U.S. president George H. W. Bush leads a coalition to oust him.
1994	For the first time since the 1910s, the United States imports more oil than it produces. Deputy energy secretary Bill White describes the situation as "the biggest trade problem we have."
2001	Vice President Dick Cheney's energy task force releases its National Energy Policy. "Our increased dependence on foreign oil profoundly illustrates our nation's failure to establish an effective energy policy," the report states, recommending a renewed commitment to domestic oil, coal, natural gas, hydropower, and nuclear power.
2005	Congress passes the Energy Policy Act, including quotas and millions of dollars in subsidies, in hopes of nearly doubling U.S. ethanol production by 2012.
2007	Congress passes the Energy Independence and Security Act, which imposes stricter fuel efficiency standards on vehicles and orders a 766 percent increase from 2007's targeted biofuel production by 2022. By 2008, ethanol had become a $32 billion business in the United States.
2008	Oil prices hit a record $148 a barrel.
2008, September	At the Republican National Convention, Michael Steele calls on Americans to "reduce our dependency on foreign sources of oil and promote oil and gas production at home. In other words: Drill, baby, drill!"
2010	A weak economy and more efficient vehicles cause oil imports to fall below half of U.S. consumption for the first time in thirteen years.
2011	U.S. shale gas production reaches 5 trillion cubic feet, five times its 1990 level, reigniting hopes for homegrown energy.

While energy policy can be introduced in myriad circumstances and for multiple purposes, establishing a national-level policy in the United States is a thorny task, mainly because the stakes are substantial. Yet many countries have made an effort. The International Energy Agency tracks such policies for every member country.[8] Many of these countries, however, are smaller, less complicated, and more efficient in their energy use than the United States.

But there are other factors at work. One of the principal obstacles to the establishment of national energy policy in the United States—not unique to this country, just more meaningful here—is the hesitation by leaders to pick (or legislate) winners and losers from the multiple possible energy futures. Most commercial energy lobbyists concentrate on what is best for their constituency and only sometimes on what is necessary for the common good.

The most effective and motivated lobbyists are those who aim at maintaining the status quo, that is, business as usual. Many organizations take this stance, as when President Trump declared in May 2018 his intention to shore up the nuclear and coal industries. He proposed subsidizing them with federal monies, even though closer adherence to strict market-based decisions continued to favor letting both continue to decline.[9]

The path to most changes in direction is littered with obstacles, even if the changes promise improvements such as cleaner air and water. For example, the Clean Air Act has become a key element in some energy policy debates, such as over emissions of mercury and carbon. And sidestepping the Clean Water Act provided a loophole for hydraulic fracturing.

One of the most common topics of consideration, debate, compromise, and lobbying is the uneven financial considerations that are often involved. Not everyone can achieve their personal goals, of course, because not everyone has the same financial deep pockets. The same is true when it comes to policy. For example, consider the lobbying industry. In simple terms, lobbyists are (usually) paid advocates for particular positions on policies and associated legislation. They represent the interests of those who pay them (although sometimes significant lobbying is conducted by committed and unpaid citizens).

Looking at the money involved will give us a sense of the influence of lobbyists, including who is winning and who is losing. Take, for example, the fossil fuel industry's position on climate change. With more than $2 billion spent lobbying against climate-change-related legislation in the United States from 2000 to 2016, the fossil fuel industry, transportation companies, and utilities outspent environmental groups and the renewable energy industry by a factor of ten (Figure 7.2).[10]

Another hurdle to the promulgation of a successful comprehensive energy policy is the complexity it commonly involves, including national laws, treaties, agency directives, and political considerations. To provide an inkling of this

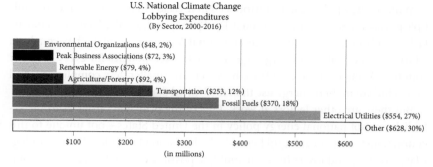

Figure 7.2   Fossil fuel industries expenditures in climate-related lobbying efforts. *Source:* http://askmagazine.org/fossil-fuel-industries-outspend-in-climate-related-lobbying-efforts/.

intricacy, consider that the energy policy of sovereign nations must include at least one and usually more of the following measures:[11]

- Statement of national policy regarding energy planning, energy generation, transmission, and usage
- Legislation on commercial energy activities (trading, transport, storage, etc.)
- Legislation affecting energy use, such as efficiency standards and emission standards
- Instructions for state-owned energy sector assets and organizations
- Active participation in, coordination of, and incentives for mineral fuels exploration and other energy-related research and development policy command
- Fiscal policies related to energy products and services (taxes, exemptions, subsidies)
- Energy security and international policy measures, such as:
  - International energy sector treaties and alliances
  - General international trade agreements
  - Unique relations with energy-rich countries, including military presence and domination

To picture the challenges of establishing national policies, imagine that you are personally assigned the task of organizing a discussion on the topic. Could you do it? How would you organize the dialog? Here are some of the most salient topics:[12]

- What is the extent of energy self-sufficiency for this nation?
- What is the location of future energy reserves?
- How will future energy resources be consumed (e.g., among sectors)?
- What fraction of the population suffers from energy poverty?
- What are the goals for future energy intensity?
- What environmental externalities are acceptable and forecast?
- What form of "portable energy" (such as motor vehicle fuels) is predicted?
- How will energy-efficient hardware such as hybrid vehicles and household appliances be encouraged?
- How will national policy drive municipal, state, and municipal functions?
- What are specific mechanisms (e.g., taxes, incentives, manufacturing standards) in place to implement the complete policy?
- What future consequences will there be for national security and foreign policy?

Another complication in crafting national energy policy is the risk of a mismatch between supply and demand. To better understand this concern, consider that your utility company must continuously keep supply and demand in close balance. They cannot (yet) store significant amounts of electricity, and they can never rest in carefully matching sources of supply with locations of demand. Maintaining such balances requires constant monitoring of equipment, plus accurate weather forecasts of such variables as temperature, cloud cover, humidity, winds, and snow cover. It also means that they must be constantly aware of the price and reliability of the electricity available to them—both within its system and outside—plus the amount of available transmission capacity and the fluctuating demands of its customers. Add into the mix the fact that some of the utility's customers—particularly those with installed solar—feed electricity back into the system. Maintaining reliability is a complicated task for utility companies.

Miscalculations and the vagaries of weather can have serious consequences. Having more demand than supply, even for a brief period, can result in life-threatening shortages. In some circumstances, such shortages could even threaten national security, as when military threat assessment capabilities suddenly lose power. Utility companies typically have policies in place for such events, including protocols for servicing critical users such as emergency rooms, intensive-care units, and people on dialysis machines.

On occasions when utility reserve safety margins are insufficient to meet unexpected demand, other policies are in place to issue public appeals to reduce electricity use voluntarily. These actions were necessary during the wildfires in

San Diego County in 2014, when 1.4 million customers were asked to cut back on their demand.[13]

If such voluntary reductions fall short, options become much more limited. Utilities might initiate rolling brownouts or even a blackout, where customers are cut off, usually for short periods, until the emergency conditions pass. Often these are tolerable if inconvenient outages; for example, timers, cable boxes, emergency equipment, and alarms may have to be reset. In extreme conditions, such as the lethal threat from forest fires ignited by transmission equipment, utility companies might suspend service to millions of customers. Such action was taken preemptively in 2019 in California, and it caused disruptions lasting up to several weeks.

In response to the cumulative burden of such inconveniences, the Public Technology Institute offers technical assistance to local governments through the Local Government Energy Assurance Planning (LEAP) program, which helps communities develop or refine existing energy assurance plans.[14] Bottom line: no one wants to have more demand than supply, even though there are policies that might be in place to mitigate the outages. Nevertheless, even with all the options, policies, and years of preparation, nothing was enough to help Puerto Rico in the fall of 2017 when Hurricane Maria devastated the island's entire electric grid. That event demonstrated how much we all have come to presume we can get all the energy we need when we need it. It also drove home how much we rely on the smooth operation of the vast energy infrastructure that has grown to maturity in the last century. We have also been learning that—despite extensive procedures, training, redundancies, and policies—disruption can always render us vulnerable.

Years of wrangling, bargaining, and deal-making about national energy policies seldom found consensus among all parties. But two were put in place in the mid-1970s, and they had broad possible bipartisan appeal. Each was designed to wean the United States off its dependence on imported oil and the associated vulnerability. At the time, we were importing more than 60 percent of our oil, and we fell victim to shifting political winds. In retaliation for U.S. support of Israel during the Yom Kippur War in 1973, the Organization of Exporting Countries (OPEC) placed an embargo on supplies sent to the United States from its member countries. With lightning speed, gasoline stations were overwhelmed by customers racing to fill up. By the time the embargo ended and OPEC members realized the power of their organization, the per-barrel price of crude oil was up 400–500 percent.

With the possibility of more embargo threats hovering over us, we increased the flexibility of our sources, thereby reducing our dependence on foreign suppliers. We also passed legislation to raise automobile mileage standards by establishing 55 miles per hour as the maximum highway speed.[15] Over the years, we improved oil recovery techniques, such as the effectiveness of hydraulic fracturing (fracking). These measures have had the desired effect: domestic supplies have gone up, and oil imports have dropped. Consequently, we have reduced the amount of money we are sending abroad for such oil by tens of billions of dollars per year. Everyone seems pleased, particularly the U.S. oil companies, which are reaping enormous new profits.

In addition to improving the efficiency of how we use oil, we also established a policy of stockpiling "black gold" in the Strategic Petroleum Reserve (SPR). President Ford set the SPR into motion when he signed the Energy Policy and Conservation Act on December 22, 1975. The legislation reinforced the stated policy of the United States to establish a reserve of up to 1 billion barrels of petroleum. We are still not quite there, but the SPR has since grown to around 700 million barrels of oil stored in four facilities in Louisiana and Texas.

While fracking and the SPR have helped the U.S. supply reduce its vulnerability, the demand for oil makes up a thin slice of the energy policy pie in the United States. Much more can be done to establish a meaningful national energy policy, but the prospects of achieving that goal do not look bright. The hesitation comes from the understandable position that all citizens and organizations view energy balances and positions differently.

When differing opinions and interests are rampant, non-governmental organizations (NGOs) can be excellent sources for neutral recommendations. One of the most respected is Resources for the Future (RFF). In 2010, in cooperation with the National Energy Policy Institute (NEPI), RFF issued a report outlining principal factors and critical energy policy choices that provide us with a picture of the complications involved in setting national energy policy.[16] More than ten years later, that report still holds timely recommendations and insights into many issues that continue to complicate settling on national energy policy. They include:

- Worries about resource dependence, particularly regarding fossil fuels
- Concerns about the emission of greenhouse gas and their effects on the global climate
- National security
- Environmental costs to the atmosphere, water, and land of continuing our present path

To addresses these issues, RFF and NEPI joined in calling for critical actions. They urged we "jumpstart the transition away from fossil fuels" with "a rigorous, wide-ranging analysis of energy policy options." They called for policymakers "to select a broad portfolio of strategies that reduce . . . the use of oil and GHG emissions and take account of other environmental damages." Their analysis used the rich assortment of available studies, going beyond what others had offered to allow for a focused, consistent ranking and analysis across specific policy options and a range of dimensions.

The report emphasized the importance of creating *specific* government policy instruments that drive changes in private markets. It stressed the vital role of two metrics. First, while keeping financial costs in mind, it called for reduced oil consumption. Second, it called for a reduction in tons of CO_2 emitted. With these two metrics in mind, it examined thirty-five policy scenarios, including four crosscutting policy options, against a reference case.

A hallmark of this report is its attention to what's known as "welfare costs." This emphasis assumes fundamental microeconomic principles in which *cost* is defined as the value of the resources that society gives up while trying to achieve a given reduction in oil use and CO_2 emissions. The report emphasized welfare costs because to exclude them would misrepresent the overall economic burden. Including them led them to essential conclusions that continue to ring true. For example, the report concluded that no single policy would simultaneously reduce oil consumption and CO_2 emissions significantly. Instead, multiple policy actions will be needed to achieve the desired result. Importantly, however, they also cautioned against taking a "buckshot" approach "in which several uncoordinated policies are implemented that may cancel out any intended benefits or even make things worse."

The National Energy Act of 1978 (NEA78) was a significant step in energy legislation that affected both supply and demand. It included the following statutes:

- Public Utility Regulatory Policies Act (PURPA)
- Energy Tax Act
- National Energy Conservation Policy Act
- Power Plant and Industrial Fuel Use Act
- Natural Gas Policy Act

This package was soon followed by the Energy Security Act on June 30, 1980. It included seven major statutes:

- U.S. Synthetic Fuels Corporation Act
- Biomass Energy and Alcohol Fuels Act
- Renewable Energy Resources Act
- Solar Energy and Energy Conservation Act
- Solar Energy and Energy Conservation Bank Act
- Geothermal Energy Act
- Ocean Thermal Energy Conversion Act

Three additional energy bills were more national in scope. One was the Energy Policy Act of 1992.[17] This law created mandates and amended utility laws to increase clean energy use and improve overall energy efficiency in the United States. The act consists of twenty-seven titles detailing various measures designed to lessen the nation's dependence on imported energy, to provide incentives for clean and renewable energy, and to promote energy conservation in buildings.

The Energy Policy Act, signed into law on August 8, 2005, made a noble if only partially successful attempt to establish some order to national policy.[18] It provided tax incentives and loan guarantees for energy production of various types. For example, it sought to increase the use of coal as an energy source—while also reducing air pollution—by (1) authorizing $200 million annually for clean coal initiatives, (2) repealing the 160-acre (0.65 km^2) cap on coal leases, (3) allowing the advance payment of royalties from coal mines, and (4) requiring an assessment of coal resources on federal lands that are not national parks. It also called for increased oil production, and it promoted nuclear reactor construction in the United States through incentives and subsidies, including cost-overrun support up to a total of $2 billion for six new nuclear plants.[19] Renewables and efficiency received comparatively little direct attention.

One of the most recent energy policies was enacted in 2016 (Figure 7.3). While it provides further guidance to the complex issue of national energy policy, it too comes up short, especially compared to many European or Asian countries (more on that later).

Many unresolved questions about energy policy still loom over us:

- Are we or are we not going to move away from coal? Obama said yes, we are. Trump said no, we are not.
- Are we or are we not promoting renewable energy? California and Hawaii say yes. Alabama and Mississippi say no.
- Are we ready to move steadily toward higher automobile mileage standards steadily? Obama instituted goals in the range of 50 mpg. Trump wanted to cut those goals substantially, while Biden aims to maintain a high standard.
- Are we going to build new nuclear power plants? The nuclear industry says yes. Most environmental groups say no.

Should we continue trying for consensus on a comprehensive national energy policy, as several presidents have proposed or even promised, or should we allow every state to decide for itself? Can a distributed policy strategy work? What about such issues as the interstate movement of electricity and air pollution? As before, our choice seems to be between continuing to move in several directions simultaneously and hoping for a technical breakthrough that resolves all our problems at once—a real golden egg.

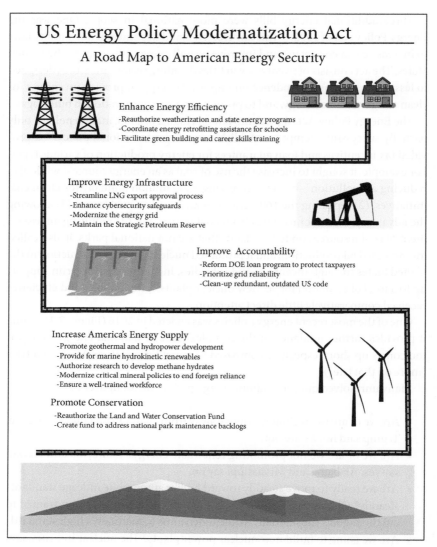

Figure 7.3 Energy Policy Modernization Act of 2016.

Energy Policies in the International Context

The geography of energy is the fundamental basis of international (and national) energy policy. The following five examples demonstrate the weighty influence of policy decisions on energy availability, energy dependency, public health, and safety, as well as national security.

France

The country's long-held pro-nuclear policy has translated into a heavy reliance on the atom for its electricity. What is behind that position? With insufficient reserves of fossil fuel to meet growing demand and a reluctance to rely on imports to make up the difference, French leaders decided several decades ago that nuclear power was the best alternative. Jean-Pierre Chausaude, of Electricité de France, noted: "In France, we have no oil, no gas, no coal, no choice. And for the French people, it was very positive to develop national energy with nuclear energy."[20]

Their policy choice sprang from the realities of the 1973 oil crisis that accompanied the Yom Kippur War.[21] When OPEC reduced oil exports and quadrupled prices, many of its dependent customers began a serious search for alternatives. On March 6, 1974, Prime Minister Pierre Messmer announced a new policy that favored nuclear power.[22] By 2019, atomic power would be generating about 72 percent of the country's electricity demand from its fifty-eight nuclear reactors. Of the total of about 380 TWh of electricity generated, 38 TWh was exported to neighboring countries.

The rest of the world watched this "experiment" in nuclear power for several dozen years, curious to learn if such a heavy reliance on one source of electricity would have any drawbacks. All seemed good until the Chernobyl accident in 1986 contaminated large parts of the surrounding area in Ukraine and Belarus, prompting second thoughts—but no stable change in policy—about France's commitment to nuclear power. The event had, however, primed the pump for what was to come next. After the Fukushima accident in 2011, France began seriously reassessing its overwhelming reliance on nuclear power. Various new policies were proposed to cut the national reliance on atomic energy back from 75 to 50 percent.[23] Subsequently, technical problems started surfacing—for example, at the Flamanville plant on the Cherbourg Peninsula—that nurtured worries and the consideration of new policies that would reduce reliance on nuclear power and accelerate the development of energy efficiency, wind, and solar. All these options carry none of the risks of nuclear power while emphasizing local solutions to energy demand.

Russia

Russia offers another telling example of the influence of the geography of energy on energy policy. It is the largest country in the world by area, and its Siberian hinterlands hold more oil and gas reserves than are needed domestically. Significantly, Russia borders fourteen countries, most of which are hungry for energy supplies. Desperate for foreign currency, Russia enthusiastically exports enormous amounts of oil and gas westward to countries such as Germany, Hungary, Poland, Finland, and the Netherlands, and eastward to countries such as China, Japan, and North Korea (Figure 7.4). From Russia's point of view, there is a double advantage. Not only does this reap financial rewards, but Russians welcome the geopolitical position of holding the importing countries dependent.

The United States

Like Russia, the United States is vast and rich in a wide range of energy resources. The unprecedented success of the United States on the world stage in many ways stems from this natural endowment, which stimulated and sustains a vast economic and energy infrastructure. So much has been invested in the development of these resources that we have accumulated massive inertia in our energy supply system, which resists shifting to more sustainable (and less messy) alternatives. Any attempts to adjust policies in a way that would put such investment in jeopardy are rarely met with open arms by those who rely in one way or another on maintaining the status quo. In other words, the fossil fuel and nuclear industries tend to resist any policies that favor renewables or thwart their interests.

Japan

Japan has famously small reserves of fossil fuels, which makes it all the more remarkable that the country has developed such a strong economy. The record-breaking earthquake and devastating tsunami that hit the northeast part of Honshu Province in 2011 tested that strength. When massive waves breached the seawall protecting the Fukushima Daiichi nuclear power plant complex, they knocked out all the backup systems, leading to the destruction of several of the units. This single event prompted a question asked around the world: is commercial nuclear power too risky? It is as yet too early to tell if the public distrust of nuclear power will be short-lived. Still, one thing is sure: the Japanese government is rapidly adopting policies that favor renewable energy.[24] As a result, solar power installations are showing up everywhere, even floating on water (Figure 7.5).

Germany

Germany has long hosted nuclear power plants within its borders. After the incidents at Chernobyl in 1986 and (more significantly) Fukushima in 2011, the

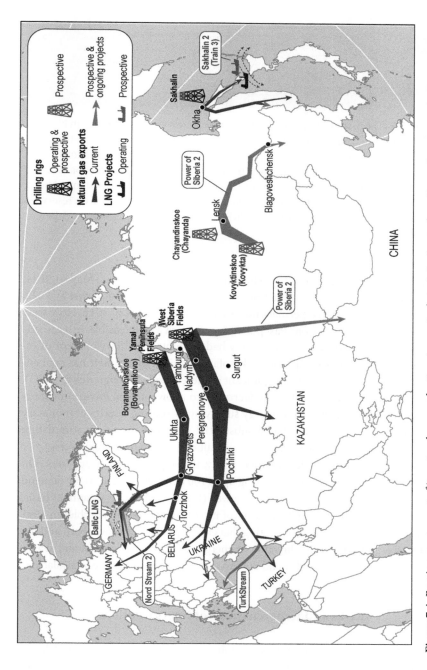

Figure 7.4 Russia exports most of its natural gas production, increasingly to Asia. *Source:* http://www.frontiere.eu/metti-sera-cena-al-kremlino-invece-del-vino-petrolio/.

Figure 7.5 The first installation of floating solar panels was ten years ago in Napa Valley, California, but the idea has yet to receive widespread acceptance in the United States. As of December 2017, the United States had seven installations of floating solar panels. Overseas, more than 100 sites have floating solar panel systems. Japan has fifty-six of the seventy largest floating solar panel systems. About 2.1 million hectares of land could be saved if solar panels were installed on reservoirs instead of on the ground, according to NREL researchers. *Source:* Photo by Adam Warren (NREL).

country changed direction. Chancellor Angela Merkel declared on May 30, 2011, that her country's seventeen nuclear power stations would shut down by 2022, if not earlier. Eight were permanently shuttered on August 6, 2011. How were they able to take such action? Past policies made such a shift feasible. In a cascade of policies over several years, Germany had promoted the development of renewable energy resources, particularly solar and wind, under their national energy transition plan. The most significant of these is called *Energiewende* (EEG), which is the planned transition by Germany to a low-carbon, environmentally sound, reliable, and affordable energy supply by instituting a series of monetary incentives.[25] This policy has had a substantial record of success in stimulating solar and (particularly) wind development, albeit at a cost to the taxpayers.

Each set of policies fueled another. Today, the impact is amply apparent. On January 1, 2018, Germany was covering up to 100 percent of its electricity demand from renewables, mostly from wind power (Figure 7.6).[26] On an annual basis, Germany rose from getting 3.1 percent of its electricity from renewables in 1991 to 38.2 percent in 2018, demonstrating that energy policies can have sharp and demonstrable impacts on the direction of energy resource development and availability, as they provide flexibility and options that would not otherwise be available.[27]

Risks and Repercussions of Energy Policies

The repercussions of energy-related policies can be so significant and long-lasting that decision-makers often tend to bend in a conservative direction.

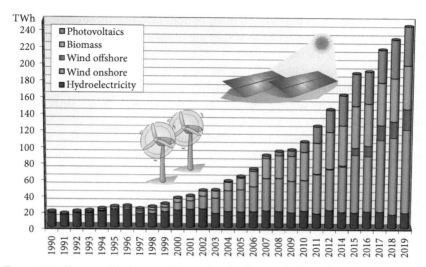

Figure 7.6 The growth of renewables under the German policy of *Energiewende* (energy transition). *Source:* https://www.volker-quaschning.de/datserv/ren-Strom-D/index_e.php. See plates for color version.

Renewable energy is excellent, they might agree, but what happens if it winds up being an insufficient source? If we put all our emphasis on developing these resources, they might reason, do we think we can build new power plants, open new mines, and drill new wells quickly enough to make up for any shortfall if it occurs suddenly? The consequences of instituting new energy policies that diverge from the present energy policies tend to put the brakes on swift and abrupt changes. It is for this reason that policymakers lean toward making a slow transition away from past practices—when they advocate such a transition at all.

Making only small changes, of course, carries its own set of risks, especially regarding the perpetuation of environmental impacts from existing energy portfolios. If we continue to rely on burning coal for electricity, won't we continue poisoning the atmosphere with carbon? Will such increased emissions of carbon bring be accompanied by unstoppable climate change, rising sea level, shifts in agricultural productivity, mass migrations, and social unrest?

And, of course, there is always a matter of economics to consider. Can we afford to support energy research with a polydirectional strategy that spreads resources so thin that it winds up limiting research development in every one of the options for the future? Should we continue spending billions of dollars trying to harness the limitless potential of fusion power to generate electricity? Or should we instead put more research money into solar energy, advanced nuclear reactors, clean coal technology, enhanced recovery strategies, geothermal energy, wave power, or biofuels? So many options exist that those entrusted with policy decisions end up acting as if they are using a compass that spins wildly, useless in pointing the way to safety.

For now, we appear likely to rely on the profit motive to drive innovation. Imagine the wealth that will be created when we succeed in providing affordable, large-scale electricity storage. The holy grail of renewable energy, such a breakthrough would release us from the bonds of intermittency and allow solar and wind power to reach their true potential. What a game-changer that will be. It might not be too far in the future.

Nuclear Energy Policy

As most people realize, nuclear power continues to hold a firm position within the debate about climate change. Generating electricity in a nuclear power plant produces no CO_2, although it is not carbon-free when the entire life cycle is considered. Over that entire cycle, nuclear power emits on average about the same amount of carbon dioxide as offshore wind installations and at a maximum about three times more, but still not as much as rooftop solar (Table 7.3). Nonetheless, the CO_2 emissions are substantially less than come from burning fossil fuels such as coal, which is about seventy-five times higher. Considered in that light, should nuclear be included as a significant contributor to future energy mixes?

Such emissions are only a part of the considerations that accompany the nuclear power debate. Others include public safety, accidents, decommissioning safety, legacy wastes from decades of nuclear research programs, long-term waste disposal, and whether or not it can compete with other resources on the price of delivered electricity. For these reasons, the future of nuclear power is on shaky ground not just in the United States but in several other countries, including, South Korea, Taiwan, and Germany.[28] And in Japan, only five of forty-three commercial reactors in Japan have been operating since the accident at Fukushima.

Several other European countries are taking a similar stance about nuclear power. In Europe, the Danish parliament passed a resolution in 1985 to prohibit nuclear power plant construction in their country.[29] Italy closed all its nuclear plants in 1990 and has no plans to build more.[30] Spain and Switzerland banned the construction of new reactors.[31] France's state-owned Électricité de France (EDF), as reliant as it is on nuclear power, is now buried in debt "thanks to its own technical missteps" and the economic and energy policies of President Emmanuel Macron.[32]

But not everyone is down on nuclear. There are about 50 nuclear reactors currently under construction in fifteen countries, notably China, India, Russia, and the United Arab Emirates.[33] About 100 power reactors with a total gross capacity

Table 7.3 Emissions of Selected Electricity Supply Technologies (g CO_2e/kWh)

Options	Direct Emissions	Infrastructure & Supply Chain Emissions	Biogenic CO_2 Emissions and Albedo Effect	Methane Emissions	Life Cycle Emissions (incl. Albedo Effect)
	Min/Median/Max	Typical Values			Min/Median/Max
Currently Commercially Available Technologies					
Coal—PC	670/760/870	9.6	0	47	740/820/910
Gas—Combined Cycle	350/370/490	1.6	0	91	410/490/650
Biomass—co-firing	N/A	–	–	–	620/740/890
Biomass—dedicated	N/A	210	27	0	130/230/420
Geothermal	0	45	0	0	6.0/38/79
Hydropower	0	19	0	88	1.0/24/2200
Nuclear	0	18	0	0	3.7/12/110
Concentrated Solar Power	0	29	0	0	8.8/27/63
Solar Photovoltaics, Rooftop	0	42	0	0	26/41/60
Solar Photovoltaics, Utility	0	66	0	0	18/48/180
Wind, Onshore	0	15	0	0	7.0/11/56
Wind, Offshore	0	17	0	0	8.0/12/35

Source: S. Schlömer et al., "Annex III: Technology-Specific Cost and Performance Parameters," in *Climate Change 2014: Mitigation of Climate Change, Contribution of Working Group III to the Fifth Assessment Report of the Intergovernmental Panel on Climate Change* (Cambridge: Cambridge University Press, 2014).

of about 110,000 MWe are on order or planned, and over 300 more are proposed. Most reactors currently planned are in Asia, with fast-growing economies and rapidly-rising electricity demand.[34]

The question remains: should the United States resume its nuclear program and move more deliberately to increase its fleet of nuclear power plants? No one in authority seems to have decided definitively. While many remain skeptical because of the technical, financial, management, regulatory, and even ethical questions surrounding commercial nuclear power, a new generation of reactors is considered by some people as a logical response to the climate threats of carbon-based fuels. They argue as well that (1) nuclear power plants have a smaller footprint because the high energy density of their fuel obviates the need for on-site fuel storage, and (2) nuclear power plants are typically quiet and unobtrusive neighbors, an advantage when compared with whirling wind turbines or large "black lakes" that solar arrays can create on the landscape.

If you remain confused about the future of nuclear power, you have a lot of company; the world is full of conflicting policies. Even in Germany, despite its clear policy to move away from the technology, some question whether other sources of electricity can pick up the slack. For its part, the United Kingdom is continuing to adhere to a policy of nuclear power plant construction, despite a vigorous anti-nuclear campaign that has been raging for decades.[35] Recent government policy decisions favor a new round of nuclear power plant construction, totaling 16 GWe of capacity that will be in operation by 2030, including Hinkley Point C, in Somerset; Wylfa, on the island of Anglesey in north Wales; and Sizewell C, which w adjacent to Sizewell A and B on the North Sea in Suffolk (Figure 7.7).[36] All continue to be controversial, and despite favorable government policy, it is still uncertain whether or not any of these will become operational, due to cost overruns and unending public resistance.

The United States remains somewhere in the middle, still unsure which policy to advocate, leaving the future of nuclear power in limbo. Of the 253 nuclear power reactors ordered from 1953 to 2008, 48 percent were canceled, 11 percent shut down prematurely, 14 percent experienced an outage lasting at least one year, and 27 percent operated without having any outage longer than a year. Thus, only about one-fourth of those ordered—or about half of those completed—are still running and have proved relatively reliable.[37] California has closed all but two units and has plans to shut those down as well within a decade. Presently, the only nuclear reactors under construction in the United States are in Georgia, and costs are now several billion dollars over their original estimates.

We in the United States do not know which new energy policies to embrace fully. As a result, nuclear power continues to receive subsidies, making for a notably uneven playing field. We do not know whether we should embrace newly designed commercial reactors that are reportedly safer.[38] We have not agreed on

Figure 7.7 Looking south toward Sizewell A (the concrete building on the left in the distance) and Sizewell B (the white dome) on the Suffolk Coast. U.K. government plans include building Sizewell C and perhaps Sizewell D units further north along the coast.

what level of risk we are willing to accept. We do not yet know how to dispose of spent nuclear fuel permanently, a question no other country has yet resolved either. We have not even decided on a firm policy regarding fuel reprocessing, with policy in this regard whipsawing over the years. For example, in October 1976, concern over nuclear weapons proliferation (notably after India demonstrated nuclear weapons capabilities using reprocessing technology) led President Gerald Ford to suspend indefinitely the commercial reprocessing and recycling of plutonium in the United States.[39] On April 7, 1977, President Jimmy Carter expanded that directive by banning the reprocessing of commercial reactor spent nuclear fuel. In both cases, the critical issue behind this policy decision was the risk of nuclear weapons proliferation by the diversion of plutonium from the civilian fuel cycle. In 1981, President Ronald Reagan lifted the ban, but because the government did not provide the substantial subsidies necessary to restart commercial reprocessing, it is not done in the United States.[40]

Complicating policy decisions about nuclear power further is the geographic dispersion of harmful radionuclides in the case of an accident. Whereas the dangers from most possible accidents at conventional fossil fuel plants tend to remain on-site, a quick consideration of the geography of nuclear power plants in western Europe reveals that abandoning the technology in one country might not protect citizens from continued operation in neighboring countries. For example, many nuclear power plants in France and Belgium are closer to London than nuclear power plants in other parts of the United Kingdom. Furthermore, the pollution plumes from the Chernobyl (and Fukushima) accidents contaminated large portions of the Northern Hemisphere.[41] Nuclear energy policies in one country can and do affect energy policy decisions in other countries.

China has indulged in no such dithering over nuclear power, embracing the technology fully. The Chinese policy decision reflected three major concerns: (1) persistent and accumulating health effects from coal-fired power plants; (2) the impact of coal power plant emissions on climate change; and (3) the inevitable rise in demand for electricity. According to its advocates, nuclear power addresses all three of these concerns. For this reason, by 2020 generating capacity was expected to reach 58 GWe, with an additional 30 GWe under construction—all of which China set out in its Energy Development Strategy Action Plan 2014–2020.[42]

Two questions continue to loom over the future of nuclear power. First, considering the need for reliable sources of electricity to support the expected increased demand in places like China, India, and the Middle East, is there any alternative to nuclear power? Second, will the growing apprehensions about global warming drive those countries that are on the fence back toward nuclear power? Some sound pro-nuclear policy decisions might make that a reality for those countries that continue to debate the issue (including the United States), but some countries have already made their choice, as we have seen. In other countries, we can expect nuclear energy policy to continue to fluctuate from one extreme to another in the coming years, influenced by world events. If there is increasing evidence and worry about global climate change, nuclear may gain popularity. If there is another serious accident, it will likely move in the opposite direction.

Renewables and Efficiency

As we have seen, energy policies are put in place to advance a preference and some favored energy resources. All lobbyists for conventional energy have long recognized this. In recent years, advocates of more significant development of renewable energy and measures promoting greater energy efficiency have come to recognize this as well. The most current and comprehensive reference source in this regard is the Database of State Incentives for Renewables and Efficiency (DSIRE).[43]

Given its progressive bent, it should be of little surprise that the state with the most incentives for development of renewable energy is California, with more than 250, while the states with the fewest include coal-rich West Virginia, with fifteen, and oil-rich Louisiana, with just under thirty. Somewhat surprising is that Texas—the state with the most dominant, productive, and richest area of oil development—has over 150 such incentives in place, plus statewide retail electricity competition. So far, the competition has produced lower electricity bills for most consumers. No doubt, these favorable policies help explain why Texas has more installed wind generating capacity than any other state, but the message does not stop there. It also infers a recognition of an expected decline in oil production and associated economic development plus the recognition that the country is beginning to transition away from fossil fuels.

As support for this transition continues gaining momentum, traditional energy companies (including utility companies) have started to push back. For example, even solar-rich Arizona has experienced resistance from the state's major utility companies, which have initiated extra charges for those who install solar modules on their roofs. For example, one Arizona company, Salt River Project, imposed a demand charge for residential solar customers. It was the first in the nation to do so. The utility initially estimated the average surcharge would be $50 a month, but it is actually closer to $100 a month. That policy adjustment erased the economic benefits of having solar, precipitating an immediate 80 percent decline in rooftop installations within its service territory. Solar advocates have sued the utility company, but as yet to no avail.[44]

Such examples signal that imposed energy policies can have competing purposes and goals. It also illustrates the difficulties of—as well as the argument for—comprehensive energy policy at the national level, instead of the mishmash of state, country, and municipal policies that currently exist. In contrast, an example of an effective national energy policy is a feed-in tariff, such as has been adopted over the years in Germany, Spain, Italy, and many other countries. Having different, overlapping, redundant, and incompatible policies among states confuses and inhibits progress toward a more sustainable energy future. Even when federal policy is put in place, each state first has to pass enabling legislation to adopt it, and then they can set the contract terms, rate of returns, and define the "avoided costs."

It is precisely in this context that energy policy can be appropriate at the national level, and the U.S. Congress has periodically stepped up to this challenge—if often with jerks and reversals—by providing periodic legislation supportive of renewable energy. Such incentives remain relatively small when compared with the multi-decade accumulation of financial subsidies approved for the fossil and nuclear industries. However, such policies have nonetheless been helping renewable energy development survive financially between the initial phases when investors underwrite new ideas and when that idea has blossomed into stand-alone profitability. It is during this interval, often called the "valley of death,"

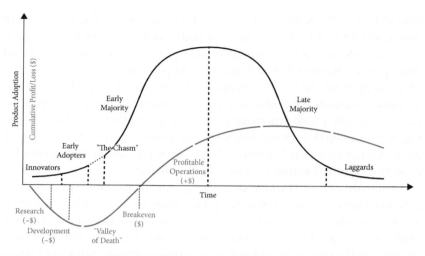

Figure 7.8 A "valley of death" often exists between the timing of the creation of initial ideas and their commercial viability. *Source:* http://neverstop.co/startco-toolkit/.

when most start-up companies fail (Figure 7.8). Supportive government policies help bridge this valley, as has been the case with the surging development of renewable energy assets.

The most successful policy aids in this regard are commonly referred to by their acronyms: PURPA, RPS, ITC, and PTC. The Public Utility Regulatory Policies Act (PURPA) was passed in 1978 as part of the National Energy Act, as we have seen. This approach aimed both at reducing demand for energy and at increasing the development of domestic energy and renewable energy resources. Even before renewable energy reached its stride in the early twenty-first century, PURPA was encouraging its development by mandating utilities to buy power from small providers—provided they could sell it at a price comparable to that coming from fossil fuels. It was one of the earliest effective policies enacted in support of alternative energy development. Its effects include:[45]

- Creating a market for power from non-utility power producers
- Increased efficiency by making use of cogeneration
- Ending promotional rate structures
- Encouraging the development of hydroelectric power
- Conservation of electric energy and natural gas

The success of PURPA has varied by resource. For example, Reuters reported that for decades, PURPA was mostly irrelevant to the wind and solar industries

2020 State Solar Power Rankings Report

How to read the report:

This chart ranks the 50 states and the District of Columbia, from best (green) to worst (red), based on their solar-friendliness. For example, Massachusetts receives the best score, while Oklahoma receives the worst.

The outermost ring (closest to each state label) shows the overall grades awarded the states. The inner rings represent factors contributing to the grades.

Grading Scale

A   B   C   D   F

Factors:

1) Overall Grade
2) Renewable Portfolio Standard (RPS)
3) Carve Out/SRECs
4) Cost of Electricity
5) Net Metering
6) Interconnection
7) Solar Rebates
8) State Solar Tax Credits
9) Sales Tax Exemption
10) Property Tax Exemption
11) Low-Income Programs

Solar Power Rocks

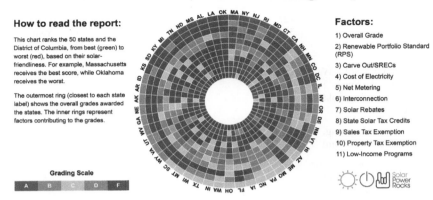

Figure 7.9 2020 State Solar Power Rankings Report. For an interactive and most current version, see https://www.solarpowerrocks.com/state-solar-power-rankings/. See plates for color version.

because such technologies were not competitive with conventional resources such as coal and uranium. That observation is no longer in effect, and this trend has been encouraging a surge of renewable power projects from developers who could rely on mandated contracts with utilities to stabilize lower costs.[46]

The effects of PURPA have also varied geographically (Figure 7.9). For example, 60 percent of the nation's current PURPA projects are in North Carolina, although the state still ranks twenty-third in installed solar capacity under PURPA. California leads the country in installed solar energy, but it still has room for improvement, as it ranks only seventh, Massachusetts is first, while solar-rich Arizona is just nineteenth.

Another successful policy inducement has been the investment tax credit (ITC).[47] These credits allow businesses to deduct from their taxes the amount that they reinvest in themselves. At the federal level, the ITC has been amended several times, most recently in February 2018. Table 7.4 shows the value of the investment tax credit for each technology by year. Expiration dates are based on when construction begins. These are policies that operate nationwide, administered by the IRS, and the states cannot interfere.

The ITC is paying dividends in the United States. It has helped the residential and commercial renewable energy sector grow by 10,000 percent since the legislation instituting the credit was enacted in 2006.[48] Solar energy is increasing faster than any other domestic energy source as manufacturing prices continue to plummet, often beating out coal and cheap natural gas. The solar energy industry created one in seventy-eight of our country's new jobs in 2019, and by

Table 7.4 Federal Business Investment Tax Credit

Technology	12/31/16	12/31/17	12/31/18	12/31/19	12/31/20	12/31/21	12/31/22	Future Years
Photovoltaics, Solar Water Heating, Solar Space Heating/Cooling, Solar Process Heat	30%	30%	30%	30%	26%	22%	10%	10%
Hybrid Solar Lighting, Fuel Cells, Small Wind	30%	30%	30%	30%	26%	22%	22%	N/A
Geothermal Heat Pumps, Microturbines, Combined Heat and Power Systems	10%	10%	10%	10%	10%	10%	N/A	N/A
Geothermal Electric	10%	10%	10%	10%	10%	10%	10%	10%
Large Wind	30%	24%	18%	12%	N/A	N/A	N/A	N/A

2020, the U.S. solar industry employed 231,474 workers. Moreover, the total of Americans employed in solar is expected to roughly double to 420,000 by the end of 2022, all while spurring roughly $140 billion in economic activity.[49] The continued success of the ITC demonstrates that long-term federal policies can drive economic growth while reducing prices and creating jobs in one of the fastest-growing sectors in the U.S. economy.[50]

The residential ITC allows homeowners to apply the credit to their personal income taxes when they purchase solar systems and have them installed in their homes. In the case of the utility credit, the business that installs, develops, and finances the project claims the credit.

A tax credit is a dollar-for-dollar reduction in the income taxes that a person or company would otherwise pay the federal government. The ITC is based on the amount of investment in solar property. Both the residential and commercial ITC are equal to 26 percent of the basis that is invested in eligible solar property that began construction through 2019. The ITC then steps down according to the following scale:[51]

- 26 percent for projects that start construction in 2020
- 22 percent for projects that start construction in 2021
- After 2021, the residential credit drops to zero while the commercial credit drops to a permanent 10 percent

A sister incentive to the ITC is the renewable electricity Production Tax Credit (PTC). It is an inflation-adjusted per-kilowatt-hour (kWh) tax credit for electricity generated by qualified energy resources and sold by the taxpayer to an unrelated person during the taxable year. The duration of the credit is ten years after the date the facility is placed in service for all facilities placed in service after August 8, 2005. Initially enacted in 1992, the federal PTC has been renewed and expanded numerous times, most recently by the American Recovery and Reinvestment Act of 2009, the American Taxpayer Relief Act of 2012, the Tax Increase Prevention Act of 2014, the Consolidated Appropriations Act of 2016, and the Bipartisan Budget Act of 2018. (The constant cycle of expirations and need for renewals tends to stymie maximum investment.)

The tax credit is phased down for wind facilities and expired for other technologies commencing construction after December 31, 2016. The phase-down for wind facilities is a percentage reduction in the tax credit amount described earlier, as follows:

- For wind facilities commencing construction in 2017, the PTC amount is reduced by 20 percent
- For wind facilities commencing construction in 2018, the PTC amount is reduced by 40 percent
- For wind facilities commencing construction in 2019, the PTC amount is reduced by 60 percent

While these policies have had success in the United States, other countries have implemented different approaches. In the United Kingdom, for example, attention has been turning toward greater energy efficiency on the demand side and more significant development of renewable energy resources on the supply side. Indeed, in response to favorable economics and government policy, renewable power—particularly wind—has been on a steeply rising curve and should continue that trajectory. Under the Renewable Energy Directive (RED) of the European Union, for example, the United Kingdom had a target for sourcing 15 percent of final energy consumption from renewable energy by 2020. The United Kingdom exceeded its 2013–14 interim target under the RED with a 6.3 percent share of renewables versus its 5.4 percent target. By 2030, wind and solar are expected to reach above 50%, more than in any other country.[52]

Elsewhere in Europe, policies have strongly supported renewable energy development. The most effective policy, the feed-in tariff (FIT)—rarely used in the United States—produced a skyrocketing effect for wind and solar energy, especially in Germany (Figure 7.10). In the simplest terms, FITs grant those who develop renewable energy a guaranteed price for each kWh generated. This price is much higher than the standard consumer rate, thus providing an economic incentive and long-term security to renewable energy producers.

In Germany, the amount of the subsidy varied by type of technology and cost of generation by those technologies. The legal underpinnings also changed

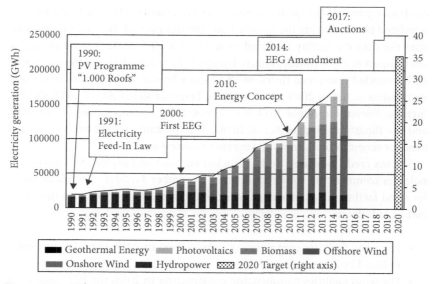

Figure 7.10 Milestones of the development of renewable electricity generation in Germany with the introduction and renewal of feed-in tariffs over the years. *Source:* Monisha Shah et al., "Clean Restructuring: Design Elements for Low-Carbon Wholesale Markets and Beyond," 2016, NREL/TP-6A50-66105, National Renewable Energy Laboratory, Golden, Colorado.

over time. Between 1991 and 1999, feed-in tariffs were prescribed through the Electricity Feed-in Law (*Stromeinspeisungsgesetz, or SEG*) at relatively moderate rates. From 2000 onward, the SEG was replaced by the Renewable Energy Sources Act (*Erneuerbare-Energien-Gesetz, EEG*), with much higher subsidy rates.[53]

Eventually, the FITs in Germany (and elsewhere) steadily declined, as they were financially unsustainable. Nonetheless, the impact of this energy policy on renewable development has been impressive. In the first half of 2018, Germany produced enough renewable energy to power every household in the country for a year. The nation's combined wind, solar, biomass, and hydroelectric power output hit a record 104 billion kWh between January and the end of June that year.[54]

Outside Europe, policies to support renewable development have also had notable success in centrally planned economies such as in China. Installed capacities of hydro, wind, and solar power in 2020 was expected to reach 350 GW, 200 GW, and 100 GW, respectively. (For comparison, the United States has about 55 GW solar photovoltaic installed.) China leads the world in the installed capacity of wind and solar power generators. In September 2016, Bloomberg reported that a 2 GW capacity solar farm with 6 million solar panels was being built in the Ningxia autonomous region, which would make it the world's largest solar plant by 2020. India is constructing solar installations that are even large, part of Prime Minister Narendra Modi's national plan.

As we've seen, national energy policies of any type, but especially in the form of economic incentives, have had a positive impact on the development of renewable energy. In individual states, net metering has had an additional positive impact on renewable energy development. In its purest form, net metering is a billing mechanism that credits solar energy system owners for the electricity they add to the grid. For example, if a residential customer has a rooftop photovoltaic system, it might generate more power than the home uses during daylight hours. Ideally, if the home is net-metered, the electricity meter will run backward to provide a credit against the electricity consumed at night or other periods where the home's electricity use exceeds the system's output. Customers are billed only for their net energy use. The benefits of net metering vary with the utility's rate plan. Arizona's Salt River Project, for example, still has net metering for residential solar, but the import/export rate is less than 4 cents/kWh, which is too low to make a big impact on the amount of time it takes to recoup the costs of the solar installation.

The largest Arizona utility company, Arizona Public Service, instituted a different approach, called net billing. The primary difference between net billing and net metering is that there are differing rates used to value the excess energy fed into the grid and energy received from the grid. Under the APS arrangement, the purchase price of electricity is about 12–13 cents per kWh, but the sale price is smaller, currently about 10 cents per kWh. Each year going forward, it drops for new installations another 10 percent or so, until it gets to the wholesale price (2.5 cents/kWh) around 2028. And you settle up every month, not just once a year. It turns out that the monthly settling-up arrangement makes net billing a better value than net metering until the difference between import and export rates exceeds 4 or 5 cents per kWh. As you can see from these two examples, different energy policies not only can make a big difference even in the same market, but are complicated enough to keep customers in the dark.

For further information on policies, you should consult the DSIRE database, the most current and comprehensive guide to the panoply of alternative energy policies. The website identifies hundreds of state-level energy policies affecting energy efficiency and renewable energy development and use. California provides an excellent example of the benefits of such policies (including incentives). California reportedly saved $75 billion between 1974 and 2012. Since 1975, the Energy Commission's building energy efficiency standards (Title 24) has saved California consumers over $30 billion. Title 24 of the California Code of Regulations, also known as the California Building Standards Code, contains the regulations that govern the construction of buildings in California. In their 2017 report, the CEC said:[55]

> In 2017, the Energy Commission demonstrated that California continues to lead the nation and the world in energy efficiency standards as the state gets more bang for the buck when managing energy usage in buildings. The Commission compared California's 2016 Building Energy Efficiency Standards to national and international standards. The comparison found California standards deliver 29 percent more energy savings than national and international standards. *That saving is equal to the electricity demand of 300,000 additional California homes annually.* Energy savings translate into financial savings for Californians and the standards developed in California get adopted around the world.

Title 24 is but one of several energy-saving policies in California. In sum, the state had saved 90 million megawatt-hours by 2016. For reference, a typical single-family household in the Phoenix area uses about 13–14 MWhr/year.

Another successful state-level policy is the Renewable Portfolio Standard (RPS), which requires a designated percentage from regulated utilities in the state to come from approved renewable energy resources.[56] This was instituted first in California in 1983 and then spread to other states. According to the most recent analysis, these standards have significantly stimulated the growth of renewable energy. By 2016, 56 percent of total U.S. retail electricity sales were from renewable sources. In Arizona, the RPS will reach 15 percent by 2025. In California, the RPS will reach 50 percent by 2030 and remain at that level in all subsequent years. Other states vary from a high of 100 percent in Hawaii by 2045 to zero percent in Alabama, Alaska, Florida, Georgia, Idaho, Kentucky, Louisiana, Mississippi, Nebraska, West Virginia (which had an RPS, but it was repealed), and Wyoming. RPS policies have been most successful at expanding the solar market. In addition to the RPS policies, voluntary renewable energy goals exist in several U.S. states, and both mandatory RPS policies and non-binding goals exist among U.S. territories (American Samoa, Guam, Puerto Rico, U.S. Virgin Islands).[57] Voluntary RPS goals are totally ineffective; however, mandatory RPS goals are essential.

While not the only type of policy that has stirred growth, RPS policies have had a considerable positive impact, especially in combination with the ITC and PTC. If supply targets are met, the approximately 300 TWh now produced by renewable energy resources in the United States would increase by half to 450 TWh by 2030. Current trends suggest this is achievable, especially as the cost of photovoltaic modules continues to decline. Hawaii's RPS goal of 100 percent by

2045 is reachable, considering how many renewable energy projects are there already.

Environmental Policy Drives Energy Policy

Because the thread of energy touches everything, it should be no surprise that policies primarily intended to affect environmental quality often have a direct effect on energy development and use. This certainty is most evident with the work of the Intergovernmental Panel on Climate Change. As we all know, the planet is warming. No question: humans are responsible for that. It is settled science. Flattening the curve and then pressing it downward is now the task at hand. Still, the practical actions we need to take are often contentious, even when the goal is to establish voluntary policies that each country agrees to pursue.[58]

The Fifth Assessment Report of the IPCC concluded that human influence on the climate system is apparent and that recent anthropogenic emissions of greenhouse gases are the highest in history. These findings informed the climate negotiations resulting in the Paris Agreement of 2015, when 197 countries committed to limiting global warming to below 2°C. According to the latest version, this agreement

> builds upon the Convention [United Nations Framework Convention on Climate Change (UNFCCC)] and for the first time brings all nations into a common cause to undertake ambitious efforts to combat climate change and adapt to its effects, with enhanced support to assist developing countries to do so. As such, it charts a new course in the global climate effort. The Paris Agreement's central aim is to strengthen the global response to the threat of climate change by keeping a global temperature rise this century well below 2 degrees Celsius above pre-industrial levels and to pursue efforts to limit the temperature increase even further to only 1.5 degrees Celsius. Additionally, the Agreement aims to strengthen the ability of countries to deal with the impacts of climate change. To reach these ambitious goals, appropriate financial flows, a new technology framework and an enhanced capacity building framework will be put in place, thus supporting action by developing countries and the most vulnerable countries, in line with their own national objectives. The Agreement also provides for enhanced transparency of action and support through a more robust transparency framework.

The IPCC's reports were also influential at the first Conference of the Parties (COP) to the Climate Convention, held in Berlin, Germany, in 1995. Attendees produced the so-called Berlin Mandate, setting out the terms for a negotiation process that would provide binding commitments by industrial countries to reduce their heat-trapping emissions after the year 2000.[59]

As of May 2018, 195 UNFCCC members have signed the agreement, and 177 have become a party to it. In a real sense this agreement is a voluntary policy, with no mechanism for enforcement.[60] Each country is free (and expected) to develop internal policies to lower its greenhouse gas emissions. Nonetheless, in June 2017, U.S. president Donald Trump announced his intention to withdraw the United States from the agreement. The earliest effective date of withdrawal for the United States under the agreement was November 2020, shortly before the end of Trump's term. President Biden signed an executive order in January 2021 to rejoin the agreement.

While the intent of the agreement is to reduce greenhouse gas emissions, the measures are de facto energy policies focused on the generation of electricity and the efficiency of our buildings and motor vehicles. For example, the Phase 2 proposals for medium- and heavy-duty vehicles, model year 2017–2018, "are expected to lower CO_2 emissions by approximately 1 billion metric tons, cut fuel costs by about $170 billion, and reduce oil consumption by up to 1.8 billion barrels over the lifetime of vehicles sold under the program. These reductions are nearly equal to the greenhouse gas emissions associated with energy use by all U.S. residences in one year."[61]

Anticipating that the planet will continue heating up and that millions and even billions of people will suffer consequences, what can we do about it? What *should* we do about it? The actions each country takes will differ according to factors such as the sources of greenhouse gas emissions, level of industrialization, opportunities for alternative energy development, types and numbers of automobiles, deforestation, use of fossil fuels, and standards of living. Such actions will mainly be linked directly to the policies they implement, agree to follow, and enforce.

On the topic of electricity use and heat production, here are some relevant questions to consider: Do you promote the use of low-carbon energy resources and avoid and/or even actively oppose further development of fossil fuels? Do you have solar modules at your house? Is your house energy efficient in terms of appliances, orientation, window treatments, and insulation? Keep in mind that California, by implementing its appliance and building codes, has saved more than $100 billion for California taxpayers.[62]

Establishing policies that provide incentives or impose penalties to promote or discourage the above questions can be useful, but they can also be challenging. Several such policies are already in place, although you might not give them much notice. For example, HOV (high-occupancy vehicle) lanes encourage carpooling by offering less congestion. But are they useful? Recent research suggests that while they reduce congestion, they also indirectly encourage urban sprawl and longer commuting distances. Bottom line: HOV lanes probably do not do much to reduce the emission of greenhouse gases.[63]

Better urban energy policies are needed. One would be to encourage denser, mixed-use zoning that minimizes the need to drive at all; this is taking place in many cities, including in Arizona. Another would be to institute a monetary penalty for driving in congested areas. In London, England, for example, the congestion charge is £11.50 daily for driving a private vehicle within the designated zone between 7:00 a.m. and 6:00 p.m., Monday to Friday (Figure 7.11). The charge rises to £14 if you wait until the next day to pay. If you don't pay by midnight on the day after you drove in the zone, you'll get a penalty charge notice. Residents receive a 90 percent discount, and registered disabled people can travel for free. Emergency services, motorcycles, taxis, and minicabs are exempt. It is reportedly a great success; between 2002 and 2014, the number of private cars coming into the zone fell by 39 percent.[64] Energy policies, by this evidence, can work if they are given careful consideration and planning.[65]

Another useful policy tool that has reduced transportation energy is rising vehicle mileage standards. As we've seen, those passed during the Obama administration would have increased the corporate average fuel economy (CAFE) target to 54.5 miles per gallon (mpg) by 2025—a significant improvement above the

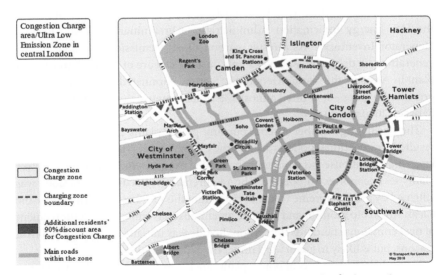

Figure 7.11 The London congestion charging zone covers 21 km² of central London. *Source:* https://tfl.gov.uk/ruc-cdn/static/cms/images/congestion-charge-ulez-map.jpg. See plates for color version.

average of 25.4 mpg in 2014. Achieving that target would be expected to reduce greenhouse gas emissions by 6 billion metric tons, while nearly doubling the fuel economy of many vehicles. But as we've also seen, changing political winds can overturn such gains by nullifying or diluting such policies.

Whether such targets are achievable is in doubt for a couple of reasons. For one thing, there's the so-called rebound effect—when drivers of more fuel-efficient cars take advantage of their lower per-mile fuel costs by driving more than they did in the past. "Typical estimates of the rebound effect are about 10%," Arthur van Benthem says. That is, "people drive about 10% more . . . when they switch to a vehicle that is twice as fuel-efficient."[66] A more effective policy would increase the cost of motor fuels by imposing higher taxes—a politically tricky step, but still just one of many that have been suggested to reduce greenhouse gases.

The last of the mitigation measures we will address here stresses the benefits of changing the policies that affect electrical generating stations. Because coal plants are significant emitters of greenhouse gases, because regulations mandating expensive emission control retrofits are accumulating, and because clean natural gas is cheaper, the long era of coal power appears to be ending in several countries where it has had a strong presence for centuries, including the United States, United Kingdom, and Germany. Between 2010 and the first quarter of 2019, U.S. power companies announced the retirement of more than 546 coal-fired power units, totaling about 102 GW of generating capacity. Plant owners intend to retire another 17 GW of coal-fired capacity by 2025, according to the U.S. Energy Information Administration's Preliminary Monthly Electric Generator Inventory.[67] As a result, carbon dioxide emissions from electricity generation are dropping. As an example, just the closure of a single large power plant in Arizona, the Navajo Generating Station, eliminated a massive source of greenhouse gases—it emitted almost 135 million metric tons of carbon dioxide between 2010 and 2017.[68]

The declinging number of coal-burning power plants in the United States resulted from national policies such as the Clean Air Act. This Act set targets for lowering greenhouse gas emissions that added expenses that made these power plants less economic to operate. The requirement for reduced emissions was affirmed by a U.S. Supreme Court decision that the EPA has authority under the Clean Air Act to monitor and regulate greenhouse gas emissions.[69] Enforcement of new emissions standards for carbon dioxide was initially resisted for political reasons, but is likely to go forward under the Biden Administration.[70]

Nevertheless, even as coal power plants are being retired in the United States, they seem to be sprouting up in places like India, Japan, and China. Currently, China has 43 percent of the coal-powered generation capacity in the world, concentrated in the eastern part of the country.[71] While its life seems to be heading toward an end, coal is not dead quite yet.

The Good and the Bad of Energy Policy

Using policy to guide behavior is a standard tool of modern society. It intends to encourage specific actions and behaviors and discourage others. Such motivation is real, for example, with time-of-use electricity rates. Those customers who concentrate their demand within off-peak times are billed at a lower unit price, while those who focus their energy use during peak times are charged at a higher rate. The result benefits both the utility company and the customer: the utility company does not have to build generating facilities to meet infrequent high-peak demand, while the customers pay less for their electricity by voluntarily shifting their use to off-peak times.

Although some policies morph into legislation, energy policies are usually intended to be temporary inducements. As such, they serve their function and then expire. This pattern has been true of the feed-in tariffs in Germany that encouraged the rapid growth of solar and wind power there in recent decades. For conventional energy resources, however, some policies have remained active so long we forget they exist. In such instances, these policies become like "force fields," effectively repelling the probes of alternative energy developers who would wish to introduce into the marketplace resources that are inherently less problematic while innately more sustainable.

This mismatch is illustrated by an analysis of subsidies to several energy technologies, which showed that fossil fuels receive the most significant portion of such incentives and have done so for many years. Between 1947 and 2015 these subsidies totaled close to $1 trillion in inflation-adjusted dollars, with two-thirds going to fossil fuels and 21 percent going to nuclear power. Incentives for other sources of energy are minuscule in comparison (Figure 7.12).

Such subsidies come in many forms. By far the largest is in the form of tax policies, accounting for 40 percent of the incentives. One example is the oil depletion allowance (ODA). The ODA is an American tax law that evolved from the policy that oil was a depleting resource. By the logic of the law, anyone in the oil business was, by definition, depleting the resource that was responsible for their livelihood. In effect, the ODA is based on the principle that depreciation can be offset against income, effectively as a capital loss. It has become controversial because, under one method of claiming the allowance, it is possible to write off more than the full capital cost of the asset.[72] It provides oil companies with a large and automatic deduction as compensation for dwindling assets in the ground, regardless of their actual expenses. And the subsidy has grown over the years. Early in its existence, it was a 5 percent deduction, but by 1926 it had grown to 27.5 percent.

Such long-standing policies often have unintended consequences. At an early stage, it could be considered sensible to provide extra incentives to those who

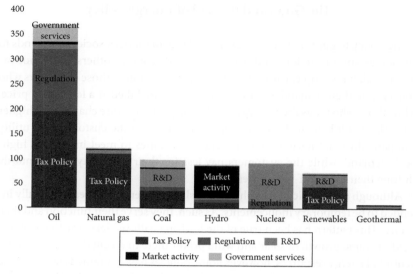

Figure 7.12 Energy incentives by resource. *Adapted from:* Roger H. Bezdek and Robert M. Wendling, "Energy Subsidies: Myths and Realities," *Public Utilities Fortnightly,* June 2012, pp. 62–67. https://www.fortnightly.com/fortnightly/2012/06/energy-subsidy-myths-and-realities. See plates for a color version.

are extracting a depleting and highly desired resource. However, the result has been that oil development has had decades of support that is mostly unavailable to other resources. This support has skewed the price signals against the development of alternative energy resources. Over the years, many have tried unsuccessfully to revoke the ODA, including President John F. Kennedy in 1963, but without success. The ODA is still in place, despite being what many people consider a legalized way to swell personal oil fortunes.[73]

Not all energy policies have been as blatantly consistent as the ODA. Renewable energy subsidies such as the ITC and the PTC—despite their success in stimulating the renewable energy industries—have suffered whiplash in Congress, repeatedly left to expire, only to be resurrected and then fade away once again. This pattern has produced insecurity within the renewable energy space, something that is apparent in the chart in Figure 7.13.

While inconsistent, these policies have had a positive impact on solar and wind power development. The question is whether either or both energy resources will become economically viable by the time these incentives phase out completely. The National Renewable Energy Laboratory (NREL) believes they will, and that the economics of solar and wind are improving to the point where both can continue to expand and compete favorably with conventional sources of generation. Elimination of the ITC is unlikely to kill continued industry growth, the NREL

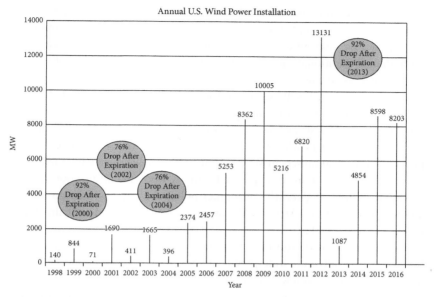

Figure 7.13 The cycle begins with the industry experiencing strong growth in development around the country while the PTC is firmly in place, and in the years leading up to the PTC's expiration. Lapses in the PTC then cause a dramatic slowdown in the implementation of planned wind projects and layoffs at wind companies and manufacturing facilities. Upon restoration, the wind power industry takes time to regain its footing and then experiences strong growth until the tax credits expire. And so on. *Source:* Union of Concerned Scientists, "Production Tax Credit for Renewable Energy," https://www.ucsusa.org/clean-energy/increase-renewable-energy/production-tax-credit#.Wv9G-0jRVPY.

predicts, because generation costs from wind and solar are now at parity with or below the costs of most other sources of generation.[74]

In other words, the policies favoring renewables have been successful. (The low costs for solar modules imported from China have helped immensely.) According to the American Wind Energy Association, these tax credits "have benefited American consumers by growing our economy, creating jobs, improving energy security, saving money for families and businesses, and supporting a new U.S. manufacturing sector . . . They have driven tens of billions of dollars into rural and Rust Belt America and brought new jobs right to the places where they are needed the most. . . . Thanks to this policy certainty, approximately 21 gigawatts (GW) of wind power capacity is now under construction or in advanced development. With the PTC phase-down, wind energy can grow to supply 10 percent of U.S. electricity by 2020 and support tens of thousands additional well-paying jobs."[75]

What does all this suggest? With the success of policies focusing on renewable energy—and their planned phase-out—one might argue against continuing tax support for conventional energy. This would seem even more reasonable given that all traditional energy resources carry hefty burdens of negative externalities such as air and water pollution and the risks of long-lived radioactive waste. Because these health, safety, and aesthetic costs are not included in what the consumer pays, the prices of these forms of energy are artificially less than they should be. Worse, as long as consumers enjoy lower prices, they do so at the expense of others who must suffer the external costs. Examples of such people include coal miners in West Virginia and northeastern China who work in hazardous conditions, disappointed tourists who encounter poor visibility in national parks, anglers who encounter polluted water few fish can tolerate, and residents in India, Africa, and Canada displaced when new hydroelectric projects are constructed.

Policies that provide subsidies to energy development of all types, conventional and alternative, are widespread outside the United States. Emerging economies in Asia, for example, account for about half of the total subsidies, while advanced economies account for nearly a quarter (Figure 7.14). In total, such

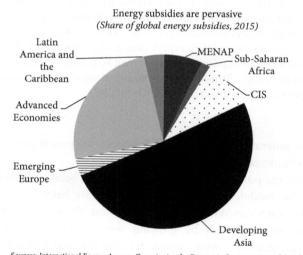

Energy subsidies are pervasive
(*Share of global energy subsidies, 2015*)

Sources: International Energy Agency; Organisation for Economic Co-operation and Development; and IMF staff estimates.
Note: MENAP = Middle East, North Africa and Pakistan; CIS = Commonwealth of Independent States

Figure 7.14 Energy subsidies are pervasive for many reasons. Often they are to favor a specific energy technology. Other times they are politically motivated. *Source:* International Monetary Fund, https://blogs.imf.org/2015/05/18/act-local-solve-global-the-5-3-trillion-energy-subsidy-problem/#post/0.

subsidies are massive. The most substantial subsidies are in China ($2.3 trillion), the United States ($699 billion), Russia ($335 billion), India ($277 billion), and Japan ($157 billion). For the European Union, subsidies total about $330 billion. The fiscal implications are mammoth as well: at $5.3 trillion, the total cost of energy subsidies—both external costs and internal (paid) costs—"exceed[s] the estimated public health spending for the entire globe. It also exceeds the world's total public investment spending. The resources freed from subsidy reform could be used to meet critical public spending needs or reduce taxes that are choking economic growth."[76]

Summing Up

Just imagine if we were to try to meet our energy needs in the absence of energy policies—that is, with no guidance as to the direction we should be heading. What would our lives look like? Could we achieve economic and personal stability, or would we devolve into chaos from the lack of clear signals? Energy policies have a place in our society because they have robust effects on our future, including effects on jobs, environmental quality, national security, geopolitics, price, sustainability, reliability, and future livability.

Winners and losers emerge regardless of which energy policies we adopt. If we decide to move away from high-carbon-emitting energy resources, for example, coal miners suffer job losses while natural gas developers prosper. If the United States weans itself from imported oil, tanker crews suffer but drilling crews prosper. If nuclear power is judged too risky in Germany, lignite miners will be in demand and wind developers will make money. Every policy option comes attached to financial costs, environmental costs, impacts on national security, and long-term consequences for future generations.

The thread of energy underlies the three converging challenges facing the United States today: economic prosperity, security, and environmental quality. Electricity continues to be a key enabler in addressing these challenges, but the status quo is not good enough. We can do better. Those who have taken up the challenge of meeting our energy needs in our overcrowded world are being increasingly encouraged to emphasize the traits of responsibility and sustainability. Energy policies that are effective and just can help us achieve that goal.

8

Geopolitics

Spiderwebs of Global Interdependency

Geopolitics of Energy

Even if you are not familiar with the term "geopolitics," it mostly explains itself. Geography + politics = geopolitics. Add in the word "energy," and you have the "geopolitics of energy." In today's world, the geopolitics of energy seem to be intertwined with just about every action taken by a country, a state, a city, and even an individual. If that does not ring true to you, answer two questions: Where are examples of international tension? Are any energy assets nearby, such as ports, reserves, pipelines, power plants, or transmission corridors? The chances are good that the answer will be yes.

Here is just a sampling of events where energy has most recently been in play:

- International claims to the Arctic Basin
- The Russian annexation of the Crimean Peninsula
- Multinational disputes over the South China Sea
- Military "tag" played daily between the navies of the United States and Iran in the Persian Gulf
- Disputes over pipeline construction between Canada and the United States
- Genocide in Sudan and South Sudan
- Ecological damage in eastern Ecuador
- Souring relations between the United States and Venezuela

While it should be no surprise that energy is tied to geopolitics, what may surprise you is how significant—even dangerous—these relationships can become. My purpose in this chapter is to illustrate these relationships, exposing how the geopolitics of energy can influence world economies, international stability and security, and the well-being and safety of everyone on the planet.

The geopolitics of energy, while common in the news today, first attracted serious public awareness several decades ago. Remember World War II from your high school history class? You might recall that after Hitler's armies invaded Poland in 1939, a significant part of his military force then headed southward. Why? At about the same time 6,000 miles to the east, Japan invaded China and

The Thread of Energy. Martin J. Pasqualetti, Oxford University Press. © Oxford University Press 2021.
DOI: 10.1093/oso/9780199394807.003.0008

soon afterward also headed southward. Again, why? What was motivating the southerly direction of both armies?

The answer is oil. Neither Germany nor Japan has ever had significant reserves of this critical energy resource. Their leaders at the time, however, understood its importance to their military ambitions. They knew that without oil, their jeeps, trucks, tanks, ships, aircraft—indeed, their entire military force—would become ineffective and eventually immobile. Without a reliable and sufficient oil supply chain, they knew, they could only manage a short war. And they did not think the war would be short. Realizing this vulnerability, Germany's immediate goal was to gain control of oil supplies near Ploiești, Romania. Years later the Allies would target these fields to cut supplies to Germany (Figure 8.1).

The Allies had maps of global oil reserves. They understood the oil dependency of several countries, including Japan. Not wanting to help Japan pursue its aggressive ambitions against China and other countries, President Franklin D. Roosevelt ordered an embargo of oil exported from the United States to Japan on July 26, 1941.

Figure 8.1 American B-24 Liberators flying over a burning oil refinery at Ploiești, 56 km north of Bucharest. Due to its role as a major supplier of oil to the Axis, Romania was a prime target of Allied strategic bombing in 1943 and 1944. *Source:* Jerry J. Jostwick, the only survivor of the 16 cameramen of the operation, August 1, 1943, National Museum file 050616-F-1234P-017.jpg, https://en.wikipedia.org/wiki/Operation_Tidal_Wave#/media/File:B-24D%27s_fly_over_Polesti_during_World_War_II.jpg.

When Japan bombed Pearl Harbor less than six months later, they made a tactical mistake. Although they sank ships and destroyed planes, they did not destroy the nearby oil storage tanks holding 10 million barrels of oil. Had they done so, U.S. retaliatory options would have been more limited. As it was, the United States was soon in hot pursuit of the Japanese navy, which was heading southward to grab military control of oil-rich Indonesia (Figure 8.2).[1] It would be several years before Japanese forces could be ousted from that region.

For a more recent example of the geopolitics of energy, one also based on the vital supply of oil, let's fast-forward to the early 1970s. At the time, the Organization of Petroleum Exporting Countries controlled most oil exports (Figure 8.3). In 1973–74, OPEC conclusively demonstrated its political power and its displeasure with countries—including the United States—that supported Israel in its Yom Kippur War with Arab states. Because OPEC countries were feeding the voracious appetites of Europe and the United States, their embargo produced quick and dramatic consequences, especially in the transportation sector. Rationing was initiated. In the United States, long lines formed around fueling stations, and fights would break out over people cutting in line. Illicit siphoning from neighbors' gas tanks became commonplace, and auto manufactures began installing locks on car fuel doors, a feature that is routine today.

Awakened to their vulnerability to future embargos, oil-importing countries initiated policies to wean themselves from OPEC supplies. In the United States, this meant opening more federal land to oil exploration and development, including wilderness areas such as the Arctic National Wildlife Refuge. Environmental organizations resisted such forays. The fight to preserve this wilderness from oil development is still ongoing, although the U.S. tax bill of 2018 included a rider to allow drilling to commence.[2] By 2020, drilling there was still being challenged in court, but by June 2021, President Biden suspended the leasing program.

The second step for the United States was to develop emergency reserves as a buffer against similar future threats from OPEC. The best-known result of this strategy was the establishment of strategic oil reserves maintained by the U.S. Department of Energy's Office of Petroleum Reserves (OPR). The OPR manages three stockpiles: the Strategic Petroleum Reserve (SPR), the Northeast Home Heating Oil Reserve (NHHOR), and the Northeast Gasoline Supply Reserve (NGSR).[3] The SPR consists of four sites (Figure 8.4). It is the largest emergency oil supply in the world, with a capacity of up to 713.5 million barrels (113,440,000 m^3). As of May 8, 2020, it held 640 million barrels.[4] As an emergency stockpile of government-owned heating oil, NHHOR was intended to meet roughly ten days' worth of demand by the northeastern states at the time. Currently, NHHOR contains a 1-million-barrel supply of ultra-low-sulfur distillate (diesel),

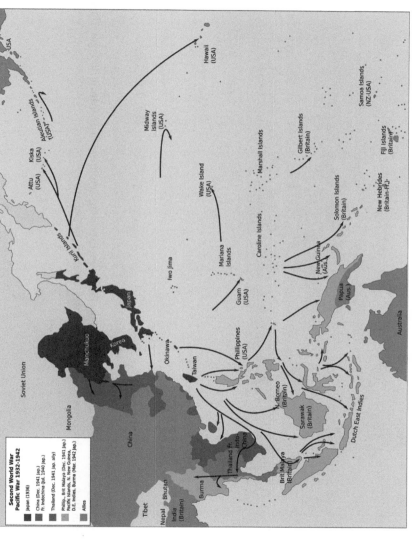

Figure 8.2 Imperial Japanese advance through the oil-rich Dutch East Indies (now Indonesia) in 1942, soon after bombing Pearl Harbor, because they had little oil of their own and needed oil to pursue their military ambitions. *Source:* https://courses.lumenlearning.com/boundless-ushistory/chapter/the-early-war-in-the-pacific/.

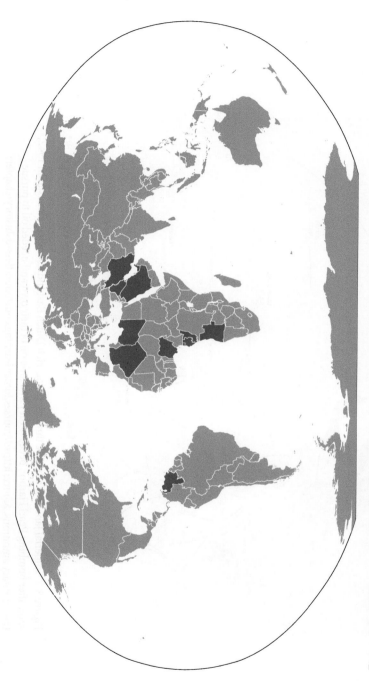

Figure 8.3 OPEC countries. The Organization of the Petroleum Exporting Countries is a permanent, intergovernmental organization, created in 1960. *Source:* http://www.opec.org/opec_web/en/about_us/24.htm.

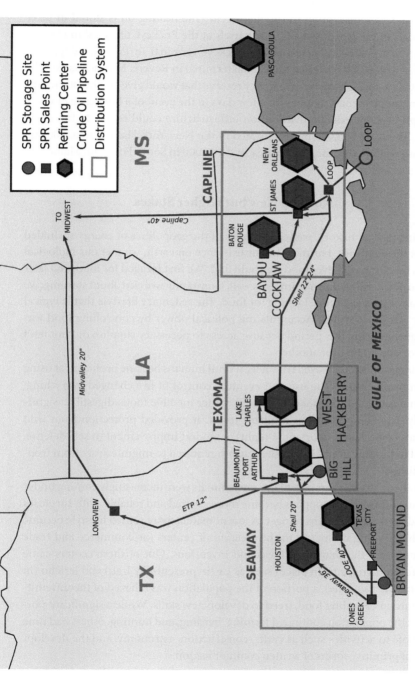

Figure 8.4 The four sites of the Strategic Petroleum Reserve are capable of storing about 714 million barrels of petroleum. Storage at the beginning of 2020 was about 640 million barrels. The Reserve contains 60 huge underground caverns. One storage cavern is large enough for Chicago's Willis Tower to fit inside with room to spare. *Source:* U.S. Department of Energy.

providing protection for homes and businesses in the northeastern United States should a disruption in supplies occur. The heating oil is stored in three terminals in the Northeast: 300,000 barrels at the Buckeye terminal in Groton, Connecticut, 300,000 barrels at the Buckeye terminal in Port Reading, New Jersey, and 400,000 barrels at the Global terminal in Revere, Massachusetts. The NGSR is a 1-million-barrel emergency reserve that would give northeastern consumers supplemental supplies for a few days in the event of a hurricane or other disruption, until existing distribution infrastructure could return to full operation. There are 700,000 barrels located in the New York Harbor area, 200,000 barrels positioned in the Boston area, and 100,000 in South Portland, Maine.

Nothing New but Higher Stakes

Looking back at history, we see examples of the geopolitics of energy sprinkled across the centuries. For most of our existence on earth, we met our biological needs by eating whatever food we could find. We first foraged for food, and later we learned to bring down animals as well. Constantly worried about starving, we were always on the move looking for food. The sedentary lifestyle that is typical today was not possible. Accumulating political power by controlling food was infrequent during this period because access to necessary supplies did not tend to favor one group over another.

This circumstance stayed true at least until humans became proficient at using and controlling fire. The use and eventual control of fire changed everything. When it was used for cooking, it could render inedible foods digestible, significantly expanding what was available to eat. It provided protection from wild animals (especially important at night), a modest improvement in self-defense, an aid in hunting animals, and the warmth required to migrate away from tropical climes.

By about 10,000 years ago, plants and animals were increasingly coming under human control. Food supplies became more assured and reliable, with surpluses providing security against hunger. A less nomadic, more settled lifestyle became possible, leading ultimately to the creation of centers for commerce and trade that nurtured the sharing of ideas and inventions. Out of these centers came the pre-industrial city. In places such as Ur (in present-day Iraq) and Jericho (in present-day Palestine), a portion of the population was relieved of the unremitting burden of finding food, freed to develop new skills. While a significant portion of the population continued farming, herding, and hunting, others had time to devote to activities such as crafts, construction, astronomy, and the development of primitive forms of written communication.

The geopolitics of energy is not new. It has existed for millennia. Probably the first example is when people started trading the energy in the food they grew for non-energy commodities that they wanted. These commodities included salt, clothing, spices, tools, and the promise of a specific service. As a result, those who controlled the supplies of energy—the most essential and fungible of commodities—gained political power. It has been that way ever since.

Whether in raw or processed form, energy is always in demand, coveted for the lifestyles it supports, the flexibility it provides, and the authority it projects. Such importance means that energy is always marketable, critical not only to the underpinning of well-being but also to the balance of global influence that we see today. It is hard to believe that major political and economic powers—such as the United States, United Kingdom, Russia, Germany, Japan, and China—could have grown to prominence without a strong foundation of energy resources upon which to project their influence.

While it is true that countries with a high standard of living tend to have a relatively high per capita energy consumption, one should not extrapolate from this correlation without caution. Better that you understand a few nuances, including that a natural abundance of energy resources does not guarantee a high standard of living, as many living in Nigeria and Venezuela can attest. Nor does it follow that countries with significant geopolitical importance must necessarily have abundant energy reserves within their borders, as resource-poor Singapore and Japan both demonstrate. Likewise, countries with vast energy reserves do not need large populations to accumulate geopolitical influence, as Bahrain and the UAE demonstrate. How then can we explain all these relationships? As Jared Diamond reminds us, these often come from generous portions of luck and history.[5] They also come from simple geographical considerations of how to move energy products from areas where they are more plentiful than needed to areas where the opposite conditions prevail.

Moving Oil

Persian Gulf countries continue as the largest concentration of oil-exporting countries (Figure 8.5).[6] The dominance of this relatively small area of the world is in part due to the fact that over 30 percent of all maritime-traded petroleum passes through the Strait of Hormuz on its way to the Indian Ocean and its destinations around the world. Such a high volume of transport helps explain why the Strait of Hormuz has become the most critical chokepoint in the world.

The strait is bounded on the north by Iran and on the south by the Musandam Peninsula of Oman. At its narrowest, it is an ample 21 nautical miles wide. However, the width of the shipping lane in either direction is only 2 miles, with

Daily transit volumes through world maritime oil chokepoints

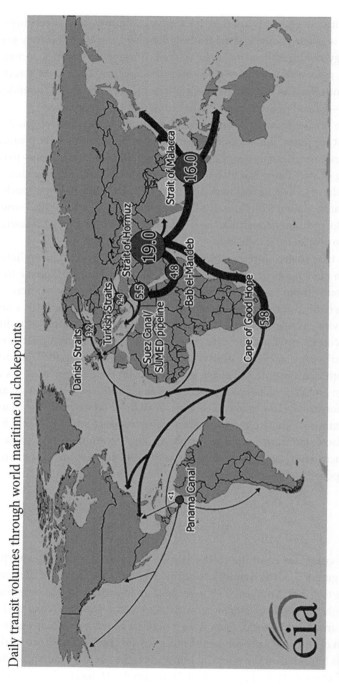

Figure 8.5 Daily transit volumes through world maritime oil chokepoints. All estimates are in millions of barrels per day, including crude oil and petroleum products. Based on 2016 data. *Source:* US EIA, https://www.eia.gov/international/analysis/special-topics/World_Oil_Transit_Chokepoints.

the two lanes separated by a 2-mile buffer zone. Several countries—including the United States—patrol these waters. The International Maritime Security Construct (IMSC), formerly known as Operation Sentinel or the Sentinel Program, is a multinational maritime effort established by the United States to ensure gulf security following Iranian seizures of commercial tankers. Members include the United Kingdom, Bahrain (where IMSC headquarters is located), the United States, Australia, Saudi Arabia, United Arab Emirates, Albania, and Lithuania. Dangerous and threatening military challenges in the region are not uncommon.[7]

The strait is deep enough and wide enough to handle the world's largest crude oil tankers, including the two-thirds of oil shipments carried by tankers in excess of 150,000 deadweight tons. Many other chokepoints are not as accommodating, however, and there is an interesting story within this context.

The era of the large crude carrier began in earnest with the closing of the Suez Canal during the Six-Day War between Israel and Egypt in 1967. The result of this closure was a long trip around the tip of Africa. Such diverted shipping routes were less profitable. For a ship traveling from the Persian Gulf, where many of the Arabian oil ports are, to Britain or western Europe, the voyage around the southern tip of Africa takes sixteen days longer and adds 4,800 miles of travel compared to going through the canal. At the time, it increased the overall cost of the voyage by as much as $20,000 (equivalent to about $150,000 in 2020).[8] To make such journeys economically feasible, they would have to carry far more oil. That was the beginning of the age of the supertanker (Figure 8.6). Geopolitics precipitated this shift in how we carry oil around the world.

Today, sea transport is massive in volume, importance, and value, making oil the most valuable commodity in world trade and critical to the maintenance of the world economy. Each ultra-large crude carrier can transport between 2 million and 3.7 million barrels of crude oil. The only U.S. port that can handle such large vessels while fully loaded is the Louisiana Offshore Oil Port, near Port Fourchon. This port currently services over 90 percent of the Gulf of Mexico's deep-water oil production. There are over 600 oil platforms within a forty-mile radius of Port Fourchon. This area furnishes 16 to 18 percent of the U.S. oil supply.

As ship sizes increased, so too did the value of their cargoes. Even at a depressed price of $50/barrel, the largest tankers can carry $185 million in oil, making them floating targets for pirates. Trying to take control of such vessels is a high-risk, high-reward gambit, sometimes ending in millions of dollars paid to the pirates in ransom, and sometimes ending with the pirates killed or captured (Figure 8.7). Such incidents—and their increasing frequency—are an example of how geography, energy, and politics are all tied into one bundle.

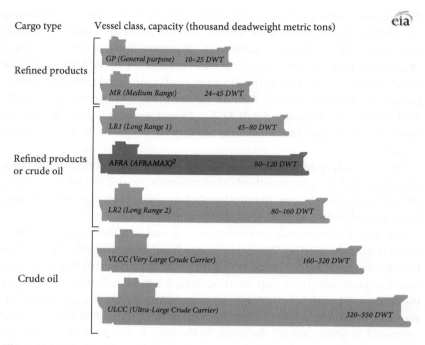

Figure 8.6 Ship size dictates whether it can squeeze through various chokepoints or must take alternative routes. A very large crude carrier (VLCC) can carry between 1.9 million and 2.2 million barrels of a West Texas Intermediate type crude oil. VLCCs are responsible for most crude oil shipments around the globe, including in the North Sea, home of the crude oil price benchmark Brent. Ultra large crude carriers (ULCCs) carry between 320 DWT and 550 DWT and cannot fit through any of three chokepoints. *Source:* U.S. EIA, based on Bloomberg and London Tanker Brokers' Panel, https://energycentral.com/c/ec/low-tanker-rates-are-enabling-more-long-distance-crude-oil-and-petroleum-product

If oil tankers represent a vivid example of the geopolitics of energy transport, oil pipelines might be the even more vulnerable. Why is this true? First, while land routes fall under the jurisdiction, control, and protection of sovereign governments, most sea transport is in international waters. Second, while ships can take many different routes from one place to another, pipeline routes are fixed and well known. Third, pipelines often cross international borders as they link areas of production with customers. These three characteristics reflect the abundance of international arrangements, debates, interruptions, and conflicts regarding energy movement.

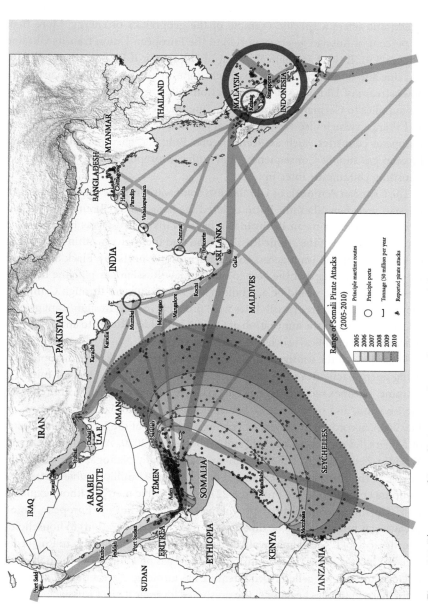

Figure 8.7 The extent of Somali pirate attacks on shipping vessels between 2005 and 2010. *Source:* Wikipedia Commons, https://en.wikipedia.org/wiki/Piracy_off_the_coast_of_Somalia.

In recent years, the clearest examples of international pipeline controversy have been in Europe, where corridors are plentiful between the great gas supplies of Russia and the hundreds of millions of gas customers in Europe (Figures 8.8).

Ukraine offers a specific example of the geopolitics of energy. More than 66 percent of Russian gas transported to the European Union transits Ukraine (Figure 8.9). If there is political unrest in Ukraine or if Russia cuts supplies to Ukraine for non-payment (as has been done in recent years), all customers downstream of Ukraine suffer. Such actions affect more than one-third of the gas supplies to Germany and more than one-quarter of the gas supplies to Italy and France (which also receive natural gas from North Africa). Some eastern European countries, such as Romania and Bulgaria, have few alternatives to Russian gas supply. Consequently, they suffer disproportionally when supplies transiting Ukraine are interrupted.

The Crimean Peninsula of Ukraine juts southward into the Black Sea. In 2014 Russian president Vladimir Putin ordered the invasion of Ukraine and the annexation of the Crimean Peninsula. What was his motive? Was he concerned about the welfare of the ethnic Russians who find themselves within the borders of Ukraine? Did he want access to the warm-water ports the Black Sea offers? Or was he thinking that the cost to build a new Russian gas pipeline between Russia and Europe would be $20 billion less if its route was through Crimea to the Black Sea? All three possibilities were probably in play. In significant geopolitical terms, might the annexation of Crimea embolden Russia to annex parts of other former republics—such as oil-rich Kazakhstan?[9]

Regardless of the motivation, the Russian invasion of Ukraine disrupted existing energy activities in several ways. Around half of the 115 coal mines in Ukraine, Europe's second-largest coal producer, halted production entirely with the invasion, and output fell 22 percent year-on-year to 5.6 million tonnes in July 2015, according to Mykhailo Volynets, the chair of the Independent Union of Miners.[10] Production had still not recovered by 2020. As another example, Valentin Zemlyansky of the Ukrainian gas company Naftogaz was quoted as saying: "[The Russians] have reduced deliveries to 92 million cubic metres per 24 hours compared to the promised 221 million cubic metres without explanation." Cuts to gas supplies risk provoking tensions within the European Union because gas might need to be diverted from industrial operations to households in order to avoid "social upheaval." Ever pragmatic, energy planners in Russia decided to increased gas trade by installing a new gas pipeline under the Baltic Sea to Germany. The so-called Nord Stream 2 has attracted a great deal of attention from many countries because of the presumed leverage it gives Russia.[11]

The examples above are just a sampling of many that illustrate the geopolitics of energy. It is easy to locate other examples. Just look for news stories on recent political unrest in Central Asia, and you will notice that Georgia as well as Dagestan, between the Black Sea and the Caspian Sea, are part of the regional

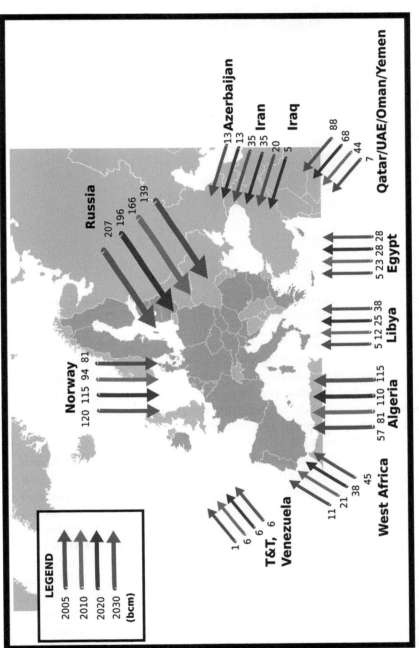

Figure 8.8 Europe receives a major portion of its natural gas supplies from surrounding countries. Such dependency poses a risk from adversaries. *Source:* https://eadaily.com/en/news/2017/03/06/syria-ukraine-gas-pipeline-wars-us-russia-alliance-inevitable. For a series of specific maps of Russian pipelines, see https://egas.com/maps.htm. See plates for color version.

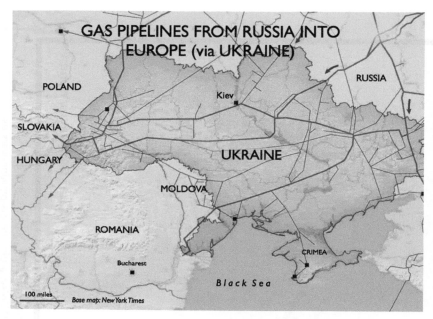

Figure 8.9 Ukraine hosts many of the most geopolitically explosive pipeline routes in the world. *Source:* http://roconsulboston.com/.

energy picture. Are you aware of the armed conflicts that have occurred in this area over the past two decades, especially in Chechnya? Apart from the possible historic ethnic and religious tensions that are common to the region, do you know what prompted these conflicts? You can get a sense of the problem just by looking at the geographical position of these political entities vis-à-vis major pipeline routes (Figure 8.10).

As if to illustrate that oil transport issues are not unique to the Eastern Hemisphere, halfway around the world the routing of pipelines attracted a great deal of attention in North America beginning in 2008 after TransCanada proposed a 1,179-mile (1,897 km), 36-inch-diameter crude oil pipeline beginning in Hardisty, Alberta, and extending south to Steele City, Nebraska. The Keystone XL pipeline was to have the capacity to transport up to 830,000 barrels of oil per day to Gulf Coast and Midwest refineries. In May 2012, TransCanada filed a new application for a presidential permit with the U.S. Department of State, a requirement for building any cross-border pipeline. The proposal fomented substantial debate for several years, mostly in the United States. The question is, why the unrest? After all, Keystone XL would not be the first oil pipeline from Canada to the United States. Nor would it be the first to transport oil to the United States from the tar sands of Alberta; rather, it would be a continuation of a trade pattern established many years ago.

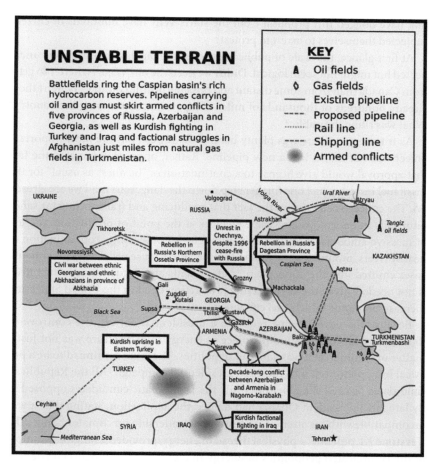

Figure 8.10 Unstable terrain. Many of the conflicts in this area of the world can be linked to their geographic position between the oil and gas resources of the Caspian Basin and the markets to the west. *Source:* Viable Opposition, "Chechnya—A Backgrounder," April 2014, http://viableopposition.blogspot.com/2013/04/chechnya-backgrounder.html. Adapted from a drawing by Joe Shoulak/San Francisco Chronicle/Polaris.

Several factors lay behind this trade pattern. First, Canada is a close neighbor, and we share history and culture. Second, the United States and Canada are on friendly terms, separated only by a mostly undefended 4,000-mile border. Third, importing oil from Canada can reduce or eliminate long-distance sea transport, ocean spills, and opportunities for piracy. Fourth, the abundance of oil in Canada—reserves are presently estimated at 170 billion barrels—ensures long-term supplies.

Given all these advantages, why would opposition and political wrangling develop over the proposed pipeline? What caused the vociferous objections that have dogged this proposal from the start? Why have hundreds of citizens subjected themselves to arrest in protest?

At first glance, the scale of public angst over the pipeline not only was unexpected but might not seem logical. Didn't we need the oil? Wasn't it better to get it from Canada than from some distant and possibly hostile country? Weren't there already hundreds of thousands of miles of pipelines crisscrossing the country? What was riling the public?

As it turns out, there was plenty behind the unrest, but the most important object was not that it was a new pipeline. Rather, opposition targeted the fact that approval would give license to a continuation of "business as usual" for the fossil fuel industry, just one further step down the dangerous road we are already on. The concern centered on the fact that producing and transporting tar sands oil results in enormous landscape changes at the point of production, the use of massive amounts of water and natural gas for processing, contamination of nearby rivers, mountains of extracted sulfur, and profuse volumes of greenhouse gases emitted into the atmosphere.[12] Opponents also argued the product itself is not needed, given the skyrocketing production of oil in the United States and increases in motor vehicle fuel efficiency.

Because the new pipeline required a presidential permit, the controversy that swirled around this unique piece of energy infrastructure was not just a matter of commerce; it was a matter of politics. The issue continued to be a political hot potato well into the 2016 presidential campaign. All the Republican candidates favored the pipeline, while the Democratic candidates opposed it. By late 2015, President Obama had denied the application, stating that it was incompatible with his attempts to foster U.S. leadership on climate change. The Keystone XL pipeline, a physical thread of energy, provided a tangible example of the complicated geopolitics of energy that can exist even between friendly neighbors.

The situation gets increasingly complicated when international bodies vie for the same resources in the same area, with each claiming sovereignty. Such is the case most recently with activities in the South China Sea. The abundance of fish, the importance of the sea for shipping, and the vast gas and oil reserves believed to exist there make the South China Sea attractive to all ten countries that lay claim to parts of it—China, Taiwan, the Philippines, Malaysia, Brunei, Vietnam, Indonesia, Singapore, Thailand, and Cambodia. Given its mighty economy and large military threaten, China's claims tend to overrun most of the others (Figures 8.11, 8.12). The sea continues to be a flashpoint for potential aggressive encounters.

The South China Sea is not the only place China is probing for energy; it is also acquiring oil rights in Sudan, Chad, Nigeria, Ecuador, Brazil, Indonesia, and countries in Central Asia. China's need for oil not only is growing but appears never-ending. It is part of a worldwide scramble for an important but finite

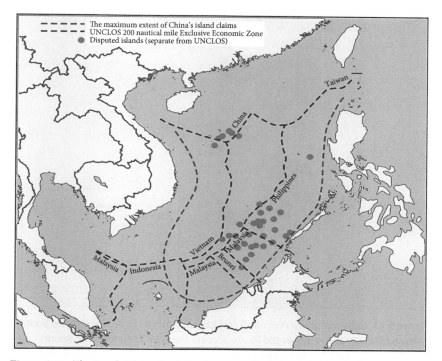

Figure 8.11 The South China Sea is the location of substantial risk of military confrontation because of overlapping territorial claims in this resource-rich area. *Source:* https://en.wikipedia.org/wiki/Territorial_disputes_in_the_South_China_Sea. See plates for color version.

resource critical to plans for economic development. These are threads of energy to keep in mind whenever you refuel your car or truck.

Politics by Other Means

On occasion, the instabilities that result from competition for oil reserves can escalate from ordinary saber-rattling to the realities of war. What is the meaning of the word "war"? Over a century ago, Carl von Clausewitz famously considered war "a mere continuation of politics by other means."[13] More recently, Michael Klare has alerted us to past, present, and possible future "resource wars," suggesting that political conflicts can result when resource demand persistently outpaces resource supply. This possibility is increasingly probable as the world becomes more crowded and competition for those resources increases.[14]

The ongoing clamor for oil has caused several notable international tensions in recent years, starting when Iraq invaded Kuwait on August 2, 1990, following

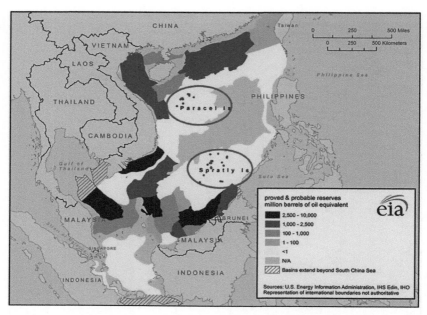

Figure 8.12 Substantial reserves of oil and gas are believed to lie beneath the South China Sea. Many countries would like to recover them for their use. The Spratly Islands and Paracel Islands are two of the most contested areas. *Source:* U.S. EIA, http://www.eia.gov/todayinenergy/detail.cfm?id=10651.

orders from Saddam Hussein. His goal was to add Kuwait's 100 billion barrels of oil reserves to the 150 billion barrels of oil reserves estimated to exist in his own country (Figure 8.13). Such an addition would have bumped Iraq up to fourth place in terms of oil reserves, after Venezuela, Saudi Arabia, and Canada.

Kuwait was likely not going to satisfy Saddam's ambitions. Most people believed that he planned to continue moving aggressively south through Kuwait into Saudi Arabia itself. The worry was that he would eventually capture the magnificent Ghawar oil field, the world's largest (Figure 8.14)—a very tempting prize for Saddam, even if its productivity has long been in dispute. (In April 2019, in the context of issuing a bond to international markets, the oil company Saudi Aramco published its profit figures for the first time since its nationalization nearly forty years ago. The bond prospectus revealed that Ghawar field can pump a maximum of 3.8 million barrels a day—well below the more than 5 million that had become conventional wisdom in the market.[15] Even this level of production is in decline now. Still, several billion more barrels remain in the field.)

Kuwait's oil fields

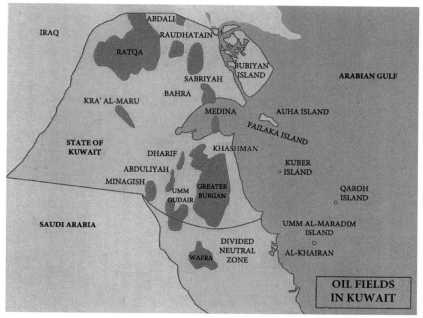

Source: Kuwait's Ministry of Oil

Figure 8.13 Oil fields of Kuwait. *Source:* https://www.markaz.com/Blog/
Markaz-Blogs/Marmore-MENA/February-2016/Kuwait-Petrochemicals.

On August 3, 1990, the United Nations Security Council called for Iraq to withdraw from Kuwait. Iraq responded by formally annexing Kuwait on August 8. Iraq's invasion and the potential threat to Saudi Arabia prompted the United States and its western European NATO allies to send troops to Saudi Arabia to deter a possible attack. Egypt and several other Arab nations joined the anti-Iraq coalition and contributed forces to the military buildup, known as Operation Desert Shield. Iraq, meanwhile, grew its occupying army in Kuwait to about 300,000 troops.

On November 29, the UN Security Council authorized the use of force against Iraq if it did not withdraw from Kuwait by January 15, 1991. By January 1991, the allied coalition against Iraq had reached a strength of 700,000 troops, including 540,000 U.S. personnel and smaller numbers of British, French, Egyptians, Saudis, Syrians, and several other national contingents (Figure 8.15).

Saddam refused to withdraw Iraqi forces from Kuwait, and this response triggered an air and ground assault by the allied coalition. The coalition offensive

Figure 8.14 The "magnificent five" contiguous fields of the super-giant Ghawar oil field. Together the string of fields is 280 km long and 40 km wide. This, the largest oil field in the world, is not far from Kuwait and was a suspected end goal of Saddam Hussein's invasion of Kuwait. *Source:* https://worldct.wordpress.com/2015/03/12/the-saga-of-saudi-aramco/.

against Iraq began on January 16–17, 1991, with a massive U.S.-led air campaign. Over the next few weeks, this sustained aerial bombardment destroyed Iraq's air defenses before attacking its communications networks, bridges and roads, government buildings, weapons plants, oil refineries, and other military targets. By mid-February, the allies had shifted their air attacks to Iraq's forward ground forces in Kuwait and southern Iraq, destroying their fortifications and tanks (Figure 8.16).

The culminating ground assault lasted just 100 hours, followed by a full capitulation by Iraqi forces, a largely destroyed infrastructure, and the loss of the majority of their military equipment. As a last act of defiance, however, retreating Iraqi troops set ablaze 600–700 oil wells along with an unspecified number of oil-filled low-lying areas, such as oil lakes and fire trenches. It was part of their "scorched earth" policy, and it created one of the largest incidents of air and water pollution in world history (Figure 8.17). Six million barrels, valued at

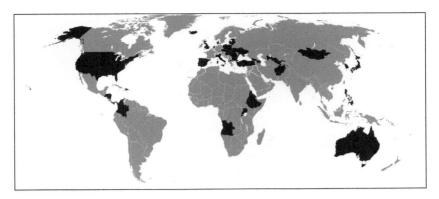

Figure 8.15 Nations that deployed coalition forces or provided support in the Gulf War. *Source:* Wikipedia, https://en.wikipedia.org/wiki/Gulf_War#/media/File:Coalition_of_the_Gulf_War_vs_Iraq.png.

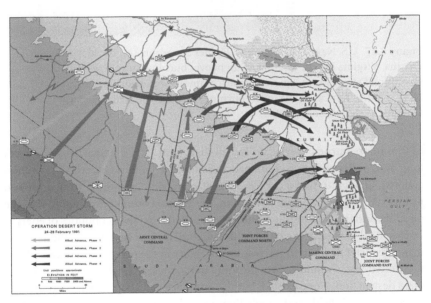

Figure 8.16 Ground troop movements, February 24–28, 1991, during Operation Desert Storm showing allied and Iraqi forces. Special arrows within the map indicate how the American 101st Airborne division moved by air and where the French 6th Light Division and American 3rd Armored Cavalry Regiment provided security. *Source:* U.S. Army Center of Military History, http://www.history.army.mil/reference/DS.jpg. See plates for color version.

Figure 8.17 U.S. Navy F-14A Tomcat, Fighter Squadron 211, Naval Air Station Oceana, Virginia Beach, Virginia, in flight over burning Kuwaiti oil wells during Operation Desert Storm. *Source:* https://commons.wikimedia.org/wiki/File:US_ Navy_F-14A_Tomcat_flying_over_burning_Kuwaiti_oil_wells_during_ Operation_Desert_Storm.JPEG.

$115 million, burned every day. The oil slick, at its largest extent, was 101 miles long by 42 miles wide and contained an estimated 8 million barrels (336 million gallons), enough to fuel a car for 214,500 years, or a jet airliner for 7.2 million miles.[16] All this—the original invasion, the ensuing bombardment, the war, the oil spills, and the choking air pollution—resulted from the geopolitics of energy.

Despite the massive use of force, the military actions against Iraq did not permanently deter Saddam from creating instability in the region. Two decades later, the United States would return, invading Iraq on a search for alleged weapons of mass destruction. None existed, but the war, once launched, dragged on for over a decade; the oil fields and supply lines were destroyed, then repaired, and then destroyed again. An estimated $1 trillion was spent just by the United States to oust Saddam Hussein and establish something resembling a democracy. The total costs of this, including all the resulting losses of blood and treasure, are still to be determined. Nevertheless, serious unrest in the area continued for years, including the turmoil created by a self-proclaimed Islamic state known as ISIS (or ISIL).

In July 2014, the ISIS terrorists began taking over oil fields in northern Iraq, capturing or destroyed refineries and pipelines. Iraq, which was still mending from decades of war, was just beginning to ramp up its oil exports again, reaching about 2.5 million barrels per day. At roughly $100 per barrel, the cessation of exports from ISIS activity amounted to over $250 million per day lost to the Iraqi economy, hindering the country's ability to return to anything resembling normalcy. Not wanting the situation to degrade any further, the United States sent limited contingents of personnel to help train the nascent Iraqi army, returning to a country that had been left on its own just three years before.

At the time of the 2003 U.S. invasion of Iraq, many people thought that securing the oil reserves was the actual goal, if not the stated one. "Of course it's about oil; we can't really deny that," said Gen. John Abizaid, former head of U.S. Central Command and Military Operations in Iraq, in 2007. Former Federal Reserve chair Alan Greenspan agreed, writing in his memoir, "I am saddened that it is politically inconvenient to acknowledge what everyone knows: the Iraq war is largely about oil." In 2007, then-U.S. senator and later defense secretary Chuck Hagel said the same thing: "People say we're not fighting for oil. Of course, we are."[17] Whether the 2003 invasion was originally about oil and, therefore, national security will be judged by historians, but what is a fact is that U.S. personnel are still there.

National Security

A persistent question at the center of most discussion about energy commerce is how it might influence national security. This single concern has saliency for all the examples provided earlier in this chapter. However, there is a related question to consider. What might happen if the present rough balance between supply and demand shifted significantly?

For a window on this question, let's look at recent events in the United States. Most dramatically, the United States is now the world's leading producer of oil. By mid-2019, U.S. oil recovery had lifted production to more than 12 million barrels per day (Figure 8.18). Imports dropped precipitously, and the U.S. Gulf Coast became a net exporter of oil.

Accompanying the welcome news about the abundance of domestic production and the customer-friendly drop in prices at the gas pump that resulted, there were also some more sour consequences. As prices dropped, 100,000 oil workers lost their jobs in the United States. In Canada, thousands more in the oil sands area of Alberta suffered similarly because profitably producing oil from the sands at lowered prices quickly became impossible. The primary reason oil production there continues at all is the billions of dollars already invested; these sunk costs

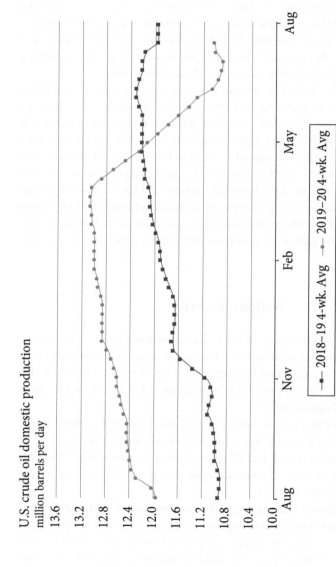

Source: U.S. Energy Information Administration

Figure 8.18 U.S. crude oil production has increased dramatically in recent years, thereby reducing dependency on imports from unstable areas of the world. *Source:* https://www.eia.gov/petroleum/weekly/crude.php.

motivate the developers to try to weather the storm, hoping that oil prices rise again soon. Such a price rise is not guaranteed, of course, as there have also been substantial increases in automobile efficiency and other improvements that have been lowering per capita demand.[18]

Lower prices may be unwelcome news to oil suppliers, particularly for exporting countries where oil provides the majority of trade revenues. Still, it can add to the national security of the countries where oil imports are a significant drain on the national treasury. Such shifts in price often have a ripple effect on geopolitical balances. For example, when our oil imports drop, so too does the political leverage held by the supplying countries. Such softened influence can have a destabilizing effect throughout the economies of those countries and even prompt a reassessment of strategies for future economic development. For example, the Saudi government has approved a comprehensive reform program, Vision 2030, which aims at diversifying the economy away from oil by that point.[19]

Such changes in the balance of trade offer greater political flexibility for government officials in other countries when it comes to relations with these nations. In turn, this flexibility affects how vigorously we must maintain our defense of tanker routes and how much we should worry about whether our oil suppliers fall under hostile influence. Imagine how the calculus would have changed within the U.S. defense establishment after Saddam invaded Kuwait if the United States and its allies had not been importing so much oil from this region. Would we all have worked together to force him to return north of the border? Similarly, the Arab oil embargo of 1973 could not have been as politically significant had the United States not been so dependent upon OPEC for a major portion of our oil imports. Would the price per barrel have quintupled after the embargo if OPEC had not been a significant supplier to western countries? Would China be flexing its muscles in the South China Sea if it felt it was able to meet all its future needs without the oil that might be there?

I am sure you can identify other examples that illustrate the geopolitics of war. As the energy geographer Peter Odell made clear in his seminal book on the subject, the control of oil mirrors world power, and it has been this way for a long time.[20] What do you think will be the geopolitical ramifications if oil reserves recede from reach through depletion or higher prices? Inevitably, there will be major shifts in world political influence as the world decreases its reliance on this "black gold."

There could be another, indirect benefit of reduced import dependence. Consider that the United States imported about 5 billion barrels of oil in 2006. By the end of 2014, this volume had declined to 3.4 billion barrels. At the price at the time of about $100 per barrel, the decline of 1.6 billion barrels equaled a savings of about $160 billion per year. This money could be reallocated to other needs. The $160 billion amounts to 150 percent of what is spent in the United States on transportation or education, 200 percent of what is spent on veterans' benefits, and six times the annual budget of the U.S. Department of Justice. In the plainest of words, lower oil prices release a lot of money, and it can have a positive impact on decisions that influence our national security.

As long as fracking and other techniques allow countries to recover more oil within their own borders, dependency on traditional oil exporters will continue to shrink. Such shrinkage, in turn, should exert downward pressure on oil prices, at least until demand rises again, if it does. Is this a good thing for everyone? How might it affect the political stability of countries where oil revenues constitute the vast majority of annual budgets? Reduced reliance on imported oil improves conditions of national security for the importing countries, such as the United States. For example, as the volume of imported oil declines, so too will the influence that the exporting countries can exert abroad. An oil embargo today would not be as effective as it was in the 1970s. Military protection of transport routes would decline, there would be less justification to invade (e.g., Iraq) or protect oil-rich countries (e.g., Kuwait), and there would be many additional benefits to our national security. Do you think all such consequences would be positive?

Consider the situation from the viewpoint not of the *importing* countries but that of the *exporting* countries. What would happen, for example, if oil exports from Saudi Arabia declined by 50 percent? Remember, the petroleum sector accounts for roughly 80 percent of that country's budget revenues, 45 percent of GDP, and 90 percent of export earnings. In other words, the country is heavily dependent on its oil exports. Moreover, more than 6 million foreign workers play an essential role in the Saudi economy, particularly in the oil and service sectors.[21]

For an answer, consider this: the drop in oil prices in 2014–2016 and then again in 2020 had quick impacts on the economies of Saudi Arabia, the UAE, Kuwait, Bahrain, and Norway (where the national economies were otherwise strong) and Venezuela, Nigeria, and Mexico (where the economies were not).[22] However, if rising demand from countries such as China does not offset the drop in prices, even the oil-rich Middle East countries will begin to suffer. Foreign nationals, who account for most of the service workers within the UAE and Saudi Arabia, might start returning home for lack of jobs and put even more strain on the economies of their home countries, which would be especially problematic if political stability there was already tenuous.

Taken together, all the impacts of such reduced demand could bring about an economic collapse, and not just in Saudi Arabia but for all the other oil-dependent economies, especially in the Persian Gulf. The negative results would quickly spread to nearby countries, including U.S. allies such as Israel. Were Israel threatened, the United States would likely become involved. It is, therefore, in the best interest of our national security to encourage economic diversification and socially conscious adjustments in oil-dependent countries so that we do not face such a situation.

Some of the richer oil-producing countries are already beginning to plan for an eventual reduction in oil revenues. For example, Dubai, which has less than 10 billion barrels in reserve (about the same as Mexico), has been developing a robust service sector, complete with opulent resorts and giant shopping malls housing top-end retailers. Even Abu Dhabi, which holds ten times more in reserves than its neighbor, has been working toward greater diversification, going so far as to build a large solar plant and what that they tout as the world's most sustainable city, Masdar.[23] In other words, these emirates are preparing for a future in which their economic health does not rely solely on oil revenues.

As if offering itself up as an example to other oil-exporting countries, Mexico is already at a stage in their history that the UAE would like to avoid. With a population of about 130 million people and a high degree of impoverishment, Mexico announced in 2015 a budget cut of over $8 billion in response to the falling price of oil and declining oil revenues.[24] Some of Mexico's historically most reliable oil fields, such as Cantarell, are in steep decline, with production dropping 40 percent in recent years. Reductions of production and revenue at this scale in a country with one-fifth the GDP per capita of the United States can result in a variety of geopolitical consequences and socioeconomic struggle.

What do you think these would be? Might these conditions influence the flow of immigrants seeking asylum in the United States? With illegal migration from Mexico to the United States already the focus of rancorous political debate, what do you think might happen if Mexico's tenuous economic stability is put under more strain as oil revenues drop? It does not take much creative thought to envision a ballooning northbound surge of desperate people. The wall that already partly separates the two countries could not stop the onslaught, and whatever geopolitical balance exists would collapse. In this case, there is a distinct thread connecting border security with the geopolitics of energy.

The Sudan Example

Substantial changes in energy demand, supply, revenue, and payment are strong forcing mechanisms affecting the political stability of Russia, Saudi Arabia, and China. These countries always receive substantial media attention, but in other countries with less direct influence on the United States, the geopolitics of energy is not as visible or recognized, despite ongoing turmoil. One such country is the African republic of Sudan.

For many years, Sudan has been suffering the consequences of internal wars that have occasionally risen above the usual clatter from the Global South. However, Sudan was barely noticed until the atrocities committed there became too horrific to ignore. Thousands died, and millions were displaced. Much of the turmoil was a manifestation of religious conflict between Arabs and non-Arabs. Still, one of the other significant factors has been oil production, processing, and transportation, and the wealth it holds for the lucky few. Sudan, a relatively small African oil producer, has been plagued by economic hardships since South Sudan seceded in 2011.

When South Sudan broke from Sudan, it took with it around 350,000 barrels per day of oil production. After the secession, the two countries were still mutually dependent, because the south has 75 percent of the oil reserves, but the north has the only current transport route for oil to get to international markets.[25] The money attached to oil development there continues to be a strong catalyst for unrest, especially because South Sudan is also one of the least developed regions on earth—only 15 percent of its citizens own a mobile phone, and few roads are paved. Ever since independence, South Sudan has been keen on using revenues from the sale of its oil to improve the standard of living for its citizens.

Several factors have thwarted this aspiration. The distribution of oil wealth between the two countries helped churn a long history of cultural disputes and distrust among the many ethnicities that cohabit in this land. Because of this geographical pattern, the two countries need a cooperative business relationship before anyone is going to reap the rewards of the oil that is underground.

Achieving such an agreement has not been easy. In January 2012, less than one year after the formation of South Sudan, the dispute between the two countries had already reached a crescendo. It centered on the surcharge that Sudan imposed on the transshipment of the oil from South Sudan to Port Sudan on the Red Sea (Figure 8.19). Sudan was demanding $32–36 per barrel, while the industry norm was about $1 per barrel. When South Sudan refused to pay this tariff, Sudan confiscated $800 million in oil as ransom. In retaliation, South Sudan closed down most of its oil operations. Neither country was profiting from the oil reserves until Sudan agreed to lower its transshipment charge, although it is still very high by global standards.

Circumstances have improved little over the past few years. Ethnic and political clashes continue despite various armistices. Valued cattle are destroyed or stolen, millions of people are displaced, and thousands are killed on all sides of the ongoing strife. Allocation of oil revenues, which could ostensibly ease the burden of the economically disadvantaged, remains central to many conflicts, disagreements, and decisions.

Although the battles over oil revenues seem primarily to be between the countries of Sudan and South Sudan, a close look reveals that these two countries are not the only actors involved. In fact, many other countries are trying to control, if not the oil revenues, then the oil reserves themselves.[26] These countries include France, Malaysia, China, Italy, and Qatar. Whether or not these countries consider the ongoing internal strife in Sudan and South Sudan of direct interest is

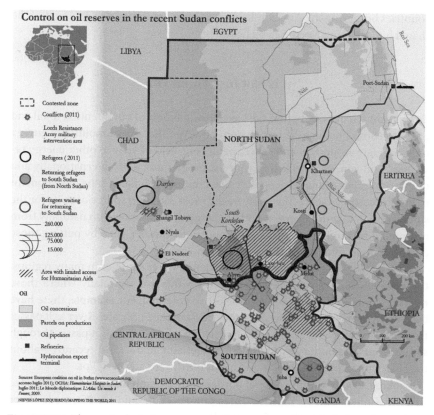

Figure 8.19 The control of oil resources in the recent conflicts in Sudan. Control of oil reserves was a major consideration in the conflicts between the south and the north of country, which resulted in the deaths of many thousands. Eventually, a new country was formed: South Sudan. *Source:* http://cartografareilpresente.org/cartoteca-africa_subsahariana-sudan-petrolio.all. Map by Nieves López Izquierdo. Used with permission.

unclear. Yet alternatives are few; the geopolitics of oil is simply part and parcel of the exploration for, discovery of, and development of oil not just in this part of Africa but in many other parts of the world as well.

As you consider the geopolitics of energy in Sudan and South Sudan, the Persian Gulf States, the South China Sea, Ukraine, Europe, the United States, Mexico, and Canada, keep in mind that these are but a few of the more visible examples. Many other examples exist, and you should take some time to investigate them in some detail. You could start with Venezuela, Nigeria, the Arctic Ocean, the coastal waters off northwestern Australia, most South American countries bordering the Amazon Basin, and even continued development of the shared but declining North Sea oil fields.[27]

It Is Not Just About Oil

To this point in our discussion, we have directed most of our attention to oil and gas. You should not get the impression, however, that the geopolitics of energy only circulate around oil and gas. Expand your field of view to other energy resources, and you will notice that your list proliferates, both internationally and domestically. For example, geopolitics are commonly linked to the development of hydroelectric projects. They may be the most common source of strife for the simple reason that the rivers—another energy thread—often run naturally from one region to another and from one country to another, thereby putting those downstream in conflict with those upstream.

With growing populations straining existing water supplies in many regions, the geopolitical consequences have been on the rise, especially where those rivers are impounded for their hydroelectric potential. For example, there have been many years of wrangling along the course of the Euphrates River as it passes from Turkey into Syria and then to Iraq; along the Mekong, as it passes through several countries on its way to the sea in Vietnam; and—most recently—about the Nile, which rises in Ethiopia and heads northward across Egypt.

Looking strictly within national borders, other examples include the Yangtze River in China, with its massive Three Gorges Dam, and the Narmada River in India, with its 3,000 dams. Like the Three Gorges project, India's enormous Narmada River Project has been promoted and subsidized for its predicted benefits to flood control, water storage, and massive electrical generation. At full scale, it is predicted to provide water to 20 to 40 million people, irrigate 1.8 to 1.9 million hectares of land, and produce 1,450 MW of power. Completion is said to be in the interest of "national development."

To many people, however, the benefits are not worth the resulting disruptions to their lives. They think of the Narmada River as the "lifeline" of the state of Madhya Pradesh. They fear that the project will flood hundreds of thousands of hectares of productive agricultural land, displace 1.5 million people, and alter centuries of cultural history. The project is the center of a heated debate between those who see it as a benefit to the national interest, especially those states through which it passes, and those who view it as "India's greatest planned environmental disaster."[28] It has pitted interest groups against one another and has created a geopolitical tug-of-war that extends beyond India's border as the developers seek international funding and cooperation.[29]

In addition to oil, gas, and hydropower, nuclear operations are among the most intriguing and dangerous examples of the geopolitics of energy. A prominent example centers on Iran. Although the matter is complicated, in essence the issues involve Iran's presumed interest in creating nuclear weapons from the various facilities they have constructed under their civilian nuclear program (Figure 8.20). Many countries, particularly Israel, Iraq, and the United States, have long worried about the threat of this proliferation. Iran, for its part, contends that such worry is unfounded, that they will not be developing nuclear weapons, and that foreign interference with their internal affairs is inappropriate, disrespectful, and insulting.

Nonetheless, many countries grouped into a coalition to impose severe financial sanctions on Iran, which have weakened the country's economy and stoked domestic unrest. Once the economic and social costs of these sanctions accumulated to unacceptable levels, Iran decided to negotiate an agreement that would allow them to pursue a civilian nuclear program while agreeing to certain conditions that would diminish the chance of the proliferation of the

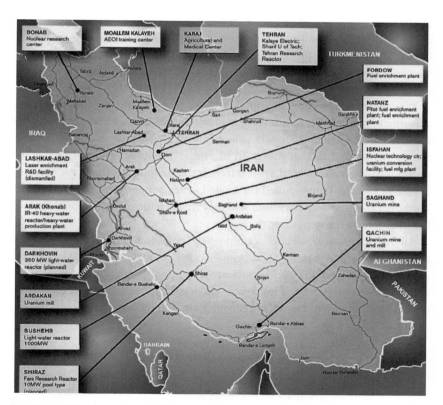

Figure 8.20 Iran's nuclear facilities provide a constant worry to those countries that prefer that Iran not develop nuclear weapons capability. *Source:* Base map courtesy of the Washington Institute. Corrected and re-drafted by Maryam Shakib.

civilian program into weapons. On July 14, 2015, the Joint Comprehensive Plan of Action (now called the Iran nuclear agreement) was formally agreed between Iran and the five permanent members of the United Nations Security Council— China, France, Russia, United Kingdom, United States—plus Germany and the rest of the European Union.

Under the arrangement, Iran agreed to eliminate its stockpile of medium-enriched uranium, cut its stockpile of low-enriched uranium by 98 percent, and reduce by about two-thirds the number of centrifuges used in the enrichment process for at least fifteen years. Iran also agreed that for the next fifteen years it would only enrich uranium up to 3.67 percent, which is substantially below weapons-grade requirements. Iran also agreed not to build any new uranium-enriching or heavy-water facilities over the same period. Uranium-enrichment activities were to be limited to a single facility using first-generation centrifuges for ten years. Other facilities would be converted to avoid proliferation risks. To monitor and verify Iran's compliance with the agreement, the International Atomic Energy Agency (IAEA) was to have regular access to all Iranian nuclear facilities. The agreement provided that in return for verifiably abiding by its commitments, Iran would receive relief from nuclear-related sanctions imposed by the United States, the European Union, and the United Nations Security Council, including the release of funds that had accumulated while the sanctions were in place.[30]

Whereas the negotiating countries believed that this was the best achievable agreement, neither every country nor every special interest viewed it with the same hope. For example, Israel's prime minister, Benjamin Netanyahu, vehemently denounced the agreement in its entirety, contending that Israel would continue to suffer an existential threat from Iran. In support of the prime minister, President Donald Trump pulled the United States out of the agreement in 2018. Washington and Tehran have both said they would return to the original deal but they disagree on the steps to get there.

All this brings up a "what if" scenario: had Iran rejected the development of a civilian nuclear power program altogether, other countries would have been less worried that Iran could develop nuclear weapons. This offers an additional example of the continuing and dangerous mixing that occurs between geopolitics and energy that can lead to critical considerations of national security.[31]

Where Does This All Lead?

Your future will be one of quick, challenging, and sometimes confusing changes. As modern energy production and trade patterns shift, so too will the geopolitics of energy. Consider the following scenario: Renewable energy (e.g., solar, wind, geothermal, biomass, and hydro) continues to grow so strongly that the proportion of demand met by fossil fuels drops significantly.[32] Coupled with great improvements in energy efficiency, the contribution made by renewable energy resources becomes dominant worldwide. How do you think this scenario would play out? What adjustments in world power would result? How would changes in the international balance of payments influence relative economic muscle? What modifications would undulate through the fuel chain, from extraction through waste disposal? How might these changes affect how and where you live? What would be the environmental responses? I pose these questions as a starting point for a hypothetical exercise you can develop for yourself.

One of the catalysts for such a transition is the perception of an existential threat from climate change. By general agreement, the global climate is changing faster than was predicted even a few years ago. The recognized culprit in this acceleration is our dominant use of fossil fuels for just about our every need. Adjustments within the energy sector—which accounts for some 60 percent of global emissions—are now considered a top priority if we are to have any hope of keeping the global temperature rise below 2°C (3.8°F). Renewable energy development is considered a large part of the solution.

UN action in this direction ramped up in Paris at the end of 2015. For the first time at any UN climate conference, "renewable energy solutions [were] showcased to governments, media and the public in a series of high profile events . . . demonstrating that renewable energy deployment is the fastest, cleanest, most reliable and most economically beneficial way to meet our climate goals."[33]

Turning to this scenario, if renewable energy does displace fossil fuels, what do you think would be the end result? To help get you thinking along these lines, consider that any of the following might occur.

- Persian Gulf countries lose political power as their markets diminish.
- Oil transport pipelines and tankers become less important.
- Critical chokepoints and transportation routes that are currently so important become redundant.
- Our system of military protection of shipping lanes shrinks, and there is a reallocation of military funding.
- Many countries become much more self-reliant in energy, and their lifestyles improve.

How will the world economy rebalance? What environmental changes—good or bad—will occur once the burning of fossil fuels declines? Will national security issues become less problematic? How will political power alter? Will the Global South gain influence? Overall, how would all the players in world geopolitics realign? These are just a few of the questions that inform the geopolitics of energy.

9

Cities

Centers of Usage

Cities as the Mirror of Energy

If you are reading this, you are very likely to live in a city. You are not alone. Not by a long shot. More than half the people in the world live in cities. In the United States, it is about 80 percent. Globally, the percentage of urban dwellers is expected to increase to 68 percent by 2050, adding another 2.5 billion people to the urban population. Ninety percent of these additional billions will be concentrated in Asia and Africa.[1] It appears that cities are the way of the world nowadays. If those predictions come to pass, city life will be a significant determinant of how much energy we need to produce, transport, and use, and we will need to be even more diligent about how energy is used in cities, how it gets to cities, and how it is made available. One can say that the urban use of energy will have increasing knock-on effects everywhere on the planet. For that reason, we need to pay attention to how it is used in cities and what we can do to make such use as efficient as possible in every city function.

We are crowding together everywhere, drawn by the conveniences, jobs, economic opportunities, and services. So thoroughly have we become dependent on secure supplies of gasoline, natural gas, diesel, and electricity that cities are entirely at the mercy of the vagaries of energy supply chains, particularly the means of transport. Highways, railroads, pipelines, transmission lines, and airplanes are like spokes on a bicycle tire converging on the concentrated hub at the center. So much must come without fail into cities to keep them going that one questions whether the presumed benefits outweigh the costs in congestion, pollution, and communicable diseases.

Numerous cities around the world are already bursting at the seams. Many of these overcrowded cities are in the Global South and have only poor access to energy supplies, particularly electricity. Do you think the world's energy supply systems can accommodate an additional 2.5 billion urbanites in the next thirty years? Can we expect that all these extra residents will be able to get all the energy they need or wish they had? Is meeting such increased demand a

The Thread of Energy. Martin J. Pasqualetti, Oxford University Press. © Oxford University Press 2021.
DOI: 10.1093/oso/9780199394807.003.0009

reasonable expectation, given that in many urban areas the infrastructure is already overtaxed or virtually absent? Cities currently consume over two-thirds of the world's energy and account for more than 70 percent of global CO_2 emissions.

Many of the most potent energy-supply challenges will be in cities of the Global South. The list that comes to mind in this regard is quite long. Think Lagos, Delhi, Djakarta, Nairobi, Caracas, Addis Ababa, Kolkata, Port au Prince, Yangon, Mumbai. These and many other cities commonly face shortfalls that exceed the needs of cities in the Global North. To heighten the worry, many if not all of these cities will need more energy per capita in the future, especially as they get air-conditioning, which will cause their electricity demand to skyrocket.

If you would like an example that is closer to home, look no further than the Global North city of Phoenix, Arizona. The population of the city proper in 2020 is about 1.5 million. The population of the surrounding metropolitan area approaches 5 million (about the population of Norway) and is growing by 85–100,000 people each year. At least another 2–3 million people are expected by midcentury, which would bring Phoenix alongside Cairo and Riyadh as the most populated desert areas in the world.[2] Do you think these cities will be able to meet the energy needs of their people in the future?

Consider Phoenix's vulnerability. Already, most of the energy it needs must be imported. Aside from a small percentage of its electricity demand being met by solar and wind installations in Arizona, every bit of the energy required to meet demands comes from outside the state. Gasoline, diesel, aviation fuel, natural gas, uranium, and electricity all come from other states. It is not hard to imagine the stress that such an increased population will place on the lengthy, fragile, and vulnerable supply infrastructure that presently serves the city (Figure 9.1).

Phoenix is a giant energy-sucking machine, somewhat like a black hole: all supplies—including energy—are drawn in, while little in the way of tangible products escapes. The city and the surrounding municipalities are attracting new residents without pause. People are attracted to its snow-free climate; resorts, parks, and other tourist destinations; nursing facilities for an aging population; second homes for residents of more northerly areas; spring baseball; education at the country's largest university; and a casual lifestyle.

Signs of this growth are everywhere in Phoenix. You can see it where saguaro cactus is carefully removed and relocated; where deserts have their remaining vegetation replaced by multistory buildings; where barriers are built across dry river channels and artificial lakes appear like a mirage; where determined

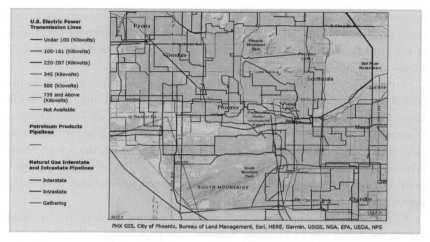

Figure 9.1 Pipelines and transmission lines converging on Phoenix. All natural gas and oil products supplying Phoenix come from outside Arizona. No roads are depicted. *Source:* U.S. EIA, https://www.eia.gov/state/?sid=AZ.

commuters fume with frustration as they try to hurry home along highways jammed with traffic; when countless new housing developments materialize in a conventional style only the Chamber of Commerce could love. How long can this go on? Water shortages are already evident around Phoenix. Will residents next suffer a scarcity of energy in the "world's least sustainable city"?[3]

Given that such an impertinent nickname is now firmly pinned to Phoenix, consider this question: why aren't civic leaders worried sick about how they will meet the needs of such a fast-growing population, especially in terms of the energy supplies that are the backbone of the entire urban enterprise, so reliant on motor vehicles and air conditioners?

Looking at any global supply-chain network reveals that regardless of the means of energy transportation—transmission lines, air, rail, pipeline, highway, or canal—the key destinations are large cities.[4] (Raise your hand if you think this is surprising.) How could it be otherwise, given that most people live and work in urban areas? Now, focus just on energy, and the picture becomes even more apparent. Considering that cities are the most significant users of energy, how will we keep all of them supplied with everything they need?

Although the pattern might eventually change with more significant development of renewable energy resources such as solar, for the foreseeable future most energy supplies are not developed within city boundaries. Instead, coal, natural gas, and oil continuously move around the world along pathways that connect scattered mines and wells with cities full of buildings, vehicles, and people. In other words, energy supplies end up in cities, the primary centers of their demand, more than any other place. Energy and cities are co-dependent. They reflect and support one another.

If cities are the mirror of energy, then continued population growth will require a significantly increased commitment of energy resources and massive expansion of the energy infrastructure that supports cities. The problem is that cities—particularly large ones—suffer from the energy consequence of their existing spatial form and functional norms, both of which tend to resist shifts toward principles of sustainability. While many organizations and funding agencies recognize this reality, progress toward modifying our approach faces the daunting challenge of overcoming the massive inertia held within the existing system and the staggering expense of actually doing so.

Try looking, really looking, at the structure and operation of the city where you live. See if you can identify where energy is coming from to keep it running. What you will discover is that cities are like a giant energy-sucking machine that might show up in an *Avenger* movie. Everything about cities screams: *A lot of energy is used here!*

Big cities need considerable, ceaseless supplies of energy in every form and for every imaginable purpose. Skyscrapers are heated and cooled day and night. High-speed elevators are on call to whiz people up and down ever-taller buildings, 24/7. Highways are like frozen rivers of concrete plied by cars and trucks burning fossil fuels. Traffic lights blink red, yellow, green; red, yellow, green. Buildings twinkle across skylines at night, messaging everyone that industrious and ambitious people are still at their desks working long after sunset. Streetlights provide visibility and security, while they also blot out urban residents' view of the stars. Various oil products, natural gas, and electricity reliably pour into our cities without fail—and, usually, without notice.

While energy *flows* are the lifeblood of modern cities, energy *development* has historically supported entire city economies. Glasgow and Manchester in the United Kingdom, Cleveland and Pittsburgh in the United States, Essen and Frankfurt in Germany, and similar cities elsewhere developed their size and complexity because they had access to concentrated forms of energy resources to support the industries that underpinned their early economies. In the past, such cities prospered because energy resources were nearby. Today, cities have slipped those bonds and—untethered to the natural distribution of energy resources— exist quite nicely even if the energy that sustains them originates hundreds or thousands of miles away.

Cities at Scale

According to the United Nations, the world population was about 7.8 billion on July 1, 2019. Cities are where most of them live nowadays, but it was not always true. In 1950, only 30 percent of the world's population lived in urban areas, a proportion that grew to 55 percent by 2018. North America is the most urbanized region, with 82 percent of its population residing in urban areas. In contrast, Asia is approximately 50 percent urban, and Africa remains mostly rural, with just 43 percent of its population living in urban areas in 2018.[5] If 70 percent of the world's population will be living in cities by the year 2050, as predicted, how will they sustain themselves?

Whenever meaningful sustainability is the goal, scale is significant. This especially applies to cities. While small cities may not provide all the services available in large cities, they often offer several advantages that make them easier to manage. Their energy needs are smaller, economies and politics simpler, supply lines less complicated, commuting times shorter, food suppliers closer, and the urban heat island (UHI) effect less pronounced.

Large cities can be much hotter than small cities, producing higher stresses on air-conditioning loads and raising the prospect of heat-related deaths.[6] This is a particularly important consideration in areas of the world where summer temperatures are commonly high, such as Phoenix, Arizona, where the UHI effect is especially apparent, making Phoenix up to 21°F hotter than nearby rural areas.[7]

We know what causes the UHI effect. Some of it results from the materials we use, some from the landscapes we plant, some from the hot exhaust of air-conditioned buildings, and some from the heat radiating away from the millions of internal combustion engines that move people and goods from place to place (Figure 9.2). Yet, despite knowing the causes of the UHI effect, we have been sluggish at making changes that would reduce it—not usually because we lack the understanding of how to make appropriate improvements, but because policies and politics thwart reasonable attempts in that direction.

Organic Cities

See if you can propose an answer to this two-part question: what is the most essential element in the creation of cities, and what supports their continued growth? The answer is the same for both parts: energy. There is no way around it: energy, in one form or another, is woven into the very fabric of modern cities. Everything from lights to limousines runs on energy. Yank out

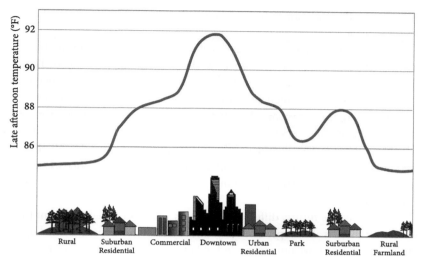

Lemmen and Warren[286]

Large amounts of concrete and asphalt in cities absorb and hold heat. Tall buildings prevent heat from dissipating and reduce air flow. At the same time, there is generally little vegetation to provide shade and evaporative cooling. As a result, parts of cities can be up to 10°F warmer than the surrounding rural areas, compounding the temperature increases that people experience as a result of human-induced warming.[313]

Figure 9.2 The urban heat island. *Source:* http://c3headlines.typepad.com/.a/ 6a010536b58035970c017744ad1afa970d-pi.

that single critical thread, and the cities you know and love would collapse. And the speed of that collapse—or at least a high degree of discomfort and inconvenience—would be quick, probably quicker than if you cut provisions of food and water.

You have certainly experienced power outages now and then, so you know how annoying they can be in the short term, and how threatening they can be if they go on for more than a couple of hours. Temperatures fluctuate out of the comfort zone, elevators halt between floors, hotel rooms lock out patrons, entertainment ceases, transport (other than bicycles) halts, chaos ensues.

The constant expansion of cities has not occurred without consequence. Imagine for a minute you are one of those responsible for the operation of São Paulo, Beijing, New Delhi, Cairo, Mexico City, or any of numerous others. You find that your city is like a monster with an endless appetite. You offer electricity, fuel, and food, but it is never "full." Worse still, every time you think you have satisfied its need, more people come, the monster grows, and its appetite expands as well. You soon realize that solving the urban problems of overcrowding, pollution, and congestion is beyond you. What do you do? Can big cities ever become sustainable?

Evidence abounds that "urban sustainability" is an oxymoron. But, just maybe, cities can become sustainable if there is enough will and dedication. The problem is that sustainable concepts and practices have seldom been integrated into the cities at their inception. Historically, most municipalities have grown and expanded organically—that is, incrementally and without much of an effective plan. In most cities, existing forms and functions interfere with sustainable ambitions. Despite the best intentions of elected officials—all of whom are temporary—it has usually been a freeway here, a building there, farmlands converted to concrete, shopping centers sprouting like weeds on any patch of empty land.

For cities of the future, different challenges will undoubtedly present themselves, but there is also hope. Critically, future cities offer unprecedented opportunities to planners wishing to amend old practices and prospect for new ones. Rather than permitting cities to evolve organically, planners will need to convince their constituents of the wisdom of incorporating sustainable practices, proactively if possible. These opportunities extend to energy systems—especially renewable energy systems—and how we can integrate them into the urban fabric.

As we work to improve the sustainability of our cities, we will quickly find that such improvements will not be easy or quick because cities are elaborate creations, perhaps the most complex entities we have ever managed to invent. The existing structure and operation of cities are intertwined with the social, economic, and political pressures that develop as cities mature. As such, cities consist not just of roads and buildings but of strands of vested interests that resist changes to a status quo that serves their needs. For this reason, few policies can alter an existing pattern in the short term. And few politicians are committed—or empowered—to adjust them in the long run.

Innovation

City life has many economic, social, and individual advantages over a pastoral existence. It offers greater personal security, more significant benefits of commercial agglomeration, and greater economies of scale. Moreover, cities are natural hubs for transportation and the exchange of goods, services, and ideas. They support a greater diversity of skills, networks, and lifestyles. The time saved from not having to carry out activities that would be necessary in rural settings frees more opportunities for invention and creativity. With all these benefits, it is no wonder most people live in cities. But they are not everywhere. Why are they where they are? What natural features favored their formation?

Ancient pre-industrial cities were often located along rivers in temperate and semi-arid climates. Waterways provided the means for irrigation, transport, and eventually renewable power to turn waterwheels. The Industrial Revolution and the availability of coal challenged this dynamic by untethering cities from such reliance on location. As coal use increased, the dominance of agrarian economies started giving way to those based on industry, especially in such parts of the world as western Europe and the eastern United States, where coal reserves were concentrated or where coal could be transported.

The capital city of London, which today receives its energy through a grid of wires and pipelines, began relying on coal as the forests of Great Britain vanished under the axes of the growing population. Carried up the Thames in ocean-going caravels that plied the stormy waters of the North Sea, the coal originated in the productive mines in the north of England near Newcastle. Other British cities such as Cardiff and Edinburgh also expanded and prospered because of coal, either by the product of its combustion or from activities associated with mining and exporting it to other cities.

Such places as Newcastle, Cardiff, and Edinburgh were the "energy cities" of their age. Centuries later, other cities would gain prominence through the extraction of energy in the form of oil. These included places like Oil City, Pennsylvania; Beaumont, Texas; and Signal Hill, California. Over the years since then, essential oil cities emerged elsewhere, including Baku, Azerbaijan; Dhahran, Saudi Arabia; Aberdeen, Scotland; Port Harcourt, Nigeria; Houston, Texas; Calgary, Alberta; Stavanger, Norway; and Kodaly, Russia.

In the United States, early cities were sited where there were double assets: natural harbors (for ease of transport and trade) and forested lands (for building materials and fuel). Eventually, some of these cities grew larger as coal was mined in nearby Appalachian coal country and could be transported to the cities using purpose-built canals and steam railroads. Such cities as Baltimore and Philadelphia were able to begin expanding their sphere of influence as access to energy supplies increased. Even inland, the availability of coal and cheap transportation facilitated the growth of cities such as Pittsburgh. Although several hundred miles west of the Atlantic Ocean, it was situated at the beginning of the mighty Ohio River and near substantial coal reserves.

Urban economies matured over time, becoming less dependent on proximity to cheap transport and raw materials. As the basis for these industrial economies shrank, people started looking elsewhere for jobs and places to live. Especially when air-conditioning became widely available, many of them moved to places where the economies were based more on services than on manufacturing. People could live just about anywhere they wanted, thanks to the ubiquitous availability of electricity. Cities, even those with little in the way of raw materials, could expand and grow to become, like Phoenix, among the largest urban areas in the country.

These changes encouraged a corresponding expansion of more elaborate, expensive, and complicated energy infrastructure. Energy supplies were no longer converging solely on a few traditional urban nodes. Instead, energy supplies could be imported wherever anyone preferred to live. In the United States, this helps explain population growth in warmer and less crowded southern and western states. The people now living in these newer cities made a calculated decision to flee the environmental and social burdens of the Rust Belt in favor of a more comfortable, safer, and more relaxed way of life offered by communities in the sunbelt. Among these is Phoenix, Arizona, which, as previously noted, is hundreds of miles from the nearest supplies of coal, oil, and natural gas. But it doesn't matter anymore; pipelines and transmission lines bring in all the energy it needs. Whether that is a sustainable model, however, is debatable.[8]

Transportation

Automobiles, owners will tell you, provide convenience, status, independence, privacy, and flexibility. While acknowledging that all this is true, we must also point out that the responsibilities attached to automobile ownership extend beyond personal expense and include high societal and environmental costs. Explaining such costs begins with the startling inefficiency of the internal combustion engine. If you own a car, consider this: how much energy does it take to transport a 180-pound person inside a 4,000-pound car? Now compare that to the amount of energy required to carry the same person an equal distance on a bicycle. Research shows that a bicycle can be up to 98.6 percent efficient in terms of converting energy expended on the pedals into forward motion.[9] Cars, on the other hand, use only 14–30 percent of the energy in the fuel to move the vehicle.[10] Virtually all that energy goes to moving the car, not the passenger.

Now imagine the aggregate cost of thousands of people choosing to drive instead of pedal. This is a choice that is happening in real time. In China, households have traded in the practicality and energy efficiency of riding bicycles for the prestige—and the expense, inefficiency, and pollution—of riding in their own car. Even as recently as 2001, few private vehicles rolled along China's streets. Instead, almost everyone rode bicycles. So prevalent were bicycles that China became known as the "kingdom of bicycles." Not many years ago, about 40 percent of Chinese cycled to work or school.

Since then, the "car revolution" has taken over, a shift that resulted in a rapid rise in the demand for transportation fuels. China's transition from bicycles to cars resulted, at least in part, from a considered political decision. As a mechanism to stimulate the economy and provide incentives to individual Chinese, the government encouraged people to own cars, even subsidizing their purchase beginning in 2009, with the expected result that in 2019 there were over 300 million registered vehicles in China, about 260 million of which are cars.[11] Such an abundance of automobiles has hurt air quality in cities such as Beijing, already one of the most polluted cities in the world because of heavy reliance on coal-fired power plants upwind.[12]

China's explosive rise in car ownership has produced yet another problem: congestion. If you ever visit Beijing, you will encounter it immediately. Congestion has increased significantly, as the city's 20 million residents and 6 million cars have effectively turned streets and highways into parking lots. A massive traffic jam that formed on August 14, 2010, extended for 100 km (62 mi) and lasted more than ten days. Many drivers were able to move only 1 km per day. In response to such calamitous events and the choking air pollution that envelops them, government officials began considering policies to return to the use of bicycles (Figure 9.3). Ironically, such a reversal is hampered by the increased use of cars, which have placed Beijing's suburbs inconveniently distant for bicycle use.

Despite Beijing's reputation for smog, it is not the world's most polluted city in terms of particulate matter. This distinction belongs to Gurugram, India, one of the fourteen cities in India that are among the top twenty most polluted cities in the world. Beijing is 122th on this list.[13]

In Beijing and Gurugram, not to mention many other places like Los Angeles (1,038th on the list of the world's most polluted cities), traffic is frequently at a standstill. There is simply no easy solution on the horizon, given the challenge of resolving air pollution and congestion simultaneously. Even if everyone in these cities converted to zero-emission vehicles overnight, congestion would remain, as would the need for more roads, parking spaces, and all the other demands of an automobile economy. In the final analysis, while cars may bring status, that comes at a high price to people everywhere. Our continued and growing reliance on automobiles makes any vision of future sustainable cities recede into the haze.

Despite apparent disadvantages, increased density does have some benefit if it is not taken too far. For example, one of the principal reasons that bicycles were once a useful mode of personal transportation in Beijing was the high population density within the city. The density around Beijing's historic Forbidden City is about 25,000 persons per km². (9,650 per sq mi) Such density also works

Figure 9.3 Traffic in China has become notorious since the encouragement of the automobile over the bicycle. *Source:* https://www.chinasmack.com/ beijing-traffic-jams-reach-new-congestion-records.

to the advantage of mass transit schemes. The Beijing subway, begun in 1969, was designed to have thirty lines, comprise 450 stations, and be 1,050 km in length by 2012, although presently it has only eighteen lines, including an airport express line. Nevertheless, it is heavily used; it is first in the world in annual ridership, with 3.21 billion rides delivered in 2013.[6] All those people are not in cars.

The California cities of San Francisco and Los Angeles offer further examples of the effects of population density on mass transit options. San Francisco, a peninsula city constrained by San Francisco Bay and the Pacific Ocean, covers about 121 km^2 (47 mi^2) with a 2020 population approaching 900,000. That means a lot of people are jammed into a rather small space. Although much less dense than Beijing, it is the second most densely settled city in the United States (after New York) with about 19,000 people per square mile. Because of this density, living without a car in San Francisco (or New York, for that matter) is not only feasible but increasingly prevalent. By contrast, Los Angeles has four times as many people (approximately 4 million people in 2020) but about ten times the land area of San Francisco (about 500 mi^2), with fewer natural barriers to expansion to the north, south, and east, giving it a population density of about 8,000 per square mile (Figures 9.4a and 9.4b). It is also—like Houston, Dallas, and Phoenix—a horizontal city, meaning that car ownership is almost manda-tory because the low population density makes mass transit costly on a person-mile basis. Car dependency also results in higher emissions of carbon dioxide. For example, compare Los Angeles at about 13.5 tons CO_2 per capita (15 percent walking, cycling, or public transport) and Mexico City at 2.5 tons CO_2 per capita (83 percent walking, cycling, or public transport) (Figure 9.5).

Sustainability

As the pace of energy use in cities continues to accelerate, several organizations are developing measures to slow it down. The quickest and least expensive ap-proach is by increasing energy efficiency. The question is: how can we measure energy efficiency in cities? For that, we turn to the American Council for an Energy-Efficient Economy (ACEEE). The ACEEE developed a City Energy Efficiency Scorecard that ranked seventy-five large cities. In their 2019 report on U.S. cities, Boston tops the list as the most energy-efficient, followed by San Francisco and Minneapolis. Tulsa and Oklahoma City rank at the bottom, suggesting a regrettable and avoidable waste of energy. The scorecard evaluated energy efficiency in each of the following categories: local government opera-tions, community-wide initiatives, buildings policies, energy and water utility policies and public benefits programs, and transportation policies. You might try using the scorecard to evaluate your own city.[14]

A second organization, the European Institute for Energy Research (EIFER), addresses the role of energy in all future decisions concerning urban develop-ment, spatial use, and settlements. To understand the role of energy in urban change, develop innovative strategies, and implement them, EIFER targets their research on five themes:

- Analyzing urban development, dynamics, and impacts on energy demand using approaches to assess and understand urban transitions
- Analyzing local energy and climate change policies
- Understanding urban governance and its interaction with energy planning, efficiency, and climate change policies
- Assessing innovative urban mobility by evaluating urban policy, technolog-ical trends, spatial simulation, and its interaction with the built environment
- Integrating concepts for sustainable energy use, considering acceptability studies, analyses of lifestyle, models for the diffusion of new technologies, analyses of the legal framework and of stakeholders, studies on governance, and the interactions between the different influence factors

(a)

(b)

Figures 9.4a and 9.4b Greater population density expands the options for mass public transit. Los Angeles population (top) is 8,000 people per mi^2 (with downtown at 4,770 per mi^2), while the population density of San Francisco (bottom) is more than twice that at 19,000 per mi^2.

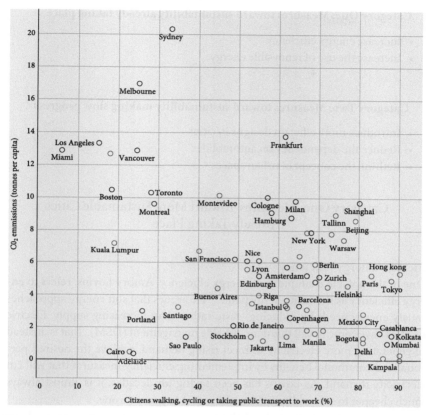

Figure 9.5 Correlation between CO_2 emissions and the share of trips made by walking, cycling, and taking public transport for global cities. Note the positions of San Francisco and Stockholm. *Source:* Environment Europe Sustainable Cities Database (http://environmenteurope.org/), 90 global cities, 2017. As cited in Stanislav E. Shmelev and Irina A. Shmeleva, "Global Urban Sustainability Assessment: A Multidimensional Approach," *Sustainable Development* 26, no. 6 (2018): 904–920.

Both the ACEEE scorecard and EIFER programmatic assessments base their analysis on the presence of various policies addressing operations, energy use, and transportation to enhance sustainable solutions. These assessments are useful when considering which strategies to adopt in each of the following categories:

Category One: Measures toward sustainability already taking place

- Increase energy efficiency
- Increase the use of renewable energy

Category Two: Measures toward sustainability making slow progress

- Introduce more efficient designs in cities
- Reduce the dependency on automobiles
- Rethink the concept of urban spaces

Category One: Measures Toward More Sustainable Cities Already Taking Place

Increase Energy Efficiency

One of the notable champions of energy efficiency, Amory Lovins, refers to energy efficiency as the "soft energy path."[15] He argues that soft energy approaches satisfy energy needs by reducing waste rather than increasing supply. Electric utilities have come to call this demand-side management (DSM). That is, instead of generating more electricity to meet rising demand, achieve the desired personal and communal benefits by implementing efficiency measures that will cut electricity demand. As Lovins has been saying for decades, it is almost always much cheaper to save a kilowatt-hour than it is to generate one.

Let us consider an example of the benefits of energy efficiency on a large scale. Say we provide $1 billion to build a 1,000 MW power plant to serve 1 million people. That would be a supply-side approach; that is, we would solve the problem of energy demand by merely increasing its supply. However, if we spent the $1 billion to design or retrofit every house and building in the city to make them more efficient, we could meet everyone's demand with a much smaller economic or environmental cost. That would be a demand-side approach. Using DSM would also avoid the disruptions of increased fuel mining and processing, fuel transport, new transmission towers, and the air and water pollution associated with the entire forty-year life of the centralized power plant that would need to be built.

Despite the logic underpinning the use of DSM to satisfy energy demand—and the long and routine acceptance of DSM throughout western Europe—it has taken several decades for the concept to filter its way into full public acceptance in the United States. Anyone who has traveled to Switzerland, Norway, or Germany, for example, has noticed that doors and windows seal tightly, lights switch off when rooms are unoccupied, air conditioners shut down when windows are opened, and incandescent lights have been replaced with compact fluorescent and LED technology. The United States, China, and many other countries are trying to catch up to these models.

Such measures reduce the energy demand in cities and make them more sustainable than cities that do not employ them. And they are paying off. For example, California has a lower per capita energy demand than any state other than Hawaii.[16] Success in California results from a robust and varied catalog of energy efficiency measures, inducements, and policies, including Title 24 (building energy efficiency) and Title 20 (commercial energy program) (Figure 9.6).[17]

Such progress toward urban sustainability is being enhanced by the acceptance of several internationally recognized "green building" certification system programs, including BREEAM and LEED. BREEAM is a suite of environmental assessment tools launched in 1990 by the U.K. group Building Research Establishment. It focuses on sustainable value and energy efficiency as outlined in its International Code for a Sustainable Built Environment. LEED (Leadership in Energy and Environmental Design) is a set of rating system certifications for the design, construction, operation, and maintenance of green buildings, homes, and neighborhoods. Both of these programs encourage greater energy efficiency in the built environment and less energy demand in cities.[18]

Increase Renewable Energy

The impacts of urban energy use do not stop at city boundaries. They extend to places sometimes thousands of miles away. As a result, if you reduce the use of

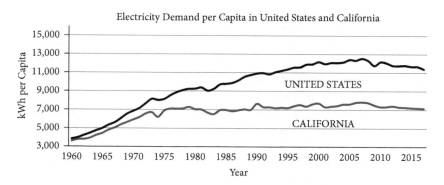

Figure 9.6 California's many initiatives to promote energy efficiency have paid off handsomely. *Source:* U.S. EIA.

conventional energy sources in cities, you also reduce the far-flung impacts of supplying that energy. We can contribute to this effort by supporting renewable energy development. Thankfully, we are witnessing rapid growth in the installation of wind and solar power worldwide. The global installed capacity totaled over 1 terawatt (TW) by the end of 2018 (Figure 9.7). That is 1 million MW. Remember that a typical house in the Phoenix area uses about 14 MWh of electricity per year, so 1 TWh would keep millions of homes fully powered for a year. For comparison, the total installed global capacity of nuclear power is about 400 GW or 0.4 TW. (Keep in mind that nuclear plants operate almost all the time, while wind turbines operate about 35 percent of the time and solar PV operate about 25 percent of the time.)

As long as cities continue growing, there will be an increasing demand for energy, especially electricity. Statewide renewable portfolio standards (RPS) encourage renewable energy development, and they have successfully driven renewable energy development to meet them.[19] High renewable targets include 50 percent in California and Oregon, 75 percent in Vermont, and 100 percent in Hawaii.[20] Already California is getting close to their goal. In May 2021, the state generating capacity was 34 per cent renewables, including hydropower. Legislation enacted in California in 2018 increased that target to 100 percent.[21] As urban growth continues, renewable energy development is expected to keep pace to help meet their growing demand, including the adoption of rooftop solar installations. By the summer of 2021, California boasted an installed solar capacity of about 26,000 MW, amounting about 32.[22] That is enough to power about 8.5 million homes.

The expansion of renewable energy is one way to reduce environmental impacts that result from urban energy demands, but there continue to be barriers to such ambitions, particularly for rooftop solar. The success of rooftop solar installations, especially in the United States, hinges on putting policies in place that encourage them. Without such policies, antiquated zoning and building codes leave property owners uncertain and reluctant to take the solar plunge.

Homeowners wishing to install rooftop solar modules may be subject to NIMBY (not-in-my-backyard) objections from neighbors and even local authorities. For instance, some homeowner associations (HOAs) resist the installations of rooftop solar out of concern that they will diminish neighborhood property values. Even in the absence of HOA restrictions, individual homeowners may cite any of a host of possible NIMBY complaints. For wind power, such complaints include interruption of sleep, intermittent noise, flickering, shadows, vibration, glare, and interference with scenic vistas.[23]

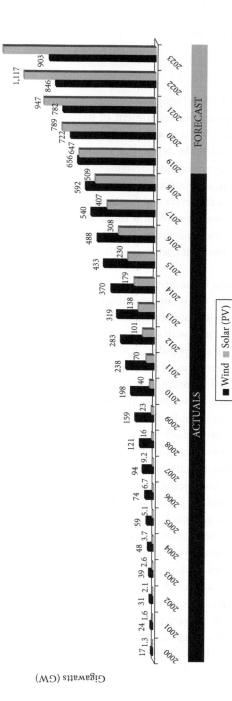

Global Cumulative Installations 2000–2023e

Figure 9.7 The quick rise in the global generating capacity of wind and solar energy is evident in this graphic, which projects to 2023. *Source:* http://www.fi-powerweb.com/Renewable-Energy.html.

Category Two: Measures Toward Sustainable Cities Making Slow Progress

Increase Efficient Urban Designs

Imagine you are given the assignment of designing or redesigning a city for greater sustainability. What would you do? Probably you can name at least a few measures that would make sense. For example, the essential elements would likely include greater walkability, designated bike lanes, curbside recycling, convenient mass transit, and—of course—sensible building codes.

Despite obstacles of inertia and bureaucracy, some cities have demonstrated a willingness to move forward, and there are many lists of "sustainable cities" where there has been good progress. The Sustainable Cities Network (https://www.sustainablestates.net/) helps keep track of progress.

Reduce Automobile Dependency

Efforts to reduce the use of automobiles have met with only marginal success, especially in cities with low population densities or no history of mass transit. However, in denser cities with mass-transit networks, getting by without a car is increasingly feasible and popular. While the majority of these networks are rail and light rail, a newer form of mass transit is being adopted in some cities: bus rapid transit (BRT) (Figure 9.8). Typically, a BRT system includes roadways that are dedicated to buses, gives buses priority at intersections where they may interact with other traffic, and includes design features to reduce delays caused by passengers getting on and off or purchasing fares. It allows greater route flexibility than a fixed-rail system, and it has already been hailed as a viable alternative in cities in Brazil, Indonesia, Australia, and Europe. As of March 2018, a total of 166 cities on six continents have implemented BRT systems servicing about 30 million passengers per day. Perhaps it will be adopted more widely in the future as an innovative approach to address increasing urban car traffic congestion.

The reality, however, is that people who can afford to buy, operate, maintain, and park their cars tend to prefer them, even though (or because?) the cars themselves rarely carry more than just the driver. Consequently, every person who drives a car contributes much more than their share of greenhouse gases than if they were to walk, carpool, use mass transit, or ride a bicycle. Each gallon of gasoline used in U.S. passenger vehicles results in about 19 pounds (8.5 kg) of GHG.[24] For a car traveling 12,000 miles per year, this means that it emits over 8,000 pounds (3,630 kg) of carbon dioxide, or more than twice its own weight. This is yet another example of why it is so important to make every effort to convince people to avoid using cars to get around, especially when distances are short and the weather is gentle.

Figure 9.8 Bus Rapid Transit (BRT). Transjakarta Zhongtong articulated buses at Harmoni Central Busway, Central Djakarta, Indonesia. *Source:* Photo by Gunawan Kartapranata, licensed under the Creative Commons Attribution-Share Alike 4.0 International license, https://commons.wikimedia.org/wiki/File:Harmoni_Central_Busway_Transjakarta_1.JPG.

Rethink the Concept of Urban Spaces

The best opportunity to design for sustainable cities lies in cities not yet built. Here, planners, architects, engineers, and builders can reconsider how to best meld the natural and built environment. Undoubtedly, one of the radical examples along that line of thinking is Arcosanti, an urban experiment designed by the visionary architect Paolo Soleri to illustrate his ideas for ecologically sound human habitats. A dynamic laboratory in the desert, Arcosanti—located 70 miles north of Phoenix, Arizona—is a demonstration of how urban architecture and the principles of ecology can work in tandem to create more efficient and livable cities. Built on 25 acres of land, the project includes thirty buildings and consumes approximately 20 percent of the energy required by conventional architecture. Arcosanti features the use of passive energy mechanisms such as the apse effect, the greenhouse effect, the chimney effect, and the heat sink effect, as well as miniaturization, elimination of the automobile, sustainable agriculture, and emphasis on building upward rather than outward (Figure 9.9).

Like many utopian communities, Arcosanti's full potential is yet to be realized. Soleri originally envisioned a compact city of 5,000 residents. Today, the community hosts between 50 and 150 residents, both short- and long-term, and the

Figure 9.9 Arcosanti, in Mayer, Arizona, an example of passive solar design within the urban experiment. Inspired and designed by architect Paolo Soleri. *Source:* Photo by Jessica Jameson. Permission courtesy of The Cosanti Foundation.

projected development is only 10 percent complete. Despite what might be called a modest beginning, Soleri's ideas have poked millions of people into being more alert to the benefits of (and, indeed, need for) sustainable urban design.[25]

Euclidean Zoning

As I hope it is apparent by now, today's cities operate with energy. Lots of energy. They cannot exist without it. This dependency is generally well understood, but it is not easy to do anything about it. The complexities of city functions and the needs of city residents create unique challenges. If you were tasked with lowering urban energy demands, would you know what to do? How would you get your city to operate better with less? How can you get them on an energy diet, but still keep them healthy and vibrant?

City planning strategies offer several possible corrective responses. This is true because many planning policies and practices exist at the intersection of energy consumption and demand. Each of the strategies we discussed above can help achieve a more sustainable city, although the degree of benefit will vary by location and age. One of the most critical considerations is the "maturity of the city"; that is, how long has it existed? How much of the existing infrastructure matches up with new technologies and conveniences? Was the city initially designed for transportation by horse and buggy? Is its pattern suited to bicycles? Is it too congested for cars? Is mass transit possible? What is the population density? Are the various functions of the city concentrated or dispersed? Do residents live above their shops?

Such considerations matter. It tends to be more difficult and costlier to retrofit an older city than to impose more energy-efficient standards on a new one. Imagine a city that meets its energy needs with dung rather than electricity, one with narrow serpentine streets rather than wide straight ones, one with an established cultural identity in contrast with one without it, one that is stagnant or shrinking compared to one that is vibrant and expanding.

The last point is particularly significant. Integrating low-energy measures in cities that are still growing, or are yet to be built, requires comprehensive policies and investments that combine the existing built environment with "smart" technologies and efficient energy distribution. In other words, land use and development decisions that result in the wiser use of energy require forethought and proactive intent. With such purpose in mind, planners can integrate efficient circulation patterns, enable convenient walkability, create bike paths, permit clustered and mixed-use housing, and discourage horizontal, low-density land use. Whatever measures you implement will have more near-term benefits in fast-growing cities than in cities that are losing population.

Density and its converse, sprawl, are critical considerations in any discussion of lower-energy cities. The density of a city strongly influences the form, efficiency, and effectiveness of its transportation network, the distances driven, access to services, and just about everything else. Low-density cities limit the ability of residents to choose low-energy transportation options such as walking and biking. Similarly, if jobs and residential areas are distant from one another, it directly affects commuter traffic volume and flows, the need for automobile use, the creation of air pollution, the need for parking facilities, and so forth. Anyone stuck in a morning or evening rush hour understands this well!

How do we overcome the problems associated with sprawl and low population density? One of the essential tools at the planner's disposal is zoning. Zoning codes are enforceable local laws intended to regulate the placement and density of development as well as permitted or prohibited uses. The first comprehensive zoning code, adopted by New York City in 1916, was designed to regulate the form of the city's expanding skyline to protect access to light and fresh air at the street level. Anyone who has visited New York City lately can judge how this has worked out.

The primary purpose of zoning is to separate incompatible land uses. Whereas industrial uses and residential areas commingled in the early twentieth century, zoning codes shifted this paradigm to keep noxious applications away from the places where people lived. Referred to as "Euclidean zoning," this single-use approach to zoning still predominates in the United States and, in many respects, continues to discourage the kind of mixed-use functions that would support a walkable and bike-friendly urban environment.[26] Except in Houston, Texas, and its suburb Pasadena, all cities with a population greater than 100,000 in the country have some form of zoning.[27]

Some communities in the United States, following the example of European cities, are now adapting their zoning codes to achieve a denser, more sustainable built environment. Form-based codes, for example, unite density with zoning laws. Instead of primarily regulating land use, form-based codes prioritize the relationship between the development and the public realm (i.e., streets and sidewalks), the massing of buildings relative to one another, and the scale of blocks and neighborhoods. In this context, they encourage mixed uses, both within buildings and between parcels.

Beyond adopting a new style of zoning code, what can cities do to adapt their existing codes to a more sustainable city? One strategy is to reconfigure codes that require a *minimum* amount of parking to a code that establishes a parking *maximum*. Parking maximums encourage shared parking between users and/ or parcels and sets minimum bicycle parking standards. Similarly, minimum lot size requirements can be modified to require or incentivize density in target areas. Earlier, we compared San Francisco, a city with natural constraints, to Los Angeles, a city with fewer topographical boundaries to continued sprawl. In Los Angeles, planning policy can help mitigate sprawling development patterns by imposing regulatory growth constraints.

Urban growth boundaries (UGBs) are another option. UGBs are regulations that direct development to areas within the boundary while significantly constraining development beyond the urban periphery. By restricting growth outside of the UGB, the region benefits from both increased density and protection of open space, agricultural land, and environmental resources. UGBs have been successfully implemented in several places, including Hong Kong, South Korea, and the United States. However, this approach is not free of criticism. Development options are constrained within UGBs, and increased demand places pressure on housing and commercial prices. In addition, UGBs do not wholly protect against urban sprawl, as development pressures can "leapfrog" beyond the protected rural zone.

At a building level, city planning can also contribute to DSM strategies. For new construction, cities can adopt more robust site and building efficiency standards within their zoning and building codes, including regulations that facilitate (and do not inhibit) passive and renewable energy options (e.g., solar photovoltaics, passive solar design). Similarly, local governments can either require or encourage the use of green development practices, such as the BREEAM and LEED standards discussed earlier.

Cities can also mandate greater efficiency in existing buildings. Weatherization programs and incentives offer one example of how local governments are striving to improve energy efficiency in cities. Weatherization assistance programs, for example, not only seek to enhance the building efficiency of older housing stock but also address financial and quality-of-life burdens borne by low-income households with minimal access to the latest technologies and home energy upgrades.

Futures

Absent existential threats like permanent traffic jams, deadly pollution, and rampant street violence, most cities will be slow to overcome the inertia of their own unsustainable, high-energy ways. This will be true both in the way we construct them and in the way we operate them. We cannot expect conditions to improve in the future as long as we continue the pattern of rural-to-urban migration, love of suburbs and private automobiles, and use of non-renewable energy to fuel transport systems, heat and cool buildings, and power factories.

All the urban considerations just discussed—including cars, highways, congestion, building design, personal comfort, landscaping, pollution, and growth trends—are tethered to cities through the commodity of energy. Cities cannot move toward a more sustainable future without considering the central role energy plays in every function, operation, enjoyment, or challenge. Furthermore, cities cannot succeed without pursuing a multi-layered approach toward sustainability without considering the energy requirements.

Cities are currently unsustainable because they are neither designed nor operated with energy efficiency as a primary guiding factor, and because they rely on carbon-rich fuels to operate. We must reduce the consumption of fossil fuels, bolster the development of renewables, and increase our attention to energy efficiency in every decision we make about our urban spaces. If we do this, we will cut the amount of energy that cities need to keep operating, make them more pleasant places to live, and accelerate our pace toward their ultimate sustainability.

10

Justice

Supply and Accountability

Energy Access

Energy access is a leading determinant influencing the life you will lead. Just as your zip code often affects the likelihood of personal success, energy decisions—even ones made *for* you (rather than *by* you)—will often control your life.[1] Such energy dependence can produce questions of justice and injustice. With stakes this high, you might ask why we permit others to make meaningful decisions on our behalf, sometimes without our consent. Is it fair when some folks flourish due to beneficial energy systems while others suffer with less? Rather than letting others make energy decisions that rule our lives, could we design energy systems that are more equitable and just, that would allow more people (or even everyone) to lead agreeable lives?

Your well-being is in large part due to access to (or lack of) the energy you want, in the form you want it, for a price you can afford. Every watt of electricity generated, every cubic meter of natural gas recovered, every barrel of oil delivered, and every piece of wood or dung collected holds embedded within it an assortment of virtual personal and environmental costs.[2] For this reason and others, energy plays a strong role in the type of person you become and the contributions you will make as a human being, family member, and citizen of the world.

My purpose in this chapter is to stretch your understanding of the role energy plays in matters that are fundamental to how you live your life and interact with others. I hope to convey to you a greater understanding of the importance energy has in your daily tasks and encounters, the opportunities that come your way, and the dreams you hope will someday come true. By the final words, you should more thoroughly appreciate the many ties between energy and how its availability triggers questions that tug at the souls of millions of people. These include matters of justice, poverty, and fairness.

Energy Justice

Health and safety are only two among a host of laudable goals of a just world in part attributable to a sustainable energy sector. Regrettably, many people across

The Thread of Energy. Martin J. Pasqualetti, Oxford University Press. © Oxford University Press 2021.
DOI: 10.1093/oso/9780199394807.003.0010

the world cannot achieve greater health and safety due in part to the energy systems on which they rely, which is a state of affairs that can be unjust. Since energy decisions substantially cause their plight, these situations exemplify "energy *in*justice." Injustice occurs when one group's benefits or harms are disproportionate to another group's (distributional injustice, also known as inequities), when such burdens are produced and suffered without consultation, consent, or compensation (procedural or representational injustice), or when one group discounts the views of and impacts on another group (recognition injustice).

The environmental justice movement brought to light that minority populations are typically more vulnerable to and less able to overcome injustice. Indigenous people are particularly vulnerable to breaches of energy justice, some of which occur when they lack adequate representation and political power. The history of energy development on the lands of the Navajo Nation records several examples of this deficit. While it can be debated whether the tribe's acceptance of an energy economy was truly voluntary or a matter of representational injustice, the resulting reality is unsettling in terms of distributional justice. The Four Corners area of the Southwest is where we find several of the most heavily polluting coal plants in the United States, such as the Four Corners Power Plant (Figure 10.1). The massive amount of air pollution they emit contributes to global warming, causes breathing difficulties at higher rates in people who live downwind, and disrupts scenic views

Figure 10.1 Four Corners power plant. A coal-burning power plant in northwestern New Mexico on the lands of the Navajo Nation, it has spewed pollution over Indian lands for decades while supplying Los Angeles and other cities with pristine electricity. *Source:* "The Price of Progress," Karen Portin, used with permission.

in the region. These negative externalities are more difficult to bear considering that one in three homes on the Navajo Nation lacks electricity; customers living hundreds of miles away get the benefit of the energy without suffering from the pollution; the fees for electricity go to well-paid energy workers and corporate ownership, and the expenses are paid by the asthma patients downwind. Therefore, harms and benefits are distributed disproportionately among these various stakeholder groups, causing injustice. Furthermore, as these power plants are being closed due to lack of economic viability, members of the Navajo Nation find it difficult to be included in some of the negotiations or must submit to the decisions of the tribal council, leading to procedural injustice.

The oil development in the rainforests of eastern Ecuador presents another claim of energy injustice suffered by indigenous people. Spills had become so numerous, damaging, and even malicious by 2011 that an Ecuadorean judge ordered the Chevron oil company to pay $18.2 billion for "extensively polluting" the Lago Agrio oil field, a few miles east of the city of Nueva Loja in the province of Sucumbios (Figure 10.2).[3] Texaco, acquired by Chevron in 2001, had been accused of doing the damage between 1964 and 1990. As is common in such cases, Chevron appealed the ruling. While Ecuador's highest judicial court upheld the verdict against Chevron in 2013, it lowered the amount of compensation to $9.5

Figure 10.2 Crude contaminates the Aguarico 4 oil pit in Ecuador, an open pool abandoned by Texaco after six years of production and never remediated. *Source:* Rainforest Action Network, photo by Caroline Bennett, used with permission.

billion. Appeals continued. Eventually, in 2014, an appeals court judge annulled the case, citing improper gathering of evidence, including bribery. The villagers received nothing. This case is often cited as one of energy injustice. While the initial polluting seems to be a harm imposed on one group by another, which is a distributional justice issue, the bribery involved in the gathering of evidence reflects a procedural violation.

These two examples of energy injustice are joined by legions of others occurring all over the world all the time. Consider the following examples:

- The Cree Indians of Ontario and Quebec lost much of their native lands when provincial authorities and utility companies constructed large dams on the rivers entering James Bay.[4]
- Radioactive contaminants from uranium mining in northern Canada leach into the local aquifer, streams, and lakes, raising concern about local health and safety.[5]
- The First Nations people in Alberta have suffered losses and damage to their land, timber, and water resources in the wake of oil sands development.[6]
- The *ejidatarios* (people who work communally owned land) of southern Oaxaca, Mexico, continue enduring negative impacts as wind turbines are installed within their *milpas* (cornfields), disrupting centuries of farming with little to no consultation or compensation (Figure 10.3).[7]

Under the scrutiny of an energy justice perspective, these cases and thousands of others, both historical and ongoing, reveal how people have been harmed by energy activities. Such infractions might occur directly, as when an energy company forces the relocation of long-time residents in order to access coal beneath their homes or to create a reservoir behind a hydroelectric dam. They might arise indirectly, as when the pollution from a power plant diminishes the value of the property you own, or when acid mine drainage kills all the fish in a lake your family frequented for generations. Often it occurs when outsiders impose their interests on local people, diminishing environmental quality, causing personal health issues, altering cultural identity, and sometimes even threatening overall human well-being and survival. These circumstances can occur when those with power take advantage of others who do not, but they can also arise inadvertently.

The *Environmental Justice Atlas*[8] includes hundreds of examples of energy injustice. These examples, which the atlas refers to as "socio-environmental conflicts," have sometimes resulted in mobilizations by local communities and social movements. Sometimes national or international networks support local communities in their fight against the negative environmental impacts of specific economic activities, infrastructure construction, or waste disposal/pollution. Public resistance to many energy projects that cause such harms has been

Figure 10.3 State police meet anti-wind protestors in Union Hidalgo, Oaxaca, Mexico, October 30, 2012. *Source:* Photo from the Asamblea de Pueblos del Istmo en Defensa de la Tierra y el Territorio.

rising, both because the harms have become more frequent and blatant but also because the issues are being recognized by people living outside the immediate area. As an example, here is a map from the *Atlas* depicting public resistance issues under the heading "fossil fuels and climate justice/energy" (Figure 10.4). Within the *Atlas* itself, one can click on any dot for further details.

One of the most prominent examples of energy injustice is the trampling of human rights in order to clear the way to build massive hydroelectric dams. Their most visible negative impact is the lands that are flooded, chiefly where these lands are occupied by settlements. A 2019 conference in Winnipeg, Canada, focused on this growing humanitarian problem in Canada, particularly the impact of the huge dams on the Cree indigenous people.[9] A 2000 report by the World Commission on Dams estimated that between 40 million and 80 million people have been displaced globally by large dams, including between 26 million and 58 million in India and China between the years 1950 and 1990, with 10 million displaced in the Yangtze Valley alone.[10] By itself, China's Three Gorges Dam, completed in 2006, required the resettlement of 1.2 million people, many of whom were forced to leave lands that had been occupied by their family for generations. While some of these individuals receive some compensation, it is not always an equivalent replacement for the home, lifestyle, or social networks they

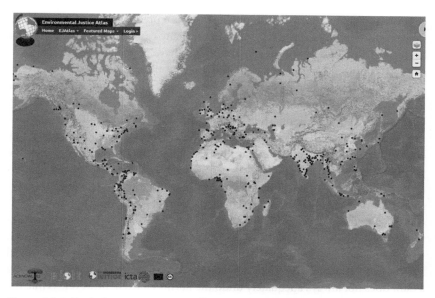

Figure 10.4 Each dot represents a conflict or an incident regarding fossil fuels and climate justice/energy recorded by the Environmental Justice Atlas. *Source:* Leah Temper, Daniela del Bene, and Joan Martinez-Alier, "Mapping the Frontiers and Front Lines of Global Environmental Justice: The EJAtlas," *Journal of Political Ecology* 22 (2015): 255–278.

previously had. Only in some cases are the displaced individuals given notice in advance with some ability to determine where they might relocate or what might be acceptable as restitution.

While physical displacement and abandonment of ancestral homes can be traumatic experiences, many examples illustrate how it can be worse. For example, consider when promises are broken—substitute land is not provided, compensation is withheld, and new facilities remain unbuilt. For instance, of the 96,000 people physically displaced for the Tarbela Dam in Pakistan, only two-thirds qualified for replacement agricultural land in Punjab and Sindh Provinces. Of these, some 20,000 people did not receive land when the amount of land provided by Sindh Province fell short of its promise.[11] The planned Sardar Sarovar Dam on India's Narmada River displaced more than 250,000 (mainly indigenous).[12] In response to the perceived injustice of such large-scale resettlement, the World Bank withdrew from the project, which was never completed. With the project incomplete, the expected benefits were not forthcoming, rendering the injustice even more intense.

The report of the World Commission on Dams chronicles dozens of other instances of energy injustice such as forced displacement, uncompensated taking of lands, lost economic opportunities, deaths from premature flooding, and outright murder of those unwilling to move. In many cases, these infractions have been foisted on people whose activities might have prevailed for centuries, including farming, fishing, and recreation. Therefore, these losses are not just immediate or short-term but greatly reset these people's lives at a community scale for gains in irrigation control, water management, and hydroelectricity that largely benefit other groups, manifesting distributional injustice. Despite the recognition of the human toll of large dams, their construction moves forward in many countries around the world, thereby illustrating that the mere recognition of the problem at a general level does not necessarily resolve issues of injustice. For example, Chinese banks and companies are involved in some 330 dams in seventy-four countries, leading to an unprecedented global dam-building boom.[13]

It is a different story in the United States, where the era of large-dam construction is over. Hundreds of smaller dams have been breached or removed. However, the largest dams remain, despite a growing chorus calling for dismantlement and habitat restoration, as with Glen Canyon Dam in Arizona and Utah.[14] Removing a dam does not always adequately undo the harms enacted, however. While the physical landscape may be rehabilitated, social injustice may still linger.

Whether we are considering human displacement, contaminated land and water, air pollution, or any other infraction of energy justice, there has been some progress at putting remedies into place. One step is mandating greater transparency, monitoring, and control over the activities of energy companies, starting with those ubiquitous fossil fuel enterprises. These mandates are examples of goals of extractive industry transparency initiatives (EITIs), voluntary, multi-stakeholder codes of conduct emerging from a collection of previous ad hoc efforts by companies, governments, and civil society.[15] Although their implementation varies from country to country, EITIs generally possess three core requirements:

1. Energy companies must disclose everything they pay to the government.
2. Institutions of the government must disclose everything they receive from energy companies.
3. Independent auditors must ensure the two sets of figures agree and that a published report is produced.

"The intent of EITIs is to track the influence and interaction among energy companies and governments, to protect citizens, but also to protect governments and investors from the 'rotten apples' that can taint the entire industry."[16] The EITI board seeks to improve the governance of the extractives sector globally. The EITIs attempt to right the injustice through this transparency, since the transparency more explicitly reveals harms and benefits, which can then be redistributed if deemed unfair. Also, they impose procedural justice through accountability. These decisions are made by representatives who are meant to be working on behalf of the people. Holding them to their word helps to ensure procedural justice. Recognition justice occurs when acknowledging and correcting this oversight allows citizens to understand the systematic picture of the situation of which they are a part and to judge whether or not it suits their interests (i.e., whether they are suffering a disadvantage to provide for another's advantage).

EITI-implementing countries number fifty-two. Several (including the United States) have withdrawn from the initiative. Others never signed up.[17]

The Example of Fuel Poverty

Some energy injustice situations are so ordinary and so commonplace that we have names for them. One of these categories is "fuel poverty." The term "fuel poverty" is often applied when a homeowner must spend more than 10 percent of their annual income to keep their living space acceptably warm. In the United Kingdom, for example, acceptable ambient temperature is set at 21°C (70°F) in the living room and 18°C (64°F) in other rooms. This situation produces distributional injustice since wealth and the means to access livable shelter can be redistributed to match everyone's needs but currently are not. Shutting off electricity without consent exemplifies a procedural injustice and demonstrates a lack of recognition of the homeowner's needs.

The drawback of this particular broad-brush standard is that it takes no account of the income level of the homeowner. For example, someone might be wealthy and still be spending over 10 percent of their income for heating. By such a definition, even Queen Elizabeth is flirting with fuel poverty despite having an annual income of over $30 million since so much is paid to heat royal buildings.[18] It is arguable that any injustice is occurring that victimizes the royals, but it may further exemplify the wealth gap (a distributional injustice) because it leads us to consider how wastefully the royal buildings might be heated, leading to the high expense. This exception is merely one demonstration that justice is often a matter of perspective. Further examples could also show that any disproportionate outcome is not always harmful or wrong. Still, while it would seem unfitting to classify the royals in this case as fuel-impoverished, they would likely wish to pay less for their heat. So the systemic failure that inappropriately prices (or values) electricity in these ways might be an unfortunate injustice of which they are victims. Such instances reflect something more closely resembling an awareness of recognition injustice, in that the lifestyles we wish to have are not properly valued (and accessible) within society. In that way, we are not properly represented because our ideals are not properly represented.

Alternatively, it is better to consider fuel poverty as occurring when the household is unable to afford the most basic levels of energy needed for routine activities. Insufficient energy can result in a variety of unpleasant outcomes, including cold and damp homes, debts on utility bills, and a reduction of household expenditure on other essential items. It can also lead to a wide range of physical and mental health illnesses, such as depression, asthma, heart disease, and unnecessary deaths in both hot and cold weather.[19] More insidiously, wide differences in energy access both reflect and worsen social inequality.

How would you personally react to news of fuel poverty or to suffering it yourself? Would you adjust? What comfort, convenience, or medical necessity might you decide you must give up? More people than you might imagine (perhaps one in five Vermonters) make a difficult choice between being warm, having shelter, being able to buy needed medicine and enough to eat.[20]

Being forced to make such choices can have serious consequences even in relatively affluent countries such as the United Kingdom, where 25 percent of families still have to choose between heating and eating.[21] They simply do not have enough money to satisfy both necessities fully. As a result, fuel poverty can result in personal costs that are more serious than discomfort alone. For example, 7,800 people die annually during winter in the United Kingdom because they cannot afford to heat their homes properly—a disturbing figure.

As a reflection of how scandalous fuel poverty has become in the United Kingdom, Ian McMillan and Ian Beesley published *The Book of Damp*, a lyrical look at fuel poverty, featuring moving anecdotal-style verse and photos of the material impact of "damp" on people's homes:

Cold makes you old
Before your time;
Cold's clutching grime
Dirties your heart
Before you can start
To live. Sharp cold,
Sharp as a long knife
Freezes your life.[22]

An average of sixty-five people die *each day* from fuel poverty in the United Kingdom, the world's fifth-largest economy. With that figure in mind, it might not surprise you to learn that fuel poverty is found elsewhere in Europe, especially the eastern European countries (Figure 10.5). You might suspect it to be even more rampant in places with colder climates, such as Russia and Scandinavia.[23] That is not true, however. The data show that roughly twice as many people per capita die in the United Kingdom than in Scandinavia and other parts of northern Europe, despite the fact that U.K. winters are typically milder. Even Siberia has lower levels of excess winter deaths.[24] So, what is happening? How can we explain why the United Kingdom has a higher incidence of death from fuel poverty than frigid Siberia? We need to look a bit deeper and ask what other factors are involved.

By definition and custom, fuel poverty is typically associated with the poor. However, other factors besides income are involved, as you will recall from my reference to Queen Elizabeth. Another factor, a critical one at that, is the price you must pay for the energy you use. High fuel prices can result in fuel poverty as readily as having insufficient funds can. Not recognizing the particular needs of distinct groups in this respect manifests a lack of representational justice, and not including them in the decisions to correct these situations would be a procedural injustice.

Another factor contributing to fuel poverty is the rate structures that utility companies use. For decades, many utilities have been charging progressively less per unit of energy as the customer used more. It amounted to an implicit incentive to consume more and more. The utility companies could do it because economies of scale lower the cost of generating each additional unit. It is like

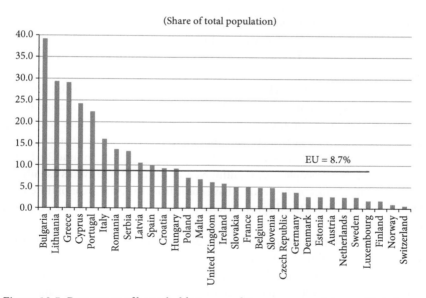

Figure 10.5 Percentage of households in several European countries unable to keep their home adequately warm. *Source:* https://en.wikipedia.org/wiki/Fuel_poverty#/File:People_who_cannot_affored_to_keep_their_home_adequately_warm._2016.svg.

when a manufacturer must charge $100,000 for a luxury car because they will not sell many of them, but another manufacturer can charge $20,000 per car because they sell millions of them. Once the assembly line is up and running, each additional car costs progressively less to manufacture. That is one of the reasons that high-volume cars, like Hondas, are less expensive than low-volume cars, like Mercedeses. It is not that a Mercedes is two or three times better than the Honda; instead, the price you are paying reflects the smaller number manufactured, marginally better performance and comfort, and added exclusivity from driving a car everyone knows is expensive.

For years, electric utility companies have been operating within this standard business model. When their baseload generating stations are running at full capacity, they produce millions of kilowatt-hours each day. The marginal cost of each additional kilowatt-hour that the power plant generates is essentially zero, but they still bill you for each one they can sell to you, pocketing most of that charge as profit. The more they sell you, the more they make. A lot more. This payment scheme neglects to recognize citizens' needs adequately, fails to encourage citizens to negotiate for lower bills when public comments do not exist, and exacerbates disproportional wealth distributions, creating a multifaceted injustice.

More recently, utility companies have been encouraged to abandon such traditional pricing schemes. Instead, they have been trending toward an inverted rate structure, making it increasingly pricey to use more electricity. Such a price signal discourages higher energy use, thereby benefitting society by lowering the environmental and other public costs that would otherwise result from higher usage.

If you happen to live in a hot climate, as I do, you might by now be thinking that fuel poverty cannot refer only to being warm during the cold season. It also applies to staying cool in the warm season, and the critical need for cool living space is increasing as heatwaves take more and more lives. For example, more than 1,000 people died of excessive heat in Karachi, Pakistan, in June 2015. It is not likely that these people were wealthy. News reports of these events usually say something like "Most of the dead . . . have come from the poor areas of the city."[25] Such a finding echoes the disproportionate impact of injustice on the poor and on minorities.

This consideration brings us to a question you might have to address someday: if you fell on hard times, how might you address fuel poverty? There are two categories of approach, the supply side and the demand side. The supply side would witness you trying to increase your income so you could afford the energy. The demand side would see you trying to decrease your demand so that you did not need any further infusions of funds. In many cases, the second option is the only option available, especially for retired, elderly, or infirm people (Figure 10.6).

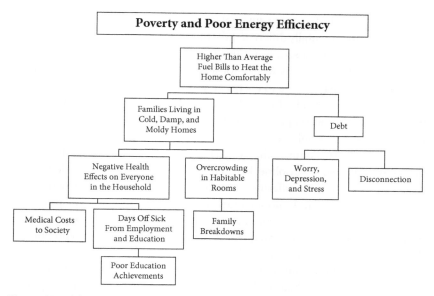

Figure 10.6 The consequences of poor energy efficiency for energy poverty.

The goal of demand-side management (DSM) is to meet energy needs through greater efficiency without sacrificing comfort, convenience, or safety. You would thus lower your energy bills as well as the environmental impacts and political entanglements that derive from your energy demand, all at the same time. DSM allows you to get more benefits for less money. It is hard to argue against such a good option, although your local utility company might subsequently increase their rates to adjust for receiving less revenue as sales decrease.

Countries such as Switzerland, Germany, Sweden, and Israel have been following a steady efficiency path for decades. Hotels in these countries rely on motion-sensor lighting in hallways and rooms to avoid unnecessary use of power when illumination is unnecessary. Windows seal tightly when closed to reduce infiltration. Air-conditioning switches off when rooms are unoccupied. Most countries fall short of such standards, including the United States. Once the United States and other countries accept energy efficiency as a standard practice, it will reduce the need for energy and the human costs necessary to provide it.

The Scourge of Energy Poverty

Another very common type of energy injustice is energy poverty. While it might seem to you to be like a lot like fuel poverty, and some people do use the terms interchangeably, the two carry different sets of connotations and can properly be considered fundamentally different. Like fuel poverty, energy poverty is a matter of energy access with a significant distinction. But while fuel poverty refers to not having sufficient funds to pay for adequate energy, energy poverty refers to having inadequate access to energy resources. In other words, it is not just a matter of being poor; it is also a matter of availability. Suffering from one affliction does not, of course, preclude anyone from bearing the burden of both, as unfortunate as that would be.

Here's one way to compare the two terms. Consider that you have the funds to buy all the coal, wood, gasoline, kerosene, or whatever fuel you needed. In that case, you would not be suffering from fuel poverty. Imagine further that even though you had the funds, no such fuels were available to buy at any price. That illustrates the essential difference between the two terms. Even though both can produce some of the same consequences—such as unhealthy living conditions—they are significantly different problems. By its very nature, energy poverty could arise from procedural injustice if you were barred from participating in decisions that would allow access had you been included in negotiations, from recognition injustice if decision-makers did not accommodate your need for energy access in their decisions, or from distributive injustice if the design of the city grid or fuel market did not provide you access. Energy poverty could also arise inadvertently, as not all territories have equal access to natural resources.

Visualize a mother raising children in Africa's Sahel, that fertile transition zone between the Sahara to the north and the savannah to the south. You would need fuel to cook meals for your family, boil water for drinking, and ward off seasonal nighttime chill. You look around and see nothing to burn. The trees that were once plentiful are all gone now, felled for firewood. With no fuel nearby, you consider your options. You might move your family to wherever fuel is available—assuming you know where that might be, and also assuming you could move there without encountering resistance from marauding thugs or residents who arrived there before you did. Alternatively, you could hire someone to find fuel for you and bring it back. The most likely scenario is that you accept that it is your responsibility alone to provide the fuel. You might wander far and wide yourself, or you might send your children, knowing that such activity displaces time for their play, studies, and opportunities for personal advancement, as well as subjects them to the possibilities of injury or assault.

Because basic human needs must take precedence, when you are faced with not having access to fuel you are likely to disregard personal responsibilities for environmental quality, thereby adding to an endless cycle that has had devastating impacts on the environment, human health, and social deprivation where you live in the first place.[26] You are part of a dynamic that offers little hope for improvement or escape. You find yourself trapped in the drudgery, the worry, and the hardship of energy poverty. Despite the efforts of many organizations such as the Global Commission to End Energy Poverty, the pattern perpetuates.[27]

The penalties of energy poverty are myriad and widespread, and even if you do not live in its grasp, imagining the reality of it can be a numbing thought. Regardless, we know that today almost 3 billion people still rely on biomass, coal, or kerosene for cooking, roughly the same number as in 2000. They have no access to clean fuels for heating and cooking. A billion of these people are in India and Africa, and more than half of China's population relies on solid fuels (coal and biomass) for cooking and heating.

These emotionless statistics mask dreadful human costs. Health impacts are among the most serious of these costs, particularly those resulting from indoor air pollution. A recent book on the fate of energy-impoverished people details these costs with gripping clarity.[28] The authors describe the lives of more than 800 million people living in rural India, where women and children must assume the responsibility of gathering household fuel (Figure 10.7).

Theirs is a life without hope for a gentler future. Each year ushers in more desperate conditions as lands continue degrading, populations continue growing, and household fuel becomes harder to find.[29] Making matters worse,

Figure 10.7 Women gathering fuelwood in Sudan. *Source:* Photo © Coordinating Committee of the Organizations for Voluntary Service / https://www.cosv.org/ ,https://creativecommons.org/licenses/by-sa/3.0/.

the low-quality fuels they burn—often animal dung—produce unhealthy smoke from traditional stoves inside poorly ventilated living quarters.

The World Health Organization estimates that between 2.8 billion and 3 billion people—most of them poor and living in low- and middle-income countries— still not only heat with solid fuels but cook with them as well. As a result, they inhale small particulate matter and other pollutants from indoor smoke. These particles inflame the airways and lungs, impairing immune response, and reduce the oxygen-carrying capacity of the blood. In poorly ventilated dwellings, exposure to soot is particularly high among women and young children, who spend the most time near the domestic hearth.

Evidence also exists that links household air pollution with low birth weight, tuberculosis, cataracts, and nasopharyngeal and laryngeal cancers. Risks from the various diseases are distributed with substantial variation across different age groups, meaning that different ages react differently to various pollutants, making it even more challenging to develop appropriate solutions. Although access to clean household fuels is improving in many places, it remains persistently low in emerging countries.

Diminished health inevitably leads to higher death rates. Largely in sub-Saharan Africa, India, and China, "an estimated 4.3 million people a year die

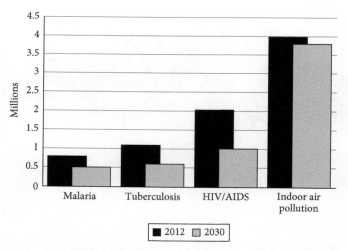

Figure 10.8 Annual deaths worldwide by cause, 2012 and 2030. Indoor air pollution is a greater risk than many diseases that attract much more public attention. *Source:* Adapted from "Energy Poverty: How to Make Modern Energy Access Universal," 2015, https://www.undp.org/content/undp/en/home/librarypage/environment-energy/sustainable_energy/energy_poverty_howtomakemodernenergyaccessuniversal.html.

prematurely from illness attributable to the household air pollution caused by the inefficient use of solid fuels (2012 data)."[30] Stunningly, people are more likely to die of indoor air pollution than from many higher-profile causes, such as malaria (Figure 10.8).

Access to Electricity and the Energy Ladder

Now, let's consider the difference in access to electricity. If you are like most people in industrialized countries, you take electricity for granted. You expect it to be available when you want it and where you want it. You give it little consideration at all, unless there is a power outage or you receive an especially large bill from the utility company. Now, think about the plight of the 1.3 billion people who have no access to electricity, mostly south of the Sahara Desert.

Ponder what your life would be like if you were in the same circumstance. It would be completely different than it is now. You would have no television or radio, no lighting for nighttime activities or security, no air-conditioning, no appliances, no tablets or computers, no cellphones, nothing at all that uses electricity. You would, if you were lucky, have access to the old standby kerosene, a liquid fuel that was fading from use in the United States in the closing years of the nineteenth century. If you could get it today, you would be using it even knowing that it comes with its own health risks, such as flesh burns, injuries, accidental fires, and poisonings from fuel ingestion. Nevertheless, you would use it because you want your children to be able to study after sundown and because you might wish to engage in small crafts and trades that require adequate lighting.

In recognizing the hardships endured by those without electricity, some organizations such as the United Nations are focusing on improving access. That is, recognition justice is leading to distributional justice. In 2012, UN secretary-general Ban Ki-moon formally launched the International Year of Sustainable Energy for All. The goals of this initiative are to ensure universal access to modern energy services for the world's poor, double the rate of improvement in energy efficiency, and double the share of renewable energy in the global energy mix. Energy—especially in the form of electricity—must be considered a basic human right.[31]

Energy poverty leads to all the aforementioned difficulties, most of which you could have imagined, but it also produces impacts that are not so obvious. For example, energy poverty places women and children at risk of injury and violence during the activity of fuel gathering, a severe problem that reflects ageism and gender-based injustices. Their particular needs go unrecognized, and harms disproportionately fall on them. In Darfur, for example, women and girls trek for hours a day in the hope of finding a few branches or roots to burn. In her essay

"Sexual Violence and Firewood Collection in Darfur," Erin Patrick describes a grim scenario:

> To avoid the midday sun, many leave in the darkness. To lessen competition, they travel alone or in very small groups. To find increasingly scarce combustible material, they may have to walk several kilometers away from the camps. In doing so, they become prime targets for the Janjaweed militia, local government or police forces and other men who act in a climate of almost total impunity.[32]

Darfur, in southwestern Sudan, has suffered war for years. Aid is sometimes available during such conflicts through the World Food Programme (WFP) of the United Nations, which is mandated to provide food during a complex humanitarian emergency. The good news ends there, unfortunately, because

> no UN agency is specifically mandated to provide fuel to cook the food. The food distributed by WFP, such as beans or whole grains, often cannot be eaten uncooked—and *no humanitarian agency considers cooking fuel its responsibility* even less so a protection concern. As a result, refugee and internally displaced women and girls must leave the relative safety of their camp or settlement to forage for firewood to cook food for their families or to sell for income. Millions of women and girls knowingly [but unavoidably] risk rape and other forms of sexual assault by trekking up to ten hours a day, several times a week in search of fuel. [Italics added][33]

Several initiatives have been established in response to this dilemma of insufficient fuels, recognizing refugees' special needs. As of May 13, 2014, the United Nations High Commissioner for Refugees committed itself to promoting safe access to fuel and energy for millions of uprooted people. One of the initial steps was to provide stoves and solar-powered lanterns (attempts to achieve distributional justice). It is a sign that those living outside the region are increasingly aware of this state of affairs, and it also reflects the increasing availability of alternative energy devices.

As economic security improves, so too will personal security and social stature because the form and quality of the energy you use reflect your position in the social hierarchy. This progression is often called the "energy ladder" (Figure 10.9). Such a ladder signals the nature of your surroundings, the status and convenience of your household, and whether you live in a rural area, near a city, or within the range of city services. For this reason, climbing the energy ladder allows you to climb the social ladder. Climbing the energy ladder helps to rectify distributive injustices by reducing your harm exposures.

Achieving electricity use is the ultimate level of the ladder. Again, the improved economics of alternative energy resources, especially solar equipment, can help people leapfrog from the drudgery and dangers of using low-quality fuels such as dung to the freedom and healthier conditions that accompany the availability and use of electricity.

As you climb the energy ladder, you are more likely to climb out of poverty, especially if you are female. The energy ladder is linked closely to economic empowerment through women's roles as household energy managers and their formal and informal networks. Their new authority helps them to restore procedural justice, which then allows them to begin to rectify recognition and distributional injustices. These networks place women in a unique position to connect with their peers, increase awareness, and deliver energy products and services. As energy users, they know what essential characteristic every energy product possesses. Similarly, "when women who are home-based micro and small-scale business owners or workers get energy access, they stand to benefit tremendously through increased productivity and lowered costs, resulting in increased incomes benefitting families, societies and local markets."[34]

Given the price women pay in the decreased health and safety that result from energy poverty, it is welcome that women are an emerging agent of change in improving such conditions. For example, ENERGIA's Women's Economic Empowerment program, running from 2014 to 2017, aimed at scaling up proven business models to strengthen the capacity of 3,000 women-led micro and small

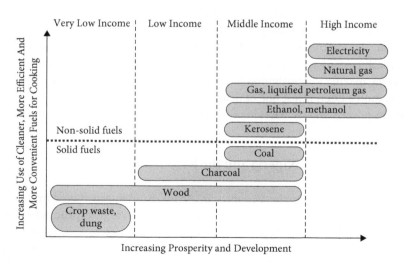

Figure 10.9 The energy ladder. *Source:* World Health Organization, "Fuel for Life," http://www.who.int/indoorair/publications/fuelforlife/en/.

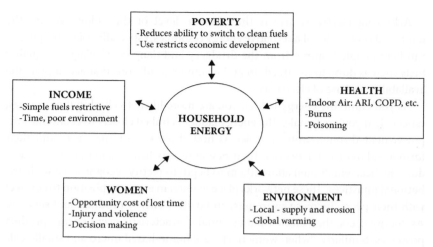

Figure 10.10 Household energy has many linkages and a host of implications. *Source:* Adapted from World Health Organization, http://www.who.int/indoorair/impacts/en/.

enterprises to deliver energy products and services to more than two million consumers.

Encouraging women to become energy entrepreneurs offers multiple develop-ment benefits. These include an expansion of economic activities for women, a diversification of productive options, and the creation of new sources of wealth and income to support family investments in education and health. In a recent article about unlocking the potential of women, Soma Dutta writes, "Women are playing a significant role in reaching energy services to the poorest and difficult-to-reach customers, who would never gain access to modern energy otherwise, thereby making a contribution to the agenda of reaching energy access to all."[35] This contribution is just one of many examples of the ties between reduction of energy poverty and increases in prosperity, health, and self-esteem (Figure 10.10).

Nigeria and the Resource Curse

Nigeria, the most populous country in Africa, is often mentioned in the con-text of energy justice; let me devote a bit of space here to explain why that is so (Figure 10.11). First, Nigeria offers compelling examples of all the afflictions we associate with energy poverty, fuel poverty, and challenges to energy justice. Central to all these considerations is Nigeria's oil reserves, perhaps as much as

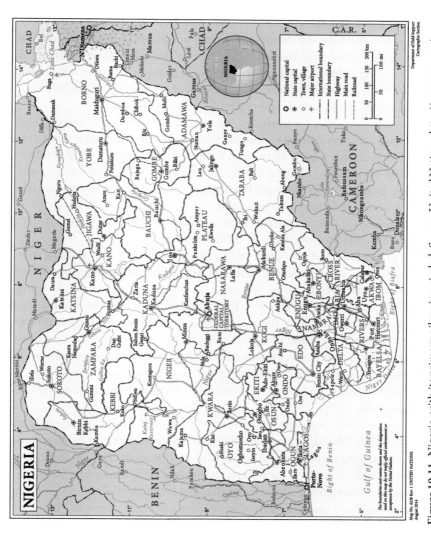

Figure 10.11 Nigeria, with its primary oil region circled. *Source:* United Nations, http://www.un.org/Depts/Cartographic/map/profile/nigeria.pdf.

35 billion barrels. Although oil development contributes only 14 percent to the national economy of Nigeria, it produces more than 95 percent of the export value. Moreover, the country's location on the eastern Atlantic Ocean enhances its value because tankers do not have to pass through any chokepoints such as the Strait of Hormuz or the Suez Canal to reach markets.

So important has oil become to Nigeria that it has lifted the country above South Africa to have the continent's largest economy. Despite such natural wealth, however, Nigerian citizens suffer because "the rising tide of oil money is not promoting stability and development, but is instead causing violence, poverty and stagnation. It also hosts a vast network of corruption that reaches deep into American and European economies."[36]

Nigeria is notorious for unrivaled corruption, environmental horrors, and ubiquitously desperate living conditions that include omnipresent risks of violence. The problems Nigeria faces result from the chaotic pattern of oil development and the ceaseless political unrest (exemplifying procedural injustice and resulting in recognition injustice) that gains momentum from the vast wealth proffered in the direction of any person, company, or country willing to clutch it. Those who succeed can accrue enormous prizes; even at low oil prices, Nigeria's oil reserves represent more than a $1 trillion treasure.

At a price of just $50 per barrel, production rates in 2019 were yielding over $100 million per day in revenue. The question is, where does this money go? By all accounts, and even by virtue of a quick on-the-ground assessment of living conditions, little of it trickles down to the majority of citizens of the country, creating distributional injustice. Because of the prevalence of pilferage and piracy, even an innocent excursion into the Niger Delta is a dangerous adventure for outsiders, as has often been chronicled in the press and in numerous books with rather scary titles such as *A Swamp Full of Dollars: Pipelines and Paramilitaries at Nigeria's Oil Frontier* and *Where Vultures Feast: Shell, Human Rights and Oil in the Niger Delta*.

While a similarly wide chasm between oil riches and quality of life exists in Ecuador, Libya, Siberia, Venezuela, China, and many other countries, Nigeria is unsurpassed in terms of the imbalance between the promise of wealth that is underground and the reality of the poverty that is rampant at the surface. Some call this injustice a "resource curse," an affliction that is common where the promise of energy wealth is matched only by an epidemic of energy poverty. First proposed by Richard Auty in 1993, the "resource curse" thesis describes countries rich in natural resources that are unsuccessful at using that wealth to boost their economies; it addresses why these countries, counterintuitively, have slower economic growth than countries without such abundance. Also known as a "paradox of plenty," the term "resource curse" colorfully describes places that, despite huge natural riches, often fare worse than places with little.[37] While there may be reasons, such as lack of technical expertise or proper equipment, that the community cannot rectify the distributional injustice that is present, there may also (or instead) be procedural corruption causing the unjust situation.

Injustice Across Time and Space

Now that you have some appreciation of the concepts of energy poverty, fuel poverty, and the resource curse, consider the injustices across territories and injustices across generations. These injustices are primarily distributional injustices and can, therefore, be described as "inequities" (called "spatial inequities" and "temporal inequities," respectively). An example of spatial inequity occurs when the electricity you use to stay cool in the summer places at risk people who mine the coal or live near a polluting power plant. A temporal inequity occurs when decisions you make today burden generations yet unborn with pollution or debt. Considering these and similar scenarios is to tacitly acknowledge that our energy decisions are not exclusive to a given time or a particular space.

If you recall from elsewhere in this book our example of uranium mining on the Navajo Nation, you will remember that although the mining took place decades ago, legacy costs to health and safety continue to this day from by-products of mining and reprocessing. One of these by-products, Plutonium 239, has a half-life of about 24,000 years and can be hazardous for up to half a million years. With that in mind, how far into the future should we extend our worry about the hazards of inadvertent or intentional exposure? The U.S. Congress has, at least for now, settled on 10,000 years, but this time estimate is a somewhat arbitrary number. Yet let's consider it in practical terms. Do you think it is possible to communicate warnings over that period of time? How would you suggest we do it? It may not be as simple as it seems.

Tackling this question fell to the U.S. Department of Energy (DOE), which was charged with developing strategies and designs to warn 400 generations of future inhabitants of the planet of the potential hazards of legacy nuclear waste. The DOE assembled four panels of experts in the early 1990s to try. They were asked to deliberate on exactly this question with regard to the Waste Isolation Pilot Plant in southeastern New Mexico, the nation's only deep geologic repository for defense-generated long-lived radioactive waste from several sites around the country (Figures 10.12 and 10.13).

All four panels had several months to develop their recommendations. They were expected to account for 10,000 years of cultural evolution, language drift, changing climate, geopolitical machinations, military conflicts, and a spectrum of other factors constrained only by the limits of human imagination. Each group

Figure 10.12 The visible buildings of WIPP in southeastern New Mexico, 30 miles east of Carlsbad. Its rather unexceptional appearance masks its purpose and what is being stored 750 meters underground. *Source:* U.S. Department of Energy.

developed scenarios and markers independently, but all shared a skepticism that the solution could be demonstrably effective. Despite all efforts to properly manage the waste today, we are foisting upon all future generations a responsibility that they never agreed to accept. Moreover, the problem could outlast language or current logic, creating an energy injustice if we continue producing these wastes (Figure 10.14).

The cousin of temporal inequity is spatial inequity, and it should be more familiar to us. It is everywhere we look, and it is not as speculative. It prompts questions such as: what effect will my energy decisions have on other people living nearby or thousands of miles away? While the answers are not always knowable, it is more likely that they can be identified in concrete terms compared to answers regarding some possible event in the future.

Why are we being asked to consider such questions nowadays? One explanation is that worldwide trade and communications have increased global interdependence at the same time as they have made it difficult to maintain ignorance of the costs of our actions outside our immediate living space. Until recently— really, for most of our time on the planet—land, water, air, and all other natural resources seemed limitless and therefore inexhaustible. We too often give little thought to the effects of dumping toxic waste into rivers or burning forests to clear them for agriculture (neglecting impacts across space, a type of recognition injustice, leading to distributional injustice). We were not paying much

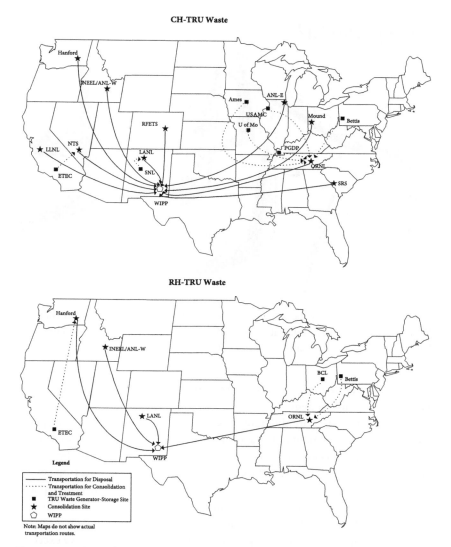

Figure 10.13 TRU (i.e., atomic number greater than 92) defense waste generating and storage sites that send wastes to WIPP. *Source:* State of Nevada, http://www.state.nv.us/nucwaste/graphics/wipp02.gif. CH-TRU: contact-handled TRU wastes. RH-TRU: remotely-handled TRU wastes.

attention, and we saw no alternative. Tacitly, we counted on nature to take care of itself and us. Before the present, when humans are living almost everywhere on the planet in sedentary communities, we moved somewhere else when environmental conditions deteriorated. Without giving it too much thought, we had

SPIKE FIELD

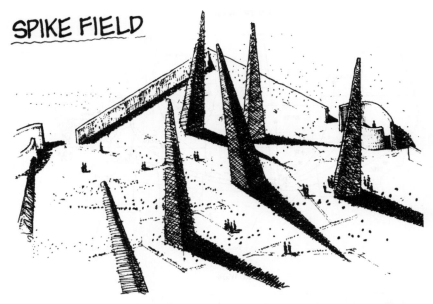

Figure 10.14 One of several warning markers developed for the WIPP site that were intended to deter inadvertent intrusion and exposure. Artwork by Michael Brill. *Source:* Sandia National Laboratory, SAND92-1382.UC-721, November 1993.

adopted a convenient way of skirting responsibility for any damage we caused. We got used to this strategy.

Today, some things are different. The spatial range of our ethical responsibility for the environmental damage we create has grown so much that we are now threatening major environmental systems, including global climate, the health of the world's oceans, and the importance of the remaining tropical rain forests. Today we realize, more than ever, that pollution from power plants, mines, refineries, and other energy-related equipment can spread hundreds or even thousands of miles from their point of origin. When this happens, everyone within the path of dispersal is subjected to some level of risk. We have exceeded the limits of built-in buffers and mechanisms that have heretofore protected us from ourselves. Injustices that were once hidden or otherwise unknown are now observable and must be managed.

It is not easy to devise a strategy to cope with such growing problems because questions of spatial inequity are often lost in the natural chaos of everyday life. In other words, as pollutants and damage accumulate, we might be unable to identify the location or origin or the responsible parties. For example, how far downwind can we reasonably expect or require operators of a coal-burning power plant to take responsibility for the impacts of their operations and accidents?

We might ask these questions to determine if unethical actions were done, to mandate restitutive actions from offenders to rectify injustices that occurred, and to try to choose ethical actions in the future. Ethical deliberations would more likely involve recognizing others, respecting them, and following proper procedures to enable their engagement. These practices would then be more likely to manifest recognition and procedural justice. By taking these considerations into account and also by choosing ethical actions, it would be more likely that just distributions occur. However, ethical actions still might lead to unjust situations or the inverse. So ethical merit also needs to be evaluated alongside justice assessments, since one does not entail the other in all cases.

We can turn again to nuclear power to provide some perspective on the spatial inequities of energy. The radioactive emissions from the 1986 accident at Chernobyl in Ukraine and the 2011 accident at Fukushima in Japan affected many thousands of people who lived hundreds or even thousands of miles from the plant (Figure 10.15). Most of these people received no direct benefits from the existence of the plant, nor were they asked if they agreed to accept the long-term health consequences that might reveal themselves many years later. They were

Chernobyl's Radioactive Cloud

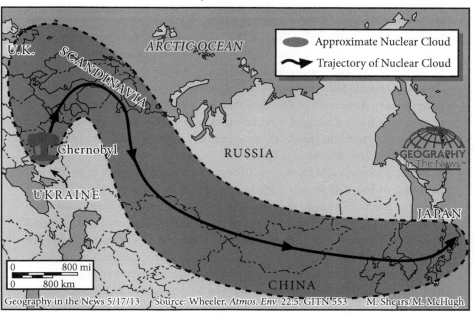

Figure 10.15 The fallout from the Chernobyl accident dispersed widely over Europe (as shown here) and also spread across the entire Northern Hemisphere. *Source:* Permission from Neal Lineback, Geography in the News LLC.

the casualties of spatial inequity in much the same way that the world community is suffering from the impacts of global climate change. So while one might not immediately jump to the conclusion that operating nuclear power plants is unethical, the resulting situation suggests that unjust harms have resulted.

Hopeful Futures

Until recently, matters of energy poverty, equity, ethics, justice, and fairness have been mostly out of view and out of mind. For the most part, these considerations have been subordinated to long-held priorities that emphasized the discovery, development, and sale of the raw materials. Matters of energy justice did not often rise to the level of concern, and when they did, it was usually on the margins of public perception. With the widespread use of the Internet, social media, and satellite imagery, it has become increasingly hard for perpetrators of injustice to hide. For this reason, such matters have begun attracting more attention from NGOs, academics, those in leadership positions, and stockholders in the energy companies themselves.

Rising public support for greater fairness, equity, and justice have obliged the energy industry to adjust their operations to a higher level of social responsibility. There are several well-known instances of this adjustment. For example, the controversy that swirled around the acidification of soils and lakes in the Adirondack Mountains of New York State prompted greater control of sulfur emissions at coal-burning power plants. Rising concern about the hazards of radioactivity resulted in substantial changes in the siting and operation of nuclear power plants. Awareness of the costs that oil development imposes on indigenous people has encouraged a reexamination of exploration techniques. And because of the resulting injustices, some political candidates are discussing banning fracking as an unethical practice.

Public awareness and more strident demands for higher levels of energy justice are growing, and we see shifts in attitudes at all levels of public discourse and government. No longer do companies have the license and the inclination to gloss over the impacts they cause without the risk of embarrassing and costly exposure. No longer are developers as likely to ignore how their decisions burden the public or the environment. Resistance to irresponsible actions is increasingly influencing which energy resources are developed, where they are developed, how they are developed, and who should be held accountable for maintaining the thread of energy justice that runs throughout the entire process. We still have much left to do in this regard, but we have made a promising start.

Estimated U.S. Energy Consumption in 2019: 100.2 Quads

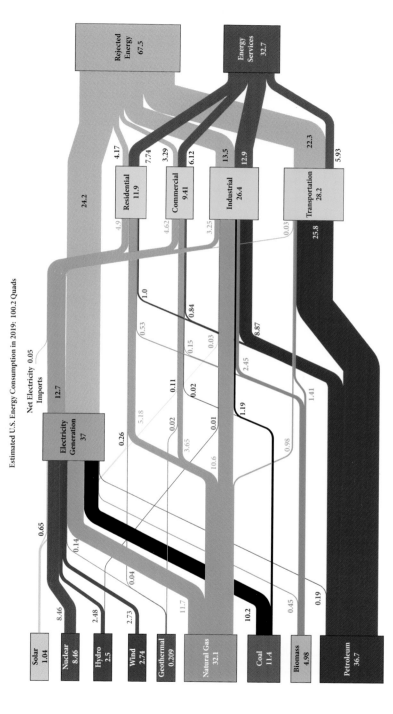

Figure 1.2 Estimated U.S. energy consumption in 2019: 100.2 quads. One quad = one quadrillion Btu. For reference, 1 pound of coal from the Kayenta coal mine in Arizona, when converted to electricity, produces about 1 kilowatt-hour. A typical household in the Phoenix area uses about 13,000 kWh per year. Source: Lawrence Livermore Laboratory, 2019, https://flowcharts.llnl.gov.

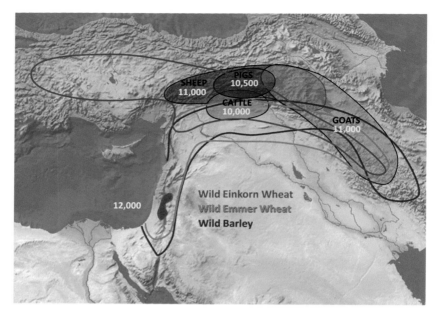

Figure 2.1 The Fertile Crescent, site of some of the earliest examples of plant and animal domestication. Drawing by Carlos Driscoll. Used with permission.

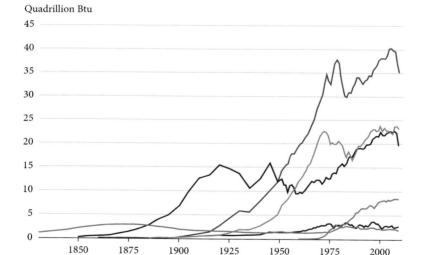

History of energy consumption in the United States, 1775–2009

Quadrillion Btu

Source: U.S. Energy Information Administration - Annual Energy Review 2009

— Petroleum   — Hydroelectric   — Coal
— Wood   — Natural Gas   — Nuclear

Figure 2.6 U.S. Energy Consumption, Per Capita (1775–2009), showing the rapid decline in the use of wood as an energy source with the rise in coal after 1850 and continuing its decline with the rise in oil use 1900. Source: U.S. Energy Information Administration

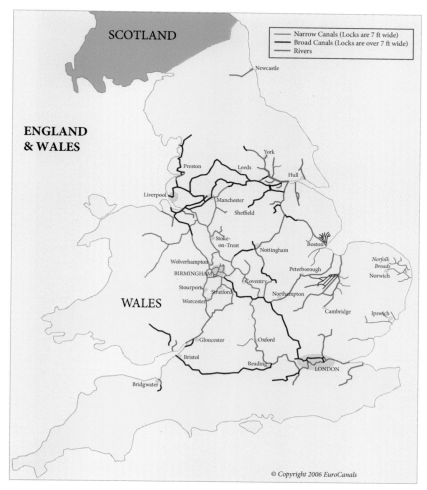

Figure 2.8 Canals, used to move heavy goods such as coal in Great Britain and elsewhere, are used today for recreation. Source: http://www.newandusedboat. co.uk/images/canal-network-map.jpg.

Figure 3.1 The World Food Programme's Hunger Map depicts the prevalence of undernourishment in the population of each country in 2016–18. From Africa and Asia to Latin America and the Near East, there are 821 million people—more than one in nine of the world's people—who do not get enough to eat. Source: https://www.wfp.org/publications/2019-hunger-map. Courtesy of World Food Programme. Used with permission.

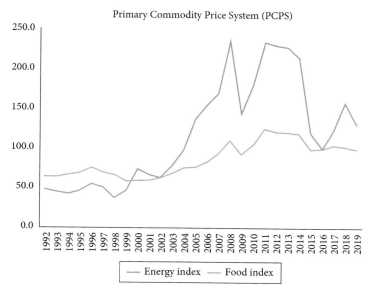

Figure 3.4 Food and fuel prices show a strong correlation, particularly after 2005. This demonstrates the importance of embodied energy in what you pay for your food. Source: IMF, as reproduced in Eric Garza, 2013, The Energy Return on Energy Invested of US Food Production. https://www.resilience.org/stories/2013-09-09/the-energy-return-of-energy-invested-of-US-food-production/

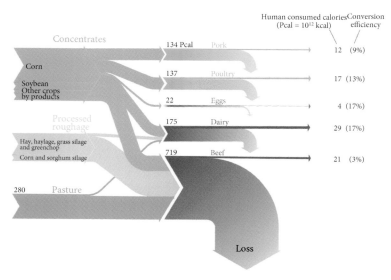

Figure 3.6 A Sankey flow diagram of the U.S. feed-to-food caloric flux from the three feed classes (left) into edible animal products (right). On the right, parenthetical percentages are the food-out/feed-in caloric conversion efficiencies of individual livestock categories. Caloric values are in Pcal (10^{12} kcal). Overall, 1187 Pcal of feed are converted into 83 Pcal edible animal products, reflecting a weighted mean conversion efficiency of approximately 7 percent. Source: A. Shepon et al., "Energy and Protein Feed-to-Food Conversion Efficiencies in the U.S. and Potential Food Security Gains from Dietary Changes," *Environmental Research Letters* 11, no. 10 (2016): 105002.

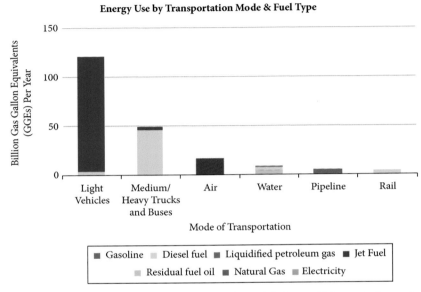

Energy Use by Transportation Mode & Fuel Type

(Legend: Gasoline, Diesel fuel, Liquidified petroleum gas, Jet Fuel, Residual fuel oil, Natural Gas, Electricity)

Figure 3.17 Energy use by transportation mode and fuel type. This graph shows the energy consumption (in gasoline gallon equivalents, GGEs) of the transportation sector by mode and fuel type for the year 2016. For the most part, each transportation mode is dominated by a different fuel type. Light-duty vehicles use the most GGEs of fuel per year, followed by medium/heavy trucks and buses. Water transportation relies mainly on residual fuel oil, a by-product of producing the light products that are the primary focus of a refinery. Pipelines are the only mode of transportation that uses non-petroleum fuels predominantly. Source: Table 2.7, Oak Ridge National Laboratory's *2019 Transportation Energy Data Book*; heat conversion factor is taken from Appendix A3 of EIA's May 2019 *Monthly Energy Review*; https:// afdc.energy.gov/data/?q=fuel+consumption++in+transportation.

Figure 4.6 Bernard Lang created this collage of coal mine images for a Paris art gallery exhibition. The original photos, and more, can be found on Bernard Lang's website, https://www.behance.net/bernhardlang.

Figures 5.10a and 5.10b Colored water indicating acidified streams resulting from coal mining near Wilkes-Barre, Pennsylvania.

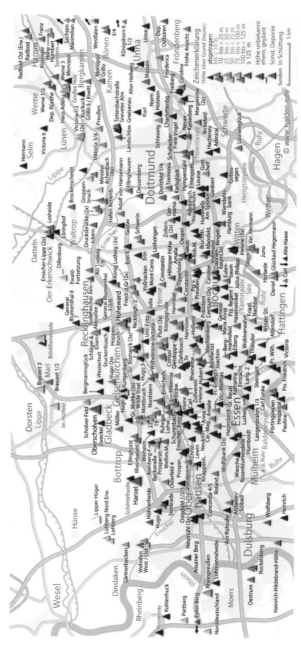

Figure 5.28 A map of the large number of waste dumps north of Essen, Germany between the Ruhr River and the Lippe River. Primarily created by coal mining, they dominate the horizon in every direction. Several disposal sites have been converted into parks and concert venues. Source: Pictures, maps, text and graphics: S. Hellmann • www.ruhrgebiet-industriekultur.de & www.halden.ruhr.

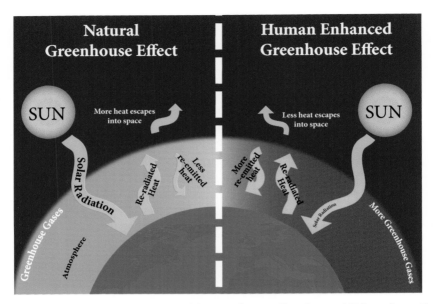

Figure 6.6 Generalized depiction of the greenhouse effect. Source: NPS.gov, https://www.nps.gov/goga/learn/nature/climate-change-causes.htm.

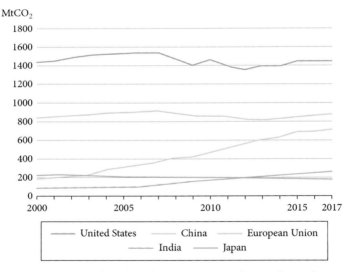

Figure 6.17 CO_2 emissions by selected countries. Note the rapid rise of emissions in China. Source: IEA, CO_2 Emissions from Fuel Combustion Highlights, 2019, https://www.epa.gov/ghgemissions/overview-greenhouse-gases

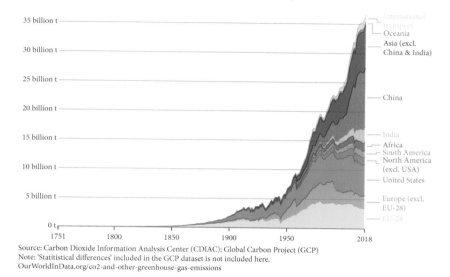

Annual total CO₂ emissions, by world region

Source: Carbon Dioxide Information Analysis Center (CDIAC); Global Carbon Project (GCP)
Note: 'Statitistical differences' included in the GCP dataset is not included here.
OurWorldInData.org/co2-and-other-greenhouse-gas-emissions

Figure 6.20 Annual total CO_2 emissions, by world region, through 2017. Source: https://ourworldindata.org/co2-and-other-greenhouse-gas-emissions#how-have-global-co2-emissions-changed-over-time.

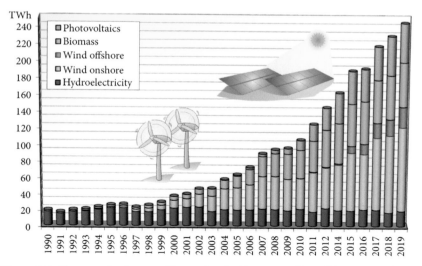

Figure 7.6 The growth of renewables under the German policy of *Energiewende* (energy transition). Source: https://www.volker-quaschning.de/datserv/ren-Strom-D/index_e.php.

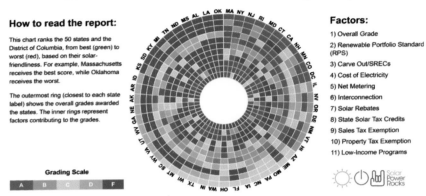

Figure 7.9 2020 State Solar Power Rankings Report. For an interactive and most current version, see https://www.solarpowerrocks.com/state-solar-power-rankings/

.

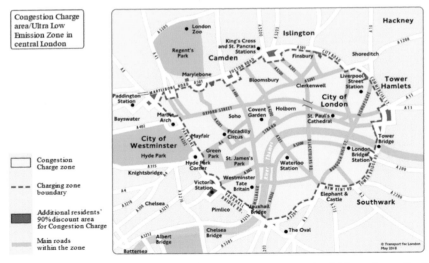

Figure 7.11 The London congestion charging zone covers 21 km² (about 8 sq mi) of central London. Source: https://tfl.gov.uk/ruc-cdn/static/cms/images/congestion-charge-ulez-map.jpg.

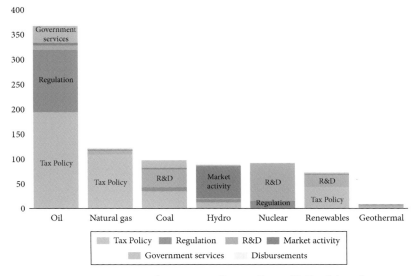

Figure 7.12 Energy incentives by resource. Source: Roger H. Bezdek and Robert M. Wendling, "Energy Subsidies: Myths and Realities," *Public Utilities Fortnightly*, June 2012, 63–67, https://www.fortnightly.com/fortnightly/2012/06/energy-subsidy-myths-and-realities.

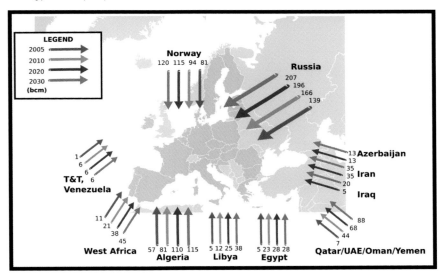

Figure 8.8 Europe receives a major portion of its natural gas supplies from surrounding countries. Such dependency poses a risk from adversaries. Source: https://eadaily.com/en/news/2017/03/06/syria-ukraine-gas-pipeline-wars-us-russia-alliance-inevitable. For a series of specific maps of Russian pipelines, see https://eegas.com/maps.htm.

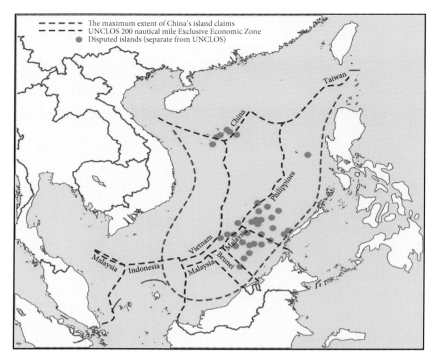

Figure 8.11 The South China Sea is the location of substantial risk of military confrontation because of overlapping territorial claims in this resource-rich area. Source: https://en.wikipedia.org/wiki/Territorial_disputes_in_the_South_China_Sea.

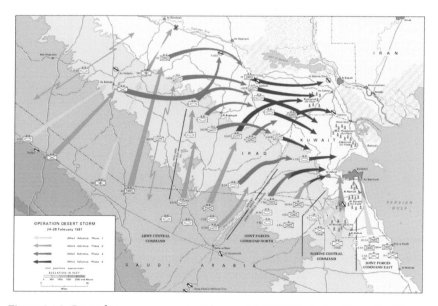

Figure 8.16 Ground troop movements, February 24–28, 1991, during Operation Desert Storm showing allied and Iraqi forces. Special arrows within the map indicate how the American 101st Airborne division moved by air and where the French 6th Light Division and American 3rd Armored Cavalry Regiment provided security. Source: U.S. Army Center of Military History, http://www.history.army.mil/reference/DS.jpg.

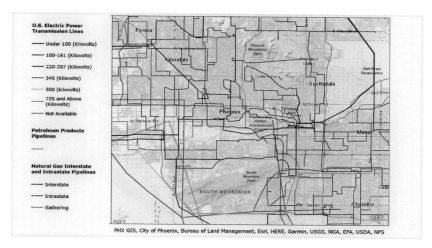

Figure 9.1 Pipelines and transmission lines converging on Phoenix. All natural gas and oil products supplying Phoenix come from outside Arizona. No roads are depicted. *Source:* U.S. EIA, https://www.eia.gov/state/?sid=AZ.

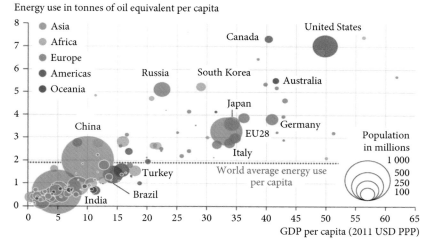

Figure 11.1 Most countries show a close relationship between GDP and energy consumption. Source: European Environment Agency, https://www.eea.europa.eu/data-and-maps/figures/correlation-of-per-capita-energy.

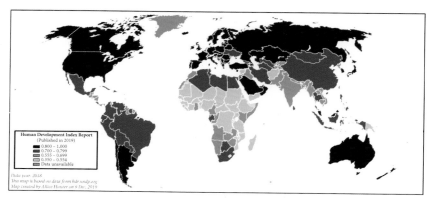

Figure 11.4 Human Development Index by country, 2018. Source: https://en.wikipedia.org/wiki/Human_Development_Index.

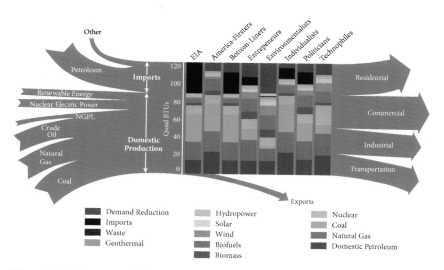

Figure 14.1 Summary of the seven perspectives' energy portfolios (year 2030). Coal has continued to fall on hard times since this figure was prepared in 2009. Source: Bruce Tonn et al., "Power from Perspective: Potential Future United States Energy Portfolios," *Energy Policy* 37, no. 4 (2009): 1432–1443.

11

Lifestyles

Security of Demand

Energy Demand and Lifestyle

When people think of the word "energy," their response often depends on which of two groups they belong. One group consists of scientists who think in terms of molecules, atoms, heat transfer, thermodynamics, chemistry, mechanics, and physics. They tend to be cerebral, well-educated, and specialized professionals, mostly affiliated with universities, government laboratories, or, occasionally, privately endowed research centers. The total number in this group is relatively small. Everybody else belongs to the vastly larger second group.

Those in the second group consider energy quite differently. For them, having access to plentiful and affordable energy supplies is the key to how they live their lives, whether they are healthy and safe, and whether their aspirations and goal are realistic or beyond their reach. Those with ready access just want to keep their lights on, their air conditioner humming, and their video games pinging and screeching. They rarely turn their attention to energy unless it becomes unavailable or too expensive. Those with insecure access tend to think about energy more often. They worry not about comfort but about survival. They must devote endless hours to gathering twigs, wood, dung, or anything else they might burn. They tend to live simple, often insecure lives. Their well-off counterparts, on the other hand, can access their energy by merely flipping a switch. Everything depends on where people are on the energy ladder.

Now that the world community has fuller access to global communication networks, it is more likely than ever before that people understand that not all energy resources are the same. For example, it is not a secret that electricity and gasoline provide more personal versatility than wood and dung. Indirectly, people probably also recognize, even if incompletely, the human dimensions of energy—that is, how energy affects their lives and the lives of their families. Do you think they all appreciate the close ties between their position on the energy ladder and their lifestyle?

One way to address this question is to try an experiment: ask some friends to identify the most critical influence on their quality of life. They will likely offer the following listing: safe water, ample healthy food, personal security, reliable

The Thread of Energy. Martin J. Pasqualetti, Oxford University Press. © Oxford University Press 2021.
DOI: 10.1093/oso/9780199394807.003.0011

health care, educational opportunities, convenient and inexpensive transportation, affordable housing, leisure opportunities, and so forth. Then ask them to explain how energy availability and consumption influence each aspiration they identified. Do you think they will be able to offer convincing explanations? Or do you suspect that many of them have not given the connections much thought? How would you try to raise their awareness? The first step is to make sure your knowledge is up to the challenge. Where might be a good place to start?

One place is to plot per capita GDP against per capita energy consumption for various countries. When you do that, you will discover a strong correlation between levels of prosperity and energy consumption (Figure 11.1). That is, for those countries where energy consumption is low, GDP per capita for that country tends to be low as well. They tend to go hand in hand; when one goes up or down, the other goes up or down. To get an initial sense of this relationship, look at the position of the United States at the top-right part of the diagram and then compare it to the position of China closer to the bottom-left part of the diagram. The United States has high GDP and high energy use. China is lower on both metrics.

Looking at China more closely, we find that as the leadership there loosened the reins on personal freedoms in recent years, there was a tremendous upswing in entrepreneurial activity, a marked rise in the accumulation of wealth, and a blatant expression of conspicuous consumption, especially noticeable in the proliferation of automobiles. The loosening of controls also produced a rapid rise in

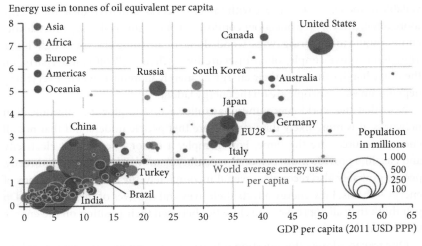

Figure 11.1 Most countries show a close relationship between GDP and energy consumption. *Source:* European Environment Agency, https://www.eea.europa.eu/data-and-maps/figures/correlation-of-per-capita-energy. See plates for color version.

GDP and, most important for our purposes, a noticeable jump in energy consumption. The Chinese were climbing up the energy ladder. But there was a price to pay, especially in terms of increased need for imported oil from Saudi Arabia and coal from Australia, along with a deadly increase in pollution.[1]

Another measure of energy consumption is kilowatt-hours per capita. For example, per capita energy consumption in Iceland is about 54,00 kWh, in Norway it is 23,000 kWh, in Bahrain it is about 20,000, in the United States it is about 13,000 kWh, and in the United Arab Emirates it is about 12,000 kWh.[2] Dubai—one of the hottest cities in the world—uses a lot of electricity just to keep cool. About 70 percent of the UAE's energy consumption is used to meet the air-conditioning needs of Dubai's glitzy resorts, a snow-sking area, soaring skyscrapers, and extravagant shopping malls (Figures 11.2 and 11.3). For example, all 12 million square feet of retail space in the Dubai Mall is cooled by a massive, energy-intensive air-conditioning system.

Even more grandiose is the planned Mall of the World, also in Dubai. It would occupy a completely enclosed and air-conditioned 48 million square feet, including a 4.35-mile promenade and connections to hundreds of hotels, all intended to attract 180 million people each year.[3] Given the energy demands of

Figure 11.2 View to the north from the observation level of the Burj Khalifa, Dubai. Expressways, clusters of tall buildings, irrigated landscaping, and the haze of its humid environment are all evident.

Figure 11.3 The Dubai Mall, a large shopping center adjacent to the world's tallest building, the Burj Khalifa, in Dubai. It requires massive air conditioning.

such projects, it is not surprising that the UAE holds the record for the world's largest ecological footprint, 57 percent of which is attributable to residential electricity consumption. If everyone lived the same way, we would need the resources of 4.5 earths to sustain us.[4]

The Human Development Index

Correlating GDP with energy consumption provides a view of one tight relationship between energy and human welfare and security. Another is the Human Development Index (HDI). The HDI is a composite index of life expectancy, education, and per capita income indicators, which are used to rank countries into four tiers of human development.[5] A country scores a higher HDI when the lifespan is higher, the education level is higher, and the gross national income per capita is higher. The HDI is expressed as a number between 0 and 1. A number close to 1 indicates very high human development. If you were to compare a world map of HDI with a world map of energy consumption, you would see that high HDI correlates with high energy use (Figures 11.4 and 11.5). For example, in 2010 Norway ranked first with an HDI of 0.938. The United States ranked

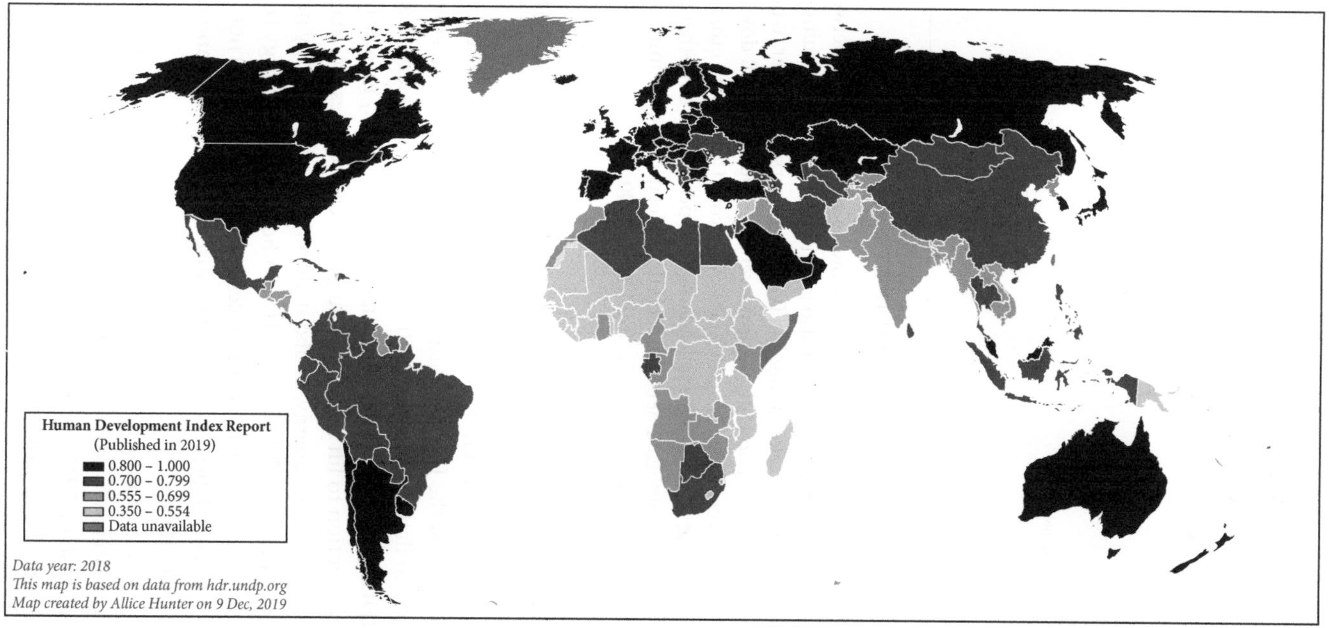
Human Development Index Report
(Published in 2019)
0.800 – 1.000
0.700 – 0.799
0.555 – 0.699
0.350 – 0.554
Data unavailable
Data year: 2018
This map is based on data from hdr.undp.org
Map created by Allice Hunter on 9 Dec, 2019

Figure 11.4 Human Development Index by country, 2018. *Source:* https://en.wikipedia.org/wiki/Human_Development_Index. See plates for color version..

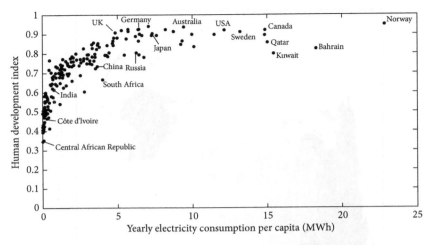

Figure 11.5 HDI vs. electricity consumption, showing a stark contrast between nations of the world. The industrialized world is well above 0.8 HDI, resulting from access to abundant energy. It takes about 3,000 kWh per person per year to be above 0.8 HDI, what we consider a good life. Getting everyone above 0.8 HDI down to about 6,000 kWh and everyone else up to 3,000 kWh would achieve an ethical and sustainable balance that is more likely to lead to global peace than any other path. *Source:* https://www.researchgate.net/figure/HDI-vs-electricity-consumption-EIA-2017-Jahan-2016-United-Nations-2014c_fig3_329246000.

fourth with an HDI of 0.902. India placed 119th with an HDI of 0.519. Sub-Saharan Africa hovered near the lower rankings.

Even modest increases in energy consumption can help improve the HDI, up to a point. Notice that the HDI increases rapidly for the first 50 gigajoules (GJ) and then begins to flatten out.[6] About 50 GJ per capita is necessary to achieve a reasonable HDI of 0.7. Above about 50 GJ, it takes progressively greater increases in energy consumption to lift the HDI. Thus, a much more respectable HDI of 0.8 requires twice as much energy, or about 100 GJ per capita. Eighty percent of the world's population is well below 0.8 HDI, indicating lower energy consumption. In other words, insufficient energy access—either because it is not available or because it is unaffordable—is a primary cause of poverty, war, and (increasingly) terrorist activities.

Bottom line: growth in the demand for energy closely follows the growth in per capita income in low- and middle-income economies, whereas high-income economies can sustain GDP growth with little if any increase in energy consumption. In either case—total energy or electricity only—each increment of improved quality of life takes an increasing amount of energy to achieve and maintain it.

Keeping in mind the premise of the energy ladder we discussed a bit earlier, it is not surprising to find a strong correlation between GDP and electricity. The more highly developed parts of the world—such as Europe and North America—have high GDP and also high electricity use. African countries are low in each category. The relative consumption is identifiable in the standard of living for that country. For example, the per capita use of electricity in the United States is about five times the world average. Canada has a high per capita use of electricity, at over six times the world average.

A more nuanced relationship between energy and the human condition is revealed in the relationship between electricity use and fertility. We find that the high consumption of electricity matches up with fewer children born per woman. This relationship suggests that there are ideal levels of electrical power and GDP necessary to build a sustainable, stable population. About this observation, Ed Caryl writes: "Those numbers would appear to be about 3,000 kWh and a GDP of about $30,000 annually per person." Caryl sees these numbers as reasonable target that would "go a long way toward improving the quality of life and stabilizing population in those areas."[7] Caryl is positing that fertility rates drop when electricity use increases. Causative factors are speculative but are most likely related to many lifestyle improvements that tend to rise together with the use of energy, electricity being the most critical driver.

Based on the relationships just described, a valuable—and recognized—early accompaniment to growing national prosperity is increasing generating capacity. China has set goals for a massive increase in generating capacity from every kind of energy resource. In recognition of the increasing environmental penalties that accompany this stance for coal-fired generation, the country's leadership has also been encouraging the aggressive development of renewables, mostly solar and wind.

This emphasis on renewables is outlined in China's five-year plan for 2016–2022. It calls for expanding investments in renewable energy, conservation, and energy efficiency as well as improving the country's integrated electricity planning and cost-based pricing decisions. As a result, China now leads the world in the scale and the speed of expansion in these sectors. To their pleasure, building more generating capacity has had the desired positive effects on standards of living in China.

Energy and Population

Another human dimension of energy consumption is the relationship between family size and GDP. Given that a fertility rate of 2.1 is the replacement level within a population, countries with a fertility rate below 2.0 will see a decline in population (if immigration is excluded). Most of Africa has a high fertility rate and low GDP (Figure 11.6). A low GDP reflects low per-capita energy consumption. If you increase access to affordable energy, the GDP will rise along with the standard of living. As family security improves, fertility rates tend to be lower. Lower fertility rates can relieve pressure on energy suppliers, but it can put additional pressure on workforce development and resources for elder care.

In contrast, a high fertility rate produces more children to help in family tasks, but per capita GDP goes down. It is a tricky relationship: which comes first, more energy or higher GDP? In truth, they tend to rise and fall together.

Several Asian countries are grappling with balancing fertility rates, GDP per capita, and energy demand. Some of these countries are prospering with low fertility rates, high GDP, and secure access to ample energy. At just 0.9 births per woman, Macao has the lowest total fertility rate in the world, followed by South Korea and Andorra, both at 1.0. Women in the United States have an average of 1.7 births in their lifetime, while women in China average 1.6; India, 2.2; and Kenya, 3.6. Niger's total fertility rate, or average births per woman over their lifetime, is 7—the highest in the world. How will those in Niger meet their energy demands?

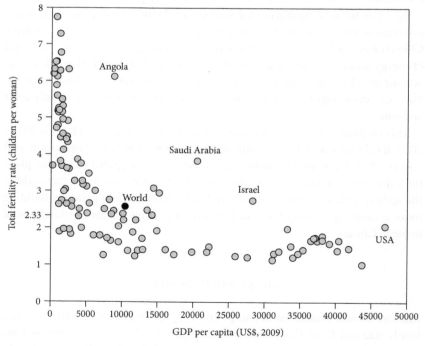

Figure 11.6 Total fertility rate trends downward as GDP per capita increases.

Asia is not alone in facing the strains of population shifts and energy demand. For example, Italian women give birth to 1.39 children on average. Germany is similar. The average across the European Union is 1.58. In response, Italy and Germany relaxed their policies about immigration, with both countries accepting millions of refugees from Africa, South Asia, and the Middle East in recent years. We can expect these patterns to be reflected in the availability, type, cost, and demand for energy.

There is a close relationship between energy demand, GDP, and population (Figure 11.7). At least as far back as we can measure, low energy use marched along in lockstep with low population growth. Once concentrated energy resources significantly replaced wood, populations rose abruptly, starting with the Industrial Revolution around 1750. Quality of life went up as well, fewer babies died (so more people reached reproductive age), and the population of the world expanded.

Coal was favored over wood for two reasons. First, the high rate of cutting was making wood scarce. Second, coal has a much higher energy density. Higher energy density means that higher transportation efficiency is possible. It also means that higher temperatures can be achieved, thus allowing a more extensive range of applications.

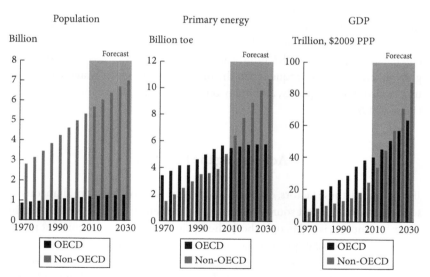

Figure 11.7 The close relationship between energy, GDP, and population. They tend to rise and fall together. *Source:* BP Energy Outlook 2035, bp.com/en/global/corporate/energy-economics/energy-outlook.html#BPstats. TOE = tons of oil equivalent. PPP = purchasing power parity.

Once energy-dense coal began substantively replacing wood, the rate of population growth started rising faster than ever before. Adding each successive billion people required shorter and short periods. World population grew from 1 billion to 2 billion in 130 years, from 2 billion to 3 billion in 30 years, to 4 billion in 14 years, to 5 billion in 13 years, then 12 more years to 6 billion, and 12 more years to 7 billion. By June 2021, the global population had risen to 7.9 billion. Considering the massive demands each additional billion people place on the earth's natural resource supply, it is difficult to imagine how this pattern can continue much longer.[8] How long do we have? When will the pattern change, how will it change, and who will bring about such change? Can we ever reach some form of equilibrium and sustainability? What will happen if we do not?

What do you think the demographic status of the world's countries will be in a few decades? Will they be represented by an inverted population pyramid, top-heavy with older people? Will they institute incentives and policies to encourage more births? Or will they relax their immigration policies? Will Singapore come to some sort of balance before it simply ages out of existence? China, which has a much lower GDP per capita, has already reacted; the one-child-per-family policy from about 1980 was abandoned in January 2016, replaced by official encouragement of two children per family. This policy was amended again to three children per family in 2021 as demographic data showed a declining birth rate.

Signs of the tension between population growth and rising energy demand are already appearing. Everyone may need energy, but will it be available when and where it is required? Will people be able to afford it? If there is not enough to go around, who will get it? Could this tension lead to what Michael Klare calls "resource wars"?[9] Some of the elements of such wars are already evident in several places (e.g., Iraq, South China Sea, Crimea, Sudan, etc.), and conditions suggest a likely increased frequency in the future. Conflicts over water supply, land use, environmental quality, and many other human needs are increasing as well (Figure 11.8).

Addressing the Problem

If we group all these factors, one way to address many problems simultaneously is to promote the cleanest sources of energy for as many people as possible. Giving the most impoverished people the highest priority will have the quickest impact because their rate of population growth is the fastest and because even a modest amount of additional energy can produce an impressive boost in HDI. Various NGOs and government agencies provide financial backing for this type of effort.

Just at Arizona State University, for example, the U.S. Agency for International Development is funding two such efforts: VOCTEC (Vocational Training

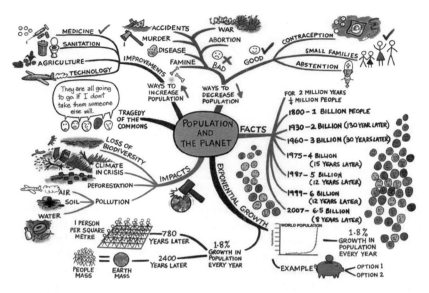

Figure 11.8 Population influences everything. *Source:* Mind Map Art, http://learningfundamentals.com.au/wp-content/uploads/Population-and-the-planet.jpg. Used with permission.

and Education for Clean Energy) and USPCAS-E (U.S.-Pakistan Centers for Advanced Studies in Energy).[10] One of the many reasons that these and similar programs promote renewable energy in developing countries is the highly favorable ratio between cost and benefit. Not only can these renewables be operational in the shortest time and for less cost than any other source, but they also have a significant and profound benefit to the quality of life of the people receiving them—and all this without the drawback of other energy sources' environmental costs and the need for elaborate supporting infrastructure. Such programs are needed in developing countries, arguably more than in any other group of countries, because that is where the greatest population increases are expected (Figure 11.9).

Summing Up

Considering all the examples just offered, we find that energy and lifestyles are completely intertwined. For most people, living comfortable, secure, and versatile lives depends on having access to reliable, affordable, high-quality sources of energy. At the opposite extreme, not having such access results in a much more

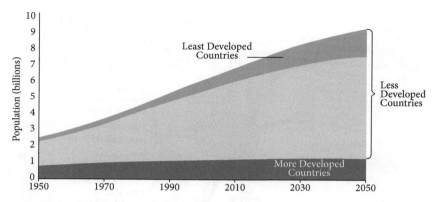

Figure 11.9 The United Nations predicts that the majority of population growth by the year 2050 will be in developing countries, with the industrialized regions remaining stable. This growth will raise the need for energy demand and produce increasing amounts of greenhouse gases. *Source:* United Nations Population Division, "World Population Prospects: The 2010 Revision," medium variant (2011), as cited in Carl Hub, "Fact Sheet: World Population Trends 2012," Population Reference Bureau, http://www.prb.org/Publications/Datasheets/2012/world-population-data-sheet/fact-sheet-world-population.aspx.

limited and stressful existence. Perhaps the most important lesson we learned is that even a small improvement in energy access—especially if it is in the form of electricity—can have a significant positive impact on how people can live. It was also apparent, however, that once an HDI of about 0.5 is reached, it takes ever more energy to lift everyone to the next level. If everyone in the world were to reach an HDI higher than 0.8, could the planet supply all the needed energy? In truth, that seems unlikely, if not impossible. If we wish to have a chance of resolving that paradox, we will need to maximize the efficiency of our energy use in all sectors of demand and maximize the development of sustainable energy resources. The third element, population stabilization, will take care of itself, even as the globe moves inevitably toward 9 billion inhabitants in the next few decades. Understanding how the thread of energy weaves in and out of lives is an essential early step.

12

Danger

Personal Risks and Hazards

First Exposure

Each year from 1978 until 2018, I would lead my students away from the chaotic urban landscapes of Phoenix and drive north to the stunning natural landscapes of the southern Colorado Plateau, four hours to the north. Within a little over two hours, each group began to sense the landscape's allure—one of emptiness, brimming with long vistas, red-walled canyons, tall mesas, and an enveloping quiet. What always escaped first impressions, however, was the setting it provided for introducing everyone to both the complexity and the hazards that accompany energy development.

Although most people today consider the open spaces and blue skies of northern Arizona ideal for relaxation and recreation, such flattering impressions are relatively recent. For several centuries it was mostly terra incognita to Europeans, not even appearing on maps. It was avoided by European settlers, who skirted either to the north or to the south. The prevailing impression was of a stark, dangerous, and empty wasteland, devoid of water, forbidding to travelers. European explorers beginning in the early sixteenth century treated the area with disinterest, if not disdain. Coronado's men, for example, sighted it in the 1520s but found it unattractive and unworthy of further attention.

For centuries, the region held its secret. It was protected by its isolation, by its bone-dry climate, and by a landscape so forbidding that Escalante and his party barely escaped with their lives after an ill-fated attempt to cross through the area from Santa Fe to California 250 years after Coronado. Its reputation as a dangerous place continued to afford silent protection for another 175 years, but by the 1950s, the Colorado Plateau had been rediscovered. Prospectors headed there in droves, determined to make their fortune by extracting long-overlooked energy riches that lay hidden underground. It was the beginning of a uranium frenzy.[1]

For purposes of a field trip, no region in the United States offers such a convenient and varied cluster of energy sites as northern Arizona. In just a single weekend, a curious visitor could tour an experiment in low-energy living in

The Thread of Energy. Martin J. Pasqualetti, Oxford University Press © Oxford University Press 2021.
DOI: 10.1093/oso/9780199394807.003.0012

Figure 12.1 Students on a field trip to northern Arizona for an energy class at Arizona State University gaze at the imposing presence of the 2250 MW Navajo Generating Station, which was decommissioned in November 2019, after more than forty years of operation.

Cordes Junction, a large transmission substation near Cameron, a retired uranium mill near Tuba City, a sub-bituminous coal mine and coal slurry pipeline high on Black Mesa, the nation's second-tallest hydroelectric dam on the Colorado River, and the hinge to such an excursion, the Navajo Generating Station, a 2,250 MW colossus located a few miles east of the city of Page.[2]

Now closed and being dismantled, no coal-fired power plant west of the Mississippi was more powerful or more imposing on the landscape than the Navajo Generating Station. At full throttle, the NGS burned a mountain of pulverized coal, weighing a total of 48 million pounds, every day. Occupying 1,700 acres of Navajo Nation land just south of Lake Powell, it struck an incongruous pose for forty-five years, jutting up atop the red Navajo sandstone and sagebrush (Figure 12.1). At night, blatantly reminding everyone of its presence, the flashing white lights on each of its 800-foot chimneys pulsed so brightly they could be seen from the International Space Station. That was true, at least, until December 2020 when, as the most noticeable stage of the plant decommissioning plan, the chimneys were demolished. For videos of this demolition, go to https://www.youtube.com/results?search_query=ngs+demolition.

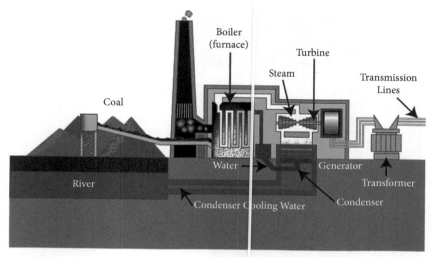

Figure 12.2 Simplified diagram of a coal-burning power plant. *Source:* Tennessee Valley Authority, https://commons.wikimedia.org/w/index.php?curid=24393916. For a video of a power plant in operation, go to https://www.youtube.com/watch?v=2IKECt4Y3RI.

Touring this energy-churning behemoth always included a stop to look inside the "business end" of all such power plants, the boiler (Figure 12.2). Through an unlatched metal door, everyone took a turn gawking into the inferno (Figure 12.3). Protected by a handheld heat shield, they witnessed a swirling 1,000-degree tornado of flame fifteen stories tall. Once their eyes adjusted to the glare, most people looking into the boiler could discern black streams of coal blasting into the blistering maelstrom from the side. After a short period, each person would hand the shield to the next in line, happy to back away from the intense heat and the remote chance that a random tongue of flame might snap out and lick their face.

Once the portal was again shut and latched, our guide reminded us that the purpose of the boiler was to, as the name suggests, boil water. To illustrate what was going on, he compared the boiler to a giant teakettle, pointing out that in the power plant, instead of whistling for your attention, the steam was forced through restrictive nozzles focused on the blades of steel turbine wheels up to 12 feet in diameter.

As with everything else at the NGS, the turbo-generators took up a lot of room. The voluminous space could comfortably accommodate three football games played simultaneously, including an occasional lofty punt (Figure 12.4).

Figure 12.3 A student gazes into one of the 800 MW boilers at the Navajo Generating Station in Page, Arizona.

Figure 12.4 Turbine bay of the Navajo Generating Station. Each of the three turbine generators are enclosed in the cowling with the tubular structure on top. This is where electricity is generated.

Encased in large cowlings, each of the turbines spun a shaft connected to a generator. This was where the magic happened. The shaft from the turbine rotated magnets around a coiled copper wire, producing one of the pillars of modern civilization—humankind's most valuable, versatile, and essential commodity, the most crucial ingredient separating the ancients from the moderns and the "haves" from the "have-nots." We call it electricity.

From the generator at NGS, a buzzing thread of energy passed through transformers, transmission lines, substations, and distribution networks that gallop off in all directions. Almost instantaneously, it reached homes in Phoenix and other cities. After it arrived, it powered computers, televisions, air conditioners, neon signs, traffic lights, light rail, insulin pumps, refrigerators, dentist drills, nail guns, and as wide an assortment of conveniences and tools as can be imagined.

The entire operation was nothing if not impressive: a ferocious fireball, crushing pressure, screaming turbines, and crackling electricity. For students on this visit—who for most of their college years had been confined within classroom walls—experiencing firsthand the creation of electricity was something they would never forget. One of their most lasting memories was the ever-present sense of raw power that triggered a profound sensory reaction. They saw it, heard it, smelled it, felt it, and even tasted it.

They also began to understand the hazards and risks that accompany every phase of energy production. Anyone working within the energy business—especially the phases of extraction, generation, and transmission—appreciates such dangers. They understand that doing their jobs can threaten their safety. They know these hazards not just intuitively but also because their employers continuously remind them to exercise sound judgment and follow all safety rules. As alert and well-warned as they are, hazards still exist, including many that are sometimes masked by distance, culture, mistakes, ignorance, mischief, or duration.

If you are like most people, you do not work in the energy industry. For you, the risks and hazards of energy development seem distant. Nevertheless, they are not absent from your experience. You know, for example, that mixing water and electricity can produce undesirable results. You know you should not put your finger in a light socket, light a cigarette while fueling your car, or enter a house when you detect the strong odor of natural gas. Apart from these examples, the risks and hazards of energy are probably not part of your everyday experiences. Knowing this—and mindful of their legal liability—energy companies provide reminders wherever the public might come across any part of the energy infrastructure (Figure 12.5). But such warnings are, at best, just superficial indicators of the hazards of energy. Let's look deeper. The purpose of this chapter is to lay bare the risks and hazards that are inherent within the energy industry. Such

Figure 12.5 Working in the energy industry, or even being around any part of the energy industry, involves inherent risks, as this sample of warning signs suggests.

dangers are relevant to those who work in the energy business, those who live nearby, and those who use energy. In other words, there are risks and hazards that we all accept in the bargain we make to get the energy we need.

Risks and Hazards of Coal

Fossil fuels—to one degree or another—hold what we call "energy density." If they did not, they would be of little utility to us, at least in modern applications. That they are concentrated forms of energy explains why we employ them to run our cars and generate our electricity. To illustrate what energy density means, recall your visit to a science museum, such as the Exploratorium in San Francisco. Remember the stationary bicycle hooked up to a lightbulb? If you have ever tried to pedal fast enough just to get a 100-watt lightbulb to flicker to life, you know the effort quickly exhausts you. If you were fit and could keep up that effort for ten hours, you would produce one kilowatt-hour (kWh) of electricity.

How much power is a kilowatt-hour? To meet the monthly energy demands of a typical single-family suburban house in the United States requires about 1,000 kWh of electricity or about 33 kWh per day. Putting those figures into terms of human endurance, producing that amount of power would require that thirty-three world-class athletes pedal their bicycles ten hours per day, every day. Even elite Tour de France riders would fall far short of accomplishing this feat. Now consider a coal-burning power plant. Burning one pound of coal at a power plant produces the same kilowatt-hour in a fraction of a second. Burning one ton of coal produces 2,000 kWh. Burning coal at a rate of 1,000 tons per hour (which was NGS's highest capacity when it was operational) would create 2 million kWh, enough to meet the needs of about 1,800 houses for a year. Generating at that capacity for 24 hours would meet the electricity demand of 44,000 homes for a year, roughly 100,000 people. That capacity is possible only because coal has a high energy density. (Some resources are even denser, including oil and uranium.)

Hazards to Workers

These facts reinforce three points for you to consider: (1) the concentrated nature of energy resources increases their value to us; (2) our practice of developing and using concentrated forms of energy exposes us to more danger than using human or animal muscle; (3) handling such concentrated types of energy resources demands respect for the hazards they create.

The hazards of handling such concentrated energy resources exist at every step in the fuel chain. Whether you participate in extraction, processing, transportation, conversion, transmission, or waste disposal, you are exposing yourself to the hazards each step holds. Dangers are always present. By working with these resources, you assume risks that are still hovering nearby, often just out of sight and out of mind.

Awareness of the hazards and risks that accompany handling energy resources is especially pertinent to employees in the energy industry. Such dangers range from the obvious to the unexpected. For example, when miners remove coal from underground, they are disrupting the native geologic stability, thereby placing themselves in harm's way. To protect themselves, miners create structural support against collapse, maintain proper ventilation to avoid the buildup of dangerous gases, and operate all equipment with caution and skill.

Still, it is a dangerous job. Working underground holds inherent risks. Every day when underground miners start their shifts, they expose themselves to the possibility of crushing cave-ins, catastrophic explosions, chemical contamination, equipment malfunctions, and electrocution. Yet miners do their jobs anyway, just as generations have before them, driven to make a living to support themselves and their families while providing the energy resources to society that makes modern life possible. And they sometimes pay the ultimate price.

Mindful of the hazards that accompany their professional activities, miners take personal safety seriously. Not only do they pay close attention to self-preservation, they heavily rely on their employers to operate the mines within the mandatory protocols of regulatory agency oversight. Such regulations—and their enforcement—vary from country to country, but in the United States they have been effective in reducing injuries and deaths. In the early twentieth century, coal mining deaths in the United States averaged 2,000 annually. By 2019, that figure had dropped to less than a dozen (Table 12.1).

While deadly accidents have become infrequent, they still do happen. For example, Monday afternoon, April 5, 2010, a spark from a piece of mining equipment ignited the coal dust in the Upper Big Branch coal mine in Montcoal, West Virginia. The explosion killed twenty-nine miners and injured two others (Figure 12.6).[3] Investigators later faulted the operators for failure to properly maintain ventilation systems, thus allowing methane levels to increase to dangerous levels leading to the explosion. It was the worst coal mining accident in the United States in forty years.

Largely because of implemented safety measures and regulatory oversight, fatalities have declined in the United States. However, numbers remain high in other countries, most prominently in China (Table 12.2). As recently as 2009, 2,631 Chinese coal miners died, compared to 18 the same year in the United States. Because China produces 3.5 times more coal than the United States, one should normalize the statistics to make the comparison more accurate. One method is to consider them in terms of the death rate per million tons of coal produced. In 2007 these numbers were 1.485 fatalities per million tons of coal mined in China and 0.029 in the United States, a rate that is less than 2 percent that in China. Yet another way to measure mining fatalities on equal terms is to consider the deaths per unit of energy, such as a terawatt-hour. The rate in China is 90 deaths per TWh of electricity generated from coal, while in the United States the number is 15. To look at the statistics in a different light, several times more coal miners die in China each year than die in commercial aircraft accidents in the entire world.

One must accept that coal mining is harsh and dangerous work. Try to imagine yourself traveling back in time to the middle of the twentieth century or before. You are a coal miner in England, Wales, Ukraine, or Kentucky during the winter months. You typically leave for work in the dark and work all day, never seeing the sky. Your workplace usually has shallow ceilings and impossibly narrow, cramped conditions. Your whole body is typically bathed in sweat and coal dust all day, every day. It covers every exposed part of your body, and you cannot avoid inhaling coal dust every time you take a breath. This is your

Table 12.1 Coal Mining Fatalities in the United States, 1900–2019

Year	Miners	Fatalities	Year	Miners	Fatalities	Year	Miners	Fatalities	Year	Miners	Fatalities
1900	448,581	1,489	1930	644,006	2,063	1960	189,679	325	1990	168,625	66
1901	485,544	1,574	1931	589,705	1,463	1961	167,568	294	1991	158,677	61
1902	518,197	1,724	1932	527,623	1,207	1962	161,286	289	1992	153,128	55
1903	566,260	1,926	1933	523,182	1,064	1963	157,126	284	1993	141,183	47
1904	593,693	1,995	1934	566,426	1,226	1964	150,761	242	1994	143,645	45
1905	626,045	2,232	1935	565,202	1,242	1965	148,734	259	1995	132,111	47
1906	640,780	2,138	1936	584,582	1,342	1966	145,244	233	1996	126,451	39
1907	680,492	3,242	1937	589,856	1,413	1967	139,312	222	1997	126,429	30
1908	690,438	2,445	1938	541,528	1,105	1968	134,467	311	1998	122,083	29
1909	666,552	2,642	1939	539,375	1,078	1969	133,302	203	1999	114,489	35
1910	725,030	2,821	1940	533,267	1,388	1970	144,480	260	2000	108,098	38
1911	728,348	2,656	1941	546,692	1,266	1971	142,108	181	2001	114,458	42
1912	722,662	2,419	1942	530,861	1,471	1972	162,207	156	2002	110,966	28
1913	747,644	2,785	1943	486,516	1,451	1973	151,892	132	2003	104,824	30
1914	763,185	2,454	1944	453,937	1,298	1974	182,274	133	2004	108,734	28
1915	734,008	2,269	1945	437,921	1,068	1975	224,412	155	2005	116,436	23

Continued

Table 12 Continued

Year	Miners	Fatalities	Year	Miners	Fatalities	Year	Miners	Fatalities	Year	Miners	Fatalities
1916	720,971	2,226	1946	463,079	968	1976	221,255	141	2006	122,975	47
1917	757,317	2,696	1947	490,356	1,158	1977	237,506	139	2007	122,936	34
1918	762,426	2,580	1948	507,333	999	1978	255,588	106	2008	133,828	30
1919	776,569	2,323	1949	485,306	585	1979	260,429	144	2009	134,089	18
1920	784,621	2,272	1950	483,239	643	1980	253,007	133	2010	135,500	48
1921	823,253	1,995	1951	441,905	785	1981	249,738	153	2011	143,437	20
1922	844,807	1,984	1952	401,329	548	1982	241,454	122	2012	137,650	20
1923	862,536	2,462	1953	351,126	461	1983	200,199	70	2013	123,259	20
1924	779,613	2,402	1954	283,705	396	1984	208,160	125	2014	116,010	16
1925	748,805	2,518	1955	260,089	420	1985	197,049	68	2015	102,804	12
1926	759,033	2,234	1956	260,285	448	1986	185,167	89	2016	81,485	8
1927	759,177	2,231	1957	254,725	478	1987	172,780	63	2017	82,843	15
1928	682,831	2,176	1958	224,890	358	1988	166,278	53	2018	82,699	12
1929	654,494	2,187	1959	203,597	293	1989	164,929	68	2019	81,361	12

Source: U.S. Department of Labor, https://arlweb.msha.gov/stats/centurystats/coalstats.asp.

Figure 12.6 Upper Big Branch Miners Memorial, Whitesville, West Virginia, in honor of the 29 miners killed in a mining explosion in 2010. The memorial is forty-eight feet long and nine feet tall at its highest peak. The serrated top of the memorial emulates the profile of the Appalachian Mountains. Etched on the front are the silhouettes of twenty-nine life-size miners standing shoulder to shoulder. Engraved above the silhouettes is the seal of West Virginia and, below, the inscription "Come to me all you who labor, and I will give you rest." *Source:* Governor Earl Ray Tomblin, Creative Commons.

experience, year after year, and—to say the least—it does not lead to healthy outcomes.

Instead, such chronic exposure leads us to consider one of the hazards of coal mining that does not involve catastrophic accidents but which is ultimately deadlier. I am referring to exposure to the fine and abrasive coal particles that float around all coal mines, even surface mines. In the early days of coal mining, little could be done to reduce such dust, despite the common suspicion that breathing it in all the time could not be good for any miner.

Inhaling such dust has been slowly suffocating miners for centuries, but it is not the only threat they face. As coal mining moves deeper through sandstone, workers are also breathing silica dust. Silica dust is even more hazardous than coal dust.[4] Despite improvements in mine air quality, coal dust and silica dust continue to take their toll.

Table 12.2 Coal Mining Fatalities in China and the United States, 2000–2009

Year	Number of Coal Mining Fatalities			Death Rate per Million Tons		
	China	United States	Ratio, China/U.S.	China	United States	Ratio, China/U.S.
2000	5,798	38	153	6.096	0.040	152
2001	5,670	42	135	5.070	0.040	127
2002	6,995	27	259	4.640	0.028	166
2003	6,702	30	223	4.170	0.031	135
2004	6,027	28	215	3.080	0.027	114
2005	5,938	23	258	2.810	0.021	134
2006	4,746	47	101	2.041	0.040	51
2007	3,786	33	115	1.485	0.029	30
2008	3,215	29	111	1.182	–	–
2009	2,631	18	146	0.892	–	–

Source: Guo Wei-ci and Chao Wu, "Comparative Study on Coal Mine Safety Between China and the U.S. from a Safety Sociology Perspective," *Procedia Engineering* 26 (2011): 2003–2011.

We commonly call the specific medical sickness that results from long-term breathing of coal dust black lung disease. More precisely, it is known as coal workers' pneumoconiosis (CWP). The American Lung Association offers this description: "Pneumoconiosis is a general term given to any lung disease caused by specks of dust that are breathed in and then deposited deep in the lungs causing damage. Pneumoconiosis is usually considered an occupational lung disease. It includes asbestosis, silicosis, and coal workers' pneumoconiosis" (Figure 12.7).[5] "In this type of disease, the lung is damaged (in this case, by coal dust); the walls of the air sacs are inflamed, and the lung stiffens from scarring of the tissue between the air sacs."[6] There is no known treatment for pneumoconiosis.

Mining accidents receive more attention because they are usually dramatic. But a look at the data reveals a clearer picture of the relative danger. For example, while 28 mining fatalities were recorded in the United States in 2004, more than 700 people died from CWP. The numbers have not improved. The National Institute for Occupational Safety and Health (NIOSH) estimates more than 76,000 miners have died since 1968 from the disease, and more than $45 billion in federal compensation benefits have been paid out to coal miners disabled by black lung and those who survived them.[7]

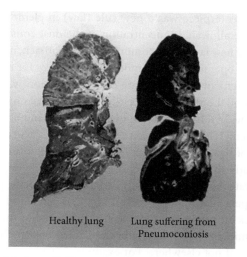

Healthy lung   Lung suffering from
Pneumoconiosis

Figure 12.7 A healthy lung (left) compared to a lung from a coal miner (right).
Source: U.S. EPA.

For many years, black lung disease was a silent killer; few outside the industry were even aware of its existence. By the 1960s, however, it started gaining national attention. In 1969, *Time* magazine reported:

> While catastrophic cave-ins are relatively rare in the U.S., about 260,000 former miners suffer from another killer: black lung disease. One reason could be that mine operators routinely circumvent air-safety standards established in 1969 to regulate the amount of coal dust in the mines. Now the government is cracking down. In Virginia last week, 13 coal companies and 15 people pleaded guilty to falsifying air samples to understate levels of coal dust. The companies face fines of as much as $500,000 each. The individuals could be imprisoned for five years and fined $250,000. Prosecutors say many more guilty pleas are forthcoming.[8]

The continuing infractions resulted from the passage of the Coal Mine Health and Safety Act and its amendments (1969 and 1997), which mandated a comprehensive set of measures to prevent CWP. According to the Centers for Disease Control and Prevention, "Enactment was followed by a marked reduction in the prevalence of CWP. in long-term coal miners. In the period 1970–1974, about 32% of miners with 25 or more years of tenure in coal mining who participated in a national x-ray surveillance program had evidence of CWP. By the period 1995–1999, the prevalence in this group had dropped to about 4%."[9]

By 2005–2006, after years of lower incidence, the prevalence of black lung increased again, to 9 percent. In 2009 the agency launched a campaign to end black lung.[10] Its centerpiece was a new rule (law) implemented on August 1, 2014, lowering the allowable concentration of the dust considered responsible for black lung. The new law took a comprehensive approach, including increased sampling by mine operators, use of modern technology for real-time sampling results, the requirement for immediate corrective action when excessive dust levels are found, and the ability to make a determination of noncompliance based on a single sample. In a further attempt to educate the public about the magnitude of the problem, NIOSH produced a video called *Faces of Black Lung II*.[11]

The number of deaths declined after the implementation of the rule, but a 2018 NIOSH study showed a resurgence of the incurable respiratory illness. It was the highest rate recorded in roughly two decades, with the peak of incidence being in the Appalachian states (Figure 12.8). Ten percent of U.S. veteran coal miners have black lung disease.[12] On the more positive side, the root cause of the problem—coal mining—is in decline in the United States. This should reduce the problem here, if not elsewhere. For example, in China, the death rate from CWP is over 6,000 per year.[13] Globally, CWP resulted in 25,000 deaths in 2013— down from 29,000 deaths in 1990.[14]

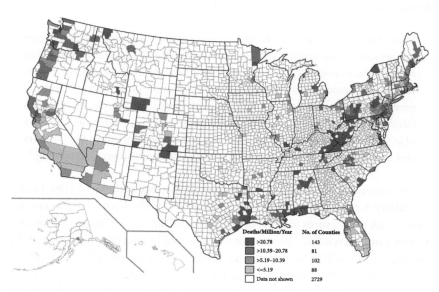

Figure 12.8 All pneumoconioses: age-adjusted death rates (per million population) by county, U.S. residents age 15 and over, 2000–2009. *Source:* CDC, https://wwwn. cdc.gov/eworld/Data/All_Pneumoconioses_Age-adjusted_death_rates_per_ million_population_by_county_US_residents_age_15_and_over_20002009/796.

Hazards to the Public

West Virginia is coal country, and it has been for centuries. The vast majority of jobs (and, therefore, residents' livelihoods) depend on the coal industry. It is also a particularly rugged part of the Appalachian Mountains. With little flat land, coal mining activities, highways, railroad tracks, and settlements all hug the banks of the many narrow stream channels. Hazards abound: mountains are being shaved off by mountain-top removal, underground mining and mine fires continue to compromise natural underground support structures, and artificial embankments tenuously impound mining waste materials. Such was the setting in West Virginia when life ended for dozens of people along Buffalo Creek in 1972.

The disaster at Buffalo Creek occurred just after 8:00 a.m. on February 26. A coal slurry impoundment burst. It was a mere four days after a federal inspector had declared it safe. Racing down the valley at 7 feet a second, 132 million gallons of black wastewater rushed through the hollow, sweeping away town after town. In short order, there were 125 dead, 1,100 injured, 1,000 cars and trucks destroyed, over 500 houses demolished, and 4,000 left homeless (Figure 12.9).

Such risky impoundments exist by the hundreds and constitute a justifiably ever-present worry.[15] One of the most notoriously hazardous circumstances continues to exist near the town of Montcoal, West Virginia, close to the location

Figure 12.9 Memorial to those lost when a coal tailings pond collapsed above Buffalo Creek, West Virginia, in 1972.

Figure 12.10 The proximity of a coal sludge impoundment to Marsh Fork Elementary School. *Source:* Photo by Lynn Willis. Used with permission.

of the Upper Branch Mine disaster in 2010. The Marsh Fork Elementary School lies directly below a coal mining impoundment. In the event of a failure of that impoundment, floodwaters would reach the school in less than three minutes (Figures 12.10 and 12.11).

The perils that coal mining exposes schoolchildren to are based not on morbid imagination but experienced tragedy. No place is better known for such heartbreak as Aberfan, a village four miles south of Merthyr Tydfil in Wales, Great Britain. The town gained infamy because of the catastrophic collapse of a coalmining sludge impoundment early on the morning of Friday, October 21, 1966. At 9:15 a.m., more than 150,000 cubic meters of water-saturated debris broke away and flowed downhill at high speed. It quickly engulfed the Pantglas Junior School, where students had arrived only minutes earlier for the last day before the half-term holiday. One hundred and forty-four people were buried alive that day, including 116 children (Figure 12.12).[16]

Despite the many standards, protocols, laws, and oversight mechanisms in place to prevent these tragic events, they continue to occur, even in the United States. To take just one example of the failure of a waste ash impoundment: a spill of millions of gallons of toxic sludge occurred in 2008 at the Kingston power plant in central Tennessee, burying some houses downstream (Figure 12.13). Thankfully it resulted only in property and environmental damage, with no fatalities, but it was a reminder that energy development imposes hazards not just on a cognizant workforce but also on an unsuspecting public.

Figure 12.11 Playground of the Marsh Fork Elementary School, located next to a coal loading silo and below a coal slurry impoundment. It is the same silo pictured in the aerial image in Figure 12.10.

Figure 12.12 Aerial photograph on the day of the Aberfan disaster in Wales, October 21, 1966. *Source:* H.M. Stationery Office, http://www.dmm.org.uk/ukreport/553-02.htm.

Figure 12.13 A waste ash pond rupture at Kingston coal power plant, Tennessee, in December 2008 sent billions of gallons of sludge downstream, burying houses. *Source:* TVA.

Coal mine fires are another common hazard. Thousands of mine fires burn in dozens of countries on every continent except Antarctica. They are ignited in various ways, including human activities such as burning trash, lightning strikes, spontaneous combustion, or nearby brush fires.[17]

Their hazards also vary by location. Sometimes the danger is small, as when they occur in lightly settled areas such as Black Mesa, Arizona (Figure 12.14). Other times coal smolders undetected for years, posing an unrecognized hazard to those living nearby. The surface might collapse, or people living there might suffer from the emission of odorless noxious gases. In still other places the odor is so noticeable that no one can miss it. Such odors gave the Powder River Basin— the location that supplies 40 percent of the nation's coal—its name, because the area smelled like burning gunpowder.[18]

Mine fires can progress from a minor inconvenience to a life-threatening problem in a hurry. When coal burns underground, it typically has little access

Figure 12.14 Student feels the heat rising from an underground coal mine fire on remote Black Mesa, on Navajo Nation lands in northeastern Arizona.

to oxygen. Once mining begins, oxygen supplies increase, and so do the dangers. At the same time, the release of dangerous, partially oxidized compounds rises to hazardous levels. Testing around the notorious mine fires of Centralia, Pennsylvania, revealed forty-five organic and inorganic chemicals, including toxins like benzene, toluene, and xylene. One of China's coal fires emits fifty-six compounds that have been detected.[19] Particulate matter and harmful gases emitted from burning coal—including sulfur, carbon and nitrogen oxides, and hydrocarbons released at the surface from gas vents, ground fissures, and the soil—cause illnesses that range from stroke to chronic obstructive pulmonary disease.[20]

Coal mine fires have burned for centuries, but they are not a thing of the past. In the United States alone, more than 100 underground fires continue burning in at least nine states, including Colorado, Kentucky, Maryland, New Mexico, Pennsylvania, and Wyoming. Specific examples include the Tiptop underground fire in Breathitt County, the Truman Shepherd fire in Floyd County, the Lotts Creek fire, and the Ruth Mullins fire, all in Kentucky.

One of the most notorious and continuing mine fires is beneath the city of Centralia, Pennsylvania. The principal activity in Centralia was mining anthracite coal. Mining continued until the 1960s, when most companies ceased operation. The mine fire started in 1962 and has not yet been extinguished. It has been burning through coal every year since then.[21]

(a)

(b)

Figures 12.15a and 12.15b Warning of public hazards in Centralia, Pennsylvania, where a coal mine fire has been burning for decades. The town was abandoned when all efforts to extinguish the fire failed. *Source:* https://www.pinterest.com/EVVphotography/a-b-a-n-d-o-n-e-d-town-centralia-pa-united-states/.

After twenty years of efforts to extinguish the fire proved unsuccessful, the continued occupation of Centralia was deemed too dangerous (Figure 12.15). Most of the residents began leaving in 1984. After the homes were sold to the Commonwealth of Pennsylvania, the structures were demolished, the ground was leveled, and trees, shrubs, and flowers were planted.[22] Today, Centralia mostly no longer exists; it is barely a ghost town, more of a tourist curiosity than anything else. The fire itself, by some estimates, could continue burning for 1,000 years.[23]

Other countries report similar and even more dramatic mine fires. One of the most persistent is the complex of fires in the coalfields of Jharia, north of the Damodar River in eastern India (Figure 12.16).[24] The Jharia mining complex is enormous, about 110 square miles, or more than twice the area of San Francisco, California. It consists of twenty-three large underground mines and nine large open cast mines. At least seventy fires are burning in the Jharia coalfield, making it the largest coal mine fire complex in the world.

First detected in 1916, many of the fires at Jharia have burned underground for over a century. If all the remaining inaccessible coal was to continuing burning

Figure 12.16 A village in Jharkhand continues to be engulfed by noxious fumes coming from a coal mine fire nearby. The Kujju-Collieries Mines in the Ramgarh district of Jharkhand has been burning continuously for twenty years, putting the lives of nearly 4,500 residents in danger. *Source:* National Public Radio, https://media.npr.org/assets/img/2015/11/23/jharia-haglund-02_custom-33bf02a12e1a6be3b92df7d150dbdfffa279813d-s1000-c85.jpg.

at the same historically average rate, the fires could last for another 3,800 years. Because of growing hazards, the Indian government is relocating entire communities, both for public safety and for economic reasons: coal worth $12 billion lies underneath, and the state government wishes to continue exploiting this resource.[25]

Another significant example is the many coal mine fires near Urumqi, the capital city of Xinjiang in western China. These fires produce the world's most polluted air.

Subsidence is an additional public hazard resulting from coal mining. Subsidence occurs after coal is removed and the supporting earth materials collapse all the way to the surface. Many examples of this phenomenon exist. Sometimes it poses little public hazard when it is in open spaces (Figure 12.17), but when it is in populated areas, it can be distinctly disruptive and even be life-threatening (Figure 12.18).

Totaling all the public hazards from coal just outlined might seem problematic, but they would not equal the scale of dangers from breathing the pollution that results from burning it. In China, for example, burning coal has the worst health impact of any source of air pollution. In 2013, it caused 366,000 premature

Figure 12.17 Coal mine subsidence above abandoned coal mines 10–15 kilometers north of Sheridan, Wyoming. These are above the Dietz coal mines, which were in operation from the 1890s to the 1920s. *Source:* Frontispiece B, U.S.G.S. Prof Paper 1164, C. R. Durned, 1976. Public domain image. https://www.osti.gov/biblio/5147087-coal-mine-subsidence-fires-sheridan-wyoming-area.

Figure 12.18 House in Wilkes-Barre, Pennsylvania, condemned and abandoned due to subsidence over a coal mine. Foundations have fallen away, and the house rests on inserted telephone poles.

deaths. Many places in India have even worse pollution; however, not all of it is from burning coal, despite India's reliance on coal for 70 percent of its electricity. Death from breathing all sources of unhealthy air in India reached 1.24 million in 2017.[26]

The public hazards of burning coal—chronic bronchitis, aggravated asthma, cardiovascular effects like heart attacks, and premature deaths—are mainly attributed to inhaling tiny particulate matter. U.S. coal power plants emitted about 200,000 tons of small airborne particles (10 micrometers or less in diameter) in 2014.[27] Other harmful emissions from coal-burning power plants should concern us as well, particularly mercury. Just 1/70th of a teaspoon of mercury deposited in a 25-acre lake can make the fish unsafe to eat. In the United States, about 42 percent of airborne mercury comes from coal plants. According to the Environmental Protection Agency's National Emissions Inventory, U.S. coal power plants emitted 45,676 pounds of mercury in 2014 (the latest year for which data are available).[28] They also emitted 3.1 million tons of sulfur dioxide, 1.5 million tons of oxides of nitrogen, 41.2 tons of lead, 9,332 pounds of cadmium and other toxic heavy metals, 576,185 tons of carbon monoxide (which causes headaches and places additional stress on people with heart disease), 22,124 tons of ozone-forming volatile organic compounds, and 77,108 pounds of arsenic. For scale, arsenic causes cancer in one out of 100 people who drink water containing 50 parts per *billion*. It is unhealthy to breathe or ingest any of these products of coal combustion. These hazards are among the many reasons 312 coal-burning power plants in the United States that existed in 2005 have been retired or are scheduled for retirement as of mid-2020.[29]

The Risks and Hazards of Oil and Gas

Hazards to Workers

The oil and gas industry, like all industries working with concentrated forms of energy, can be dangerous to those in their employ. These dangers are found everywhere along the supply chain, and just about everywhere in the world. The reason for this broad spatial extent is simple: the oil and gas business is the biggest in the world. Consider these statistics:[30]

- More than 4 billion metric tons of oil is produced worldwide annually.
- Total global annual revenue of the oil industry is more than $1.25 trillion.
- The amount paid for crude oil each day in the United States is about $1 billion.
- The value of the crude oil in a single large tanker can be as high as $100 million or more.
- The annual value of upstream capital spending (such as exploration) in the oil and gas industry has been in the range of $500 billion to $800 billion in the years up to 2018.

The massive and complex global oil and gas enterprise is supported by the effort of hundreds of thousands of skilled workers. Because oil and natural gas products are combustible, often at relatively low temperatures, working every day with these products routinely exposes industry workers to an assortment of hazards too numerous to itemize here. Suffice it to say that these workers assumed the associated risks of accompanying their industry when they signed their employment contracts. They also agreed to follow all safety procedures at all times.

Despite all the best intentions, training, and oversight, accidents do happen. When they do, the consequences can be deadly. For example, 823 oil and gas extraction workers lost their lives on the job in the United States between 2003 and 2010. That fatality rate was seven times greater than the rate for all U.S. industries (Table 12.3).[31]

Some of the risks of oil and gas production are common everywhere, but perhaps the most dangerous place to work is offshore. Despite the risks, however, offshore development continues because, to state the obvious, that is where the

Table 12.3 U.S. Worker Fatalities in All Industries Compared to Worker Fatalities in the Oil Industry

Year	All Industries, Fatalities per 100,000 Workers	Oil Industry, Fatalities per 100,000 Workers
2013	3.2	29.0
2012	3.4	31.9
2011	3.5	29.0
2010	3.6	32.4
2009	3.5	28.5
2008	3.7	25.2
2007	4.0	16.1
2006	4.2	24.6
2005	4.0	22.4
2004	4.1	25.2
2003	4.0	19.1

Source: U.S. Bureau of Labor Statistics, http://www.bls.gov/iif/oshcfoi1.htm.

reserves are. For example, the Gulf of Mexico and its littoral supply 18 percent of U.S. oil production (and half of U.S. refining capacity).[32] The 2016 Hollywood movie *Deepwater Horizon* provided a vivid portrayal of the unfiltered terror of working on a drilling rig engulfed in flames (Figure 12.19). Eleven died in the 2010 accident portrayed in the film, with many more injured. It cost the oil company BP over $50 billion.

As tragic as that incident was, other offshore accidents have resulted in greater loss of life. The Piper Alpha disaster in the North Sea in July 1988 killed 167 people. Only 61 of the workers on the oil rig survived, earning it the miserable distinction of being the deadliest offshore oil rig accident in history.[33] This example and many others are reminders of the personal risks that offshore workers assume and why they receive hazard pay.

The hazards of oil and gas development extend beyond the activities of extraction. Working in an oil refinery is, by most accounts, even more dangerous. The Environmental Protection Agency—which is the responsible regulatory agency dealing with refinery spills and toxic gas releases—says no other industry suffers as many catastrophic incidents involving hazardous chemicals as refineries. For example, in August 2012, a massive explosion occurred at Chevron's Richmond, California, refinery, easily visible from across San Francisco Bay (Figure 12.20).[34]

Figure 12.19 Deepwater Horizon explosion and fire. *Source:* U.S. Coast Guard, http://en.wikipedia.org/wiki/Deepwater_Horizon#/media/File:Deepwater_Horizon_offshore_drilling_unit_on_fire_2010.jpg.

Figure 12.20 Flames and plumes of smoke rise from the Chevron refinery in Richmond, California, after an explosion in 2012. Nearby residents were warned to stay inside, but more than 14,000 were eventually treated at nearby hospitals. *Source:* Photo by Phil McGrew. Used with permission.

The Chevron incident, while expensive to repair, was comparatively minor. Other refinery accidents have been truly catastrophic. For example, on March 23, 2005, an explosion at BP's Texas City refinery (now owned by Marathon and called Galveston Bay) injured nearly 200 and killed 15. It resulted in hefty fines and billions of dollars of lost revenues.

Although there are other hazards for workers in the oil and gas business— such as piracy—we will mention just one more here. I am referring to the increased risk of terrorist attacks, especially in unstable parts of the world. One of the deadliest attacks occurred when Islamic militants stormed the In Amenas gas processing facility, located near the Libyan border and jointly operated by Sonatrach, BP, and Statoil. The January 16, 2013, attack damaged two of the facility's three processing "trains," each of which can process 3 billion cubic meters per year (106 billion cubic feet per year). Algerian Special Forces raided the site after four days, to free the hostages. At least 29 terrorists and 39 foreign hostages were killed along with an Algerian security guard, although the actual figures are not known. Three terrorists were captured. A total of 685 Algerian workers and 107 foreigners were freed.[35] Natural gas output at In Amenas was first partially restarted at the end of February 2013 at one of the three production trains, with the second train returning to service two months later. The third train was out of service at least until 2016. Afterward, BP and Statoil withdrew their staff from In Amenas and the In Salah gas facility (located 373 miles to the West of In Amenas), setting back plans to boost output at both projects.

Hazards for the Public

Oil and gas development holds a broad range of dangers for the public, some more obvious than others. One of the less obvious resulted from the development of the Bakken Formation in the northern Great Plains. Because it was a new site for oil production, there were few existing pipelines. The only alternative was to use railroad tanker cars. This mode of oil transport has a higher potential for public interaction owing to cross traffic, weather challenges, and the consequences of human error.

The Association of American Railroads claims that far less than 1 percent of all derailments in 2017 involved crude oil tank cars, and that "more than 99.999% of all tank cars containing crude oil arrive at their destination without an accident caused by a release."[36] A 2014 article in the New York Times, however, reported that 200 "virtual pipelines . . . snake in endless processions across the horizon daily."[37] Given the amount of oil being transported by train each day, even a very small percentage of derailment incidents can be seriously problematic.

Rail incidents involving crude oil jumped nearly sixteen-fold between 2010 and 2014. There were at least six accidents—usually derailments—in the first four months of 2015, some of them forcing the evacuation of nearby towns.[38]

Concern about the rising incidents of oil rail accidents peaked in recent years in response to the carnage that occurred in Lac-Megantic, Quebec, in the early morning hours of July 6, 2013. At 1:15 a.m., an unattended freight train carrying Bakken formation crude oil rolled downhill from its parked position at Nantes, Quebec. Two boxcars and sixty-six tank cars derailed, resulting in fire and the explosion of multiple tank cars. Forty-two people were confirmed dead from the conflagration. Five more were declared missing and presumed dead. Fifty-three vehicles and forty buildings in the town's center, roughly half of the downtown area, were destroyed. All but three of the thirty-nine remaining downtown buildings needed to be demolished due to petroleum contamination.[39] Most of the fatalities were young people who had been enjoying the early morning hours in the Musi-Café nightclub, tragically unaware of the disaster rolling their way.

In 2015, the New York Times prepared a short documentary film about this growing concern, called A Danger on the Rails.[40] The film points out that one of the problems is that the flashpoint of the type of crude oil being transported is quite low. One of the commentators referred to each of the half million 30,000-gallon tank cars that traverse the country yearly as "rolling bombs."

In any debate weighing "imposed risk" and "received benefit," you should re-member that perspectives differ between the point of origin of the oil and the point of delivery of the oil. While the developer, the rail company, and the buyer all accrue benefits, those living along or near this particular thread of energy also incur risk.

Where available, underground pipelines are a popular alternative to rail transport. Based on cost per ton-mile, they are an efficient, low-cost means of transportation. They are mostly out of sight, mostly immune to the vagaries of weather, and do not have to adjust to cross traffic. These benefits do not mean, however, that they are free of risk or always out of the public eye, as is illus-trated by the opposition to the proposed Keystone XL pipeline, as discussed in Chapter 8. The purpose of the pipeline, as we saw, is to move oil from the tar sands of northeastern Alberta into the middle of the United States.[41] There were several public arguments against the pipeline, with the fundamental opposition resting on a matter of principle more than practicality. That is, building and op-erating the pipeline helped perpetuate our reliance on fossil fuels—in this case, a particularly dirty form of oil.

Public concerns about oil pipelines rise as oil production increases. As one ex-ample, just in the United States, there were more than 2,500 significant onshore incidents involving pipelines carrying crude oil and refined petroleum products in the years 1995–2014 (Figure 12.21). These incidents resulted in $3.4 billion in property damage, according to the federal Pipeline and Hazardous Materials Safety Administration.[42]

Because most oil pipelines are underground, hazards posed by leaks usually are more of an environmental concern than a personal one. For example, a spill from an oil pipeline near Santa Barbara, California, in May 2015, wreaked havoc on sea birds and mammals. In contrast, the impact on humans was mainly con-fined to inconvenience when beaches closed to public use.

Gas pipelines present a somewhat different set of public hazards. In the event of a leak, there is no liquid to spill. The gas simply dissipates into the atmosphere. In its natural condition, natural gas is invisible and odorless.[43] The industry injects an odor, something that is easy to detect, especially indoors. A small gas leak is less of a worry than a large one because there must be a concentration of between 4 and 15 percent gas in the air before it will ignite. Still, anything that can create a spark—even a flashlight—may ignite a fire when those concentrations have been reached. If gas has accumulated and especially if there is a limitless source of gas, such as from a pipeline, explosions can occur.

Natural gas explosions are not uncommon.[44] But some have been so dramatic as to attract massive public attention. One such recent incident resulted from a break in a major 30-inch natural gas transmission line a few miles west of the San Francisco International Airport in 2010. Fifty million cubic feet of natural gas

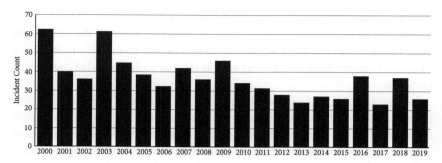

Figure 12.21 PHMSA pipeline incidents, 2000–2019. About 2.6 million miles of transportation pipelines deliver the energy products the American public needs to keep its homes and businesses running. *Source:* U.S. DOT, Pipeline and Hazardous Materials Safety Administration, https://www.phmsa.dot.gov/. For an interactive map of PG&E's gas pipeline transmission systems, see https://www.pge.com/en_US/safety/how-the-system-works/natural-gas-system-overview/gas-transmission-pipeline/gas-transmission-pipelines.page.

spewed into the air. To put that into perspective, the average home in America uses about 1,200 cubic feet per year. Fifty million cubic feet has an energy content of more than 400,000 gallons of automobile fuel or enough electricity to power 1 million homes in Phoenix for a year. The accident killed eight people, injured even more, leveled thirty-eight houses, and damaged many others (Figure 12.22a and 12.22b). The resulting fire burned for over twelve hours and was difficult to control until the gas supply was shut off, a step that took ninety minutes. People living in the affected development had no idea massive volumes of natural gas passed continuously beneath their feet. A similar unexpected mishap occurred in September 2018, when sixty gas-leak explosions occurred in the Boston suburb of Lawrence.

Natural gas hazards seem to be multiplying. The newest concern results from the remarkable success of fracking. In 2015, 300,000 fracked wells produced more than 53 billion cubic feet of natural gas per day.[45] However, although fracking has dramatically increased production of natural gas, some experts suggest that it is "time to pay attention to concerns to develop a more comprehensive understanding and assessment of the environmental impacts of hydraulic fracking."[46]

The problem is that fracking involves forcing fluids, including about 750 chemicals, underground under high pressure. About 650 of these chemicals are known carcinogens, such as benzene and toluene. Fracking techniques have been directly linked to water contamination and increased earthquakes.[47] How to avoid public risks from fracking? The general recommendation to those living

(a)

(b)

Figure 12.22a, 12.22b A leak in a gas transmission line led to a massive explosion and fire in San Bruno, California, on September 9, 2010. It focused attention on issues of public safety and pipeline maintenance. Eight people lost their lives, fifty-one people required inpatient hospitalization, and thirty-eight homes were destroyed. *Source:* Photos by Brocken Inaglory. Used with permission. http://en.wikipedia.org/wiki/2010_San_Bruno_pipeline_explosion#/media/File:Devastation_in_San_Bruno.jpg.

where there are geological formations attractive for fracking is to keep your distance—the farther away, the better.

However, such leaks are only part of the problem with fracking. Induced seismicity is another problem. Forcing fluids underground lubricates faults. Strong earthquakes can occur when these slickened faults move. The most noticeable increase in earthquakes in the United States has been in Oklahoma. By 2014, Oklahoma was experiencing more earthquakes than California, the result of a vast increase in the use of reinjection wells used in hydraulic fracturing.

Uranium

If you are like most people, you are adept at recognizing and avoiding the hazards of everyday life. You do not walk into traffic, jump off cliffs, roll around in poison ivy, or knowingly expose yourself to any of hundreds of other obvious threats to your safety. This learned behavior includes cautious use of energy. The energy hazards that are simplest to avoid are the ones that are most familiar and apparent. You avoid downed power lines, you don't stick your finger into an electric outlet, you don't step in front of a moving coal train, and you would be sensibly cautious visiting a coal mine or an oil drilling rig. However, sometimes the hazards associated with energy are not as obvious.

Take, for example, the hazards that accompany our use of uranium. Uranium has become important in the generation of electricity, especially in the United States, western Europe, and Japan. The 442 commercial reactors that were operating as of June 18, 2018, produced about 10 percent of electricity worldwide and about 20 percent in the United States.[48] And more are under construction, particularly in the fast-growing Asian economies such as in China.[49]

Demand for uranium fuel is therefore expected to rise markedly soon. Including military applications, uranium production totaled 62,071 tons in 2016. Two-thirds of that comes from Kazakhstan, Canada, and Australia (Figures 12.23 and 12.24). Precisely assessing the hazards of nuclear fuel is complicated by three interlaced characteristics of the resource: its radioactivity, its carbon-free nature, and its potential for use in weapons of mass destruction. Any assessment is complicated even further by the relatively short experience we had had with uranium compared with fossil fuels; that is, we might yet discover that nuclear fuels are more dangerous—or, possibly, less dangerous—than we once thought.

Uranium Processing

Uranium requires elaborate processing before it can be used as a fuel. Unlike oil, gas, coal, wood, direct solar energy, or the natural heat of the earth, uranium does not volunteer itself in forms useful for power generation or military purposes. It is rather like copper in this regard. It must be heavily processed before it is ready for use. Such processing leaves behind substantial solids and other

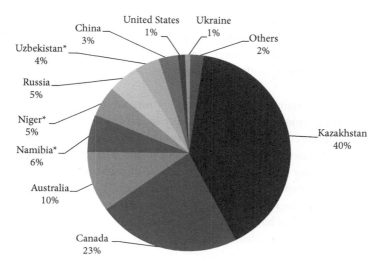

Figure 12.23 Global uranium production in 2016: 62,071 tonnes of uranium (tU). *Source:* https://read.oecd-ilibrary.org/nuclear-energy/uranium-2018_uranium-2018-en#page58.

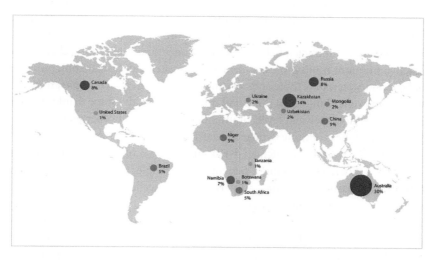

Figure 12.24 Global distribution of identified reserves at a price of approximately US$130 per kg of uranium as of January 1, 2017. *Source:* Nuclear Energy Agency and International Atomic Energy Agency, "Uranium 2018: Resources, Production, and Demand," 2018.

waste products. For example, a short ton (2,000 lbs or 907 kg) of ore yields 1 to 5 pounds (0.45 to 2.3 kg) of uranium, depending on the concentration required. The remnants, called tailings, weigh almost as much as the ore and therefore accumulate quickly. Such residue can retain up to 85 percent of the ore's original radioactivity.

Conventional uranium mining is a risky business. It holds the risk of mine collapse and the unique hazards of exposure to radioactive particles and radon gas. Two categories of human illness can result from such exposure: cancers (of all kinds) and genetic mutations. The mechanisms and severity of biological harm vary in many ways, including the length of exposure and the source of the radioactivity. Exposure to uranium can damage internal organs, cause congenital disabilities, increase the risk of leukemia, and mutate human DNA and chromosomes. Studies have shown that exposure to radon, uranium, and uranium decay elements can also cause cancer of the bone marrow, stomach, liver, intestine, gallbladder, kidney, and skin.[50] These are hazards both to workers and to the public.

Public exposure to the human threats of uranium accompanies several activities, some of which come from "legacy" hazards. Many of the most immediate legacy hazards emanate from abandoned sites of uranium mining and milling. In the United States, these sites are concentrated in the western states, mostly away from urban areas but commonly on or near tribal lands. On the Navajo Nation's lands alone, nearly 4 million tons of uranium ore were extracted between 1944 and 1986. Such mining left behind 500 abandoned uranium mining claims, four inactive uranium milling sites, a former dump site, widespread contamination of land and water, plus thousands of other mine features such as pits, trenches, holes, and homes built from mine and mill site materials. Some drinking water sources now hold elevated levels of uranium, radium, and other metals (Figure 12.25). These sites continue to pose a hazard for all who live nearby.

Sometimes we take a calculated risk and expose ourselves to radioactive materials deliberately, seeking a benefit. For example, nuclear tracers help in medical diagnostic procedures, all part of what is called "nuclear medicine," a specialized area of radiology that uses tiny amounts of radioactive materials, or radiopharmaceuticals, to examine organ function and structure. More commonly, physicians and dentists use x-rays to assess injuries. As with all such tests, you cannot feel, taste, smell, see, or hear the radiation passing through your body. Nor can you sense any possible accumulating damage from repeated exposure.

What is a safe dose? Some would argue that the less exposure, the better—that any exposure carries risk. This view is what is known as the "linear theory."[51] In essence, this theory purports that no level of exposure is safe, that the risk is proportional to the dose.

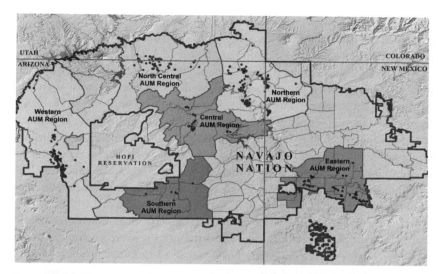

Figure 12.25 Uranium mining occurred in six major areas of the Navajo Nation. The mines are now abandoned. This map indicates many of the 521 sites mapped by the U.S. EPA, but there are likely hundreds more. Source. U.S. EPA, http://www.sric. org/nbcs/docs/2014_Lancet_Uranium%20mines_mistrust%20and%20lung%20 disease%20among%20Native%20Americans_TLRM%20News_Furlow_021914.pdf.

In the early days of the Cold War, military needs outweighed concern for the possible dangers to civilian or industry workers of uranium mining and processing. In part, this approach was a necessity of national defense. Still, it also stemmed from another factor: it often takes a long period—decades, in many cases—before the clinical effects of uranium poisoning manifest themselves. It is similar in this way to black lung disease; by the time the damage is recognized, little can be done to reverse the adverse effects.

Increased incidence of cancer is the primary concern from exposure. Radium, via oral exposure, can cause bone, head, and nasal passage tumors in humans. Radon, via inhalation exposure, can cause lung cancer. Long-term inhalation exposure to uranium and radon in humans can result in respiratory problems such as chronic lung disease. Uranium might also cause lung cancer and tumors of the lymphatic and hematopoietic tissues.[52] Radium exposure can result in acute leukopenia, anemia, necrosis of the jaw, and other effects.

In the United States, the majority of the earliest uranium mining activities took place around the Four Corners region, where Colorado, Utah, New Mexico, and Arizona meet. In the earliest days, many of the hazards associated with handling uranium were not well understood. Nowadays, uranium mining and processing have almost wholly disappeared. Today, given the several years of remediation that have been completed, the most noticeable hints of earlier uranium activities are found in the occasional innocent sign (Figure 12.26).

After several decades during which mining companies and public officials paid little attention to the possible health effects of working with uranium, attitudes began changing. Publicity, evidence, and expediency merged to create legislation aimed at reducing the hazards scattered about the areas of concentrated mining, much of it on Indian lands. Disused uranium mills were the first target for remediation. One example is near Moab, Utah, between the banks of the Colorado River and the red rock scenery of Arches National Park. The Uranium Reduction Company mill—more commonly remembered as the Atlas Minerals plant—started operating during the uranium boom of the 1950s. Moab, until then a quiet Mormon farming town, relished the economic development that accompanied the uranium frenzy that the Atlas plant represented.[53] Some local merchants even renamed their businesses in a supportive response (Figure 12.27). Reflecting a mood of self-promotion and perhaps willful ignorance, the company once declared nearby Moab "Uraniumland." It was a nickname that would be changed once health concerns increased and economic activity shifted to more benign forms. Today Moab is better known as the "Mountain Bike Capital of the World."[54]

Figure 12.26 Sign in Naturita, Colorado, center of decades of extensive uranium mining.

Figure 12.27 Motel sign in Moab, Utah, reflecting the importance of uranium mining and milling to the economic development of the town. Today, the motel name has been changed and Moab has become a mecca for mountain bikers and slick-rock enthusiasts.

Years after the Atlas plant closed in 1984, the building and tailings remained in place. Eventually, after the removal of the buildings, the site was listed as a federal Uranium Mill Tailings Remedial Action project. The company EnergySolutions contracted to remove the 16 million tons of tailings, hauling them by rail to lined disposal pits just north of the junction of U.S. Highway 191 and I-70 near the tiny town of Crescent Junction, Utah. The estimated cost for this job, scheduled for completion by 2030, is between $750 million and $1 billion.[55] (Figures 12.28 and 12.29)

Similar cleanups have taken place at other locations—almost all of which are in the Intermountain West. These include several remediated sites near Grants, New Mexico; another at Mexican Hat, Utah; one at Cane Valley, Utah; and one near Tuba City, Arizona.

Perhaps the most flagrant example of the unexpected hazards of uranium mining comes from Grand Junction, Colorado.[56] From the early 1950s to 1966, Climate Uranium Company donated to the city over 300,000 tons of uranium mill tailings—as a "public service"—for use in construction.[57] Gratefully received, it was innocently used in house foundations, sewer pipes, and many other projects. In a 2009 newspaper article, Sharon Sullivan wrote: "At the time it wasn't widely known, locally anyway, that the waste product was highly radioactive."[58] Soon afterward, Grand Junction became known as the "most radioactive town in

Figure 12.28 Uranium mill in Moab, Utah, alongside the Colorado River. The mill processed 1,400 tons of ore daily on average during its operational life between 1956 and 1984 . It sat idle for two decades before being demolished (see Figure 12.29).

Figure 12.29 Uranium mill site in Moab, Utah, in 2013 after substantial decommissioning and remediation. Sixteen millions tons of mill tailings and contaminated soil were present when the U.S. Department of Energy assumed site ownership.

America." By the time government officials and the public followed that nuclear thread to the end, it had again become clear that the public is not always aware of the hazards that accompany the energy choices they make.

Nowadays, with over half a century of scientific studies and public experience, plus voluntary and mandatory reporting mechanisms in place, the public has access to more information than ever about the hazards of radioactive materials. Greater information, however, does not necessarily lead to greater protection. For example, public dangers continue to exist on Indian lands in many other places, such as the Black Hills of South Dakota, lands of the Lakota Nation. Reportedly, the Black Hills hold

> hundreds of unsealed mines, as well as thousands of exploration wells and drill holes, some more than 200 m deep, scattered across the four-state area. Many are filled with water and there is the constant danger of leaks and spills into the surrounding creeks that can potentially contaminate underground aquifers or the larger Cheyenne and Missouri Rivers. Field studies in the years 1999–2000 found radiation doses of . . . about 200–400 times natural background radiation. . . . To this day, there has not been an adequate scientific workup of the Lakota's health problems.[59]

Legacy hazards are not confined to Indian lands or the American Southwest; similar stories abound in other countries as well. A case in point is Wismut, Germany, where past activities produced the so-called Wismut Legacy. Between 1946 and 1990, 7,163 former uranium miners at that facility died from lung cancer. For 5,237 of them, occupational exposure was recognized as the cause of the disease. The incidence of such cancers has declined in Germany, as it has in the United States, with adherence to safer working conditions and the popularity of new recovery techniques.[60]

In an odd turn of events, however, Germany experienced unintended consequences when efforts were made to reduce hazards to miners. Using advanced ventilation efforts, they lowered the doses of radiation for miners, but exposure levels for residents living near the ventilation exhaust shafts spiked. The problem continued even after the mine shut down, as long as ventilation continued. The problem was traced to the location of the exhaust. Once discovered, the solution was simple. For example, radon levels for residential areas of the town of Schlema in Saxony shrank significantly in April 1992 after the exhausts were relocated to more remote places.[61]

The rising recognition of the health risks of uranium mining has resulted in several pieces of protective legislation in many parts of the world, including the United States. The Radiation Exposure Compensation Act (RECA), for example, specified that not just underground uranium miners but downwinders (i.e., people living downwind of nuclear processing facilities) and nuclear test-site workers were eligible for compensation for harms suffered as a result of exposure to radiation. Later the act was amended to include surface miners, ore truck haulers, and mill workers. On the Navajo Nation, where much of the early uranium mining took place, the Office of Navajo Uranium Workers assisted workers and their families to apply for RECA funds.[62]

Greater ventilation and other measures in uranium mines reduced radon levels. As a result, the average annual exposure of uranium miners has fallen substantially, and this has reduced the risk of occupationally induced cancer from radon. Nevertheless, health hazards remain an issue both for those who are currently employed in uranium mines and for those who have been so employed in the past.[63] It is likely, however, that this particular health risk is also dropping, now that in-situ leaching is becoming the most popular recovery method (Figure 12.30).

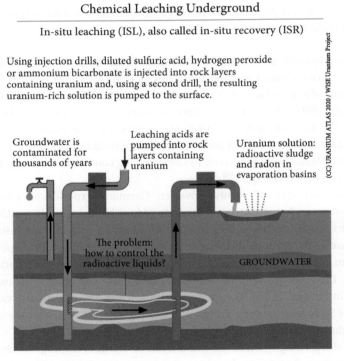

Figure 12.30 The in-situ recovery process circumvents many of the health hazards of underground uranium mining. *Source:* U.S. Nuclear Regulatory Commission, https://en.wikipedia.org/wiki/In_situ_leach#/media/File:NRC_Uranium_In_Situ_Leach.png.

Nuclear Accidents

Further along the fuel chain—after mining, milling, and enrichment—we find the nuclear power plants themselves. Although many studies show that the risks of nuclear power generation are relatively low, these risks tend to receive the most public attention, especially after accidents. Four such accidents stand out:

1. On October 10, 1957, a fire broke out at the Windscale Piles, a nuclear power plant on the northwest coast of England. It has come to be regarded as the worst nuclear accident in Great Britain's history, although the hazards at the time were not well understood. The wind carried the emitted radioactive pollution eastward across Britain, the North Sea, and on into continental Europe. Elevated levels of leukemia have been reported in clusters downwind in Cumbria.

2. About two decades later, in 1979, a small amount of radioactive particles was released after a partial meltdown of Unit #2 at the Three Mile Island nuclear plant near Harrisburg, Pennsylvania. Out of an abundance of caution and a disarrayed public information protocol, tens of thousands left the area in response to an evacuation order.

3. After the April 26, 1986, accident at Chernobyl, in Ukraine, radioactive fallout wafted over much of Europe, the Mediterranean littoral, and European Russia before it circumnavigated the Northern Hemisphere. The accident was responsible for several dozen immediate deaths. According to some sources, there will be thousands of delayed deaths.[64]

4. The March 11, 2011, accident at the Fukushima Daiichi power plant on northeastern Honshu Island, Japan, resulted from a catastrophic tsunami emanating from the Tōhoku earthquake a few miles offshore. The great wave breached the breakwater, destroying units 1–3. Substantial radioactive pollution from the accident was carried eastward over the Pacific Ocean to North America and beyond.

Table 12.4 summarizes the radioactive releases from these four accidents.

These and other less serious accidents continue to fuel the ongoing debate about the future of nuclear power in several countries, including the United States and Japan. Germany, although not the home of a nuclear accident, has chosen to shut down all its reactors.[65]

Nuclear Waste Disposal

In the United States, about 3 million packages of radioactive materials are shipped yearly, either by highway, rail, air, or water. Regulating the safety of these

Table 12.4   Radioactive Releases Compared (in TBq)

Material	Half-life	Windscale	Three Mile Island (compared to Windscale)	Chernobyl	Fukushima Daiichi (atmospheric)
Iodine-131	8.0197 days	740	Much less	1,760,000	130,000
Cesium-137	30.17 years	22	Much less	79,500	35,000
Xenon-133	5.243 days	12,000	–	6,500,000	17,000,000
Xenon-135	9.2 hours	–	25× Windscale	–	–
Strontium-90	28.79 years	–	Much less	80,000	–
Plutonium	–	–	–	6,100	–

Source: https://en.wikipedia.org/wiki/Windscale_fire.

shipments is the joint responsibility of the U.S. Nuclear Regulatory Commission (NRC) and the U.S. Department of Transportation. The NRC establishes requirements for the design and manufacture of packages for radioactive materials. The Department of Transportation regulates the shipments while they are in transit and sets standards for labeling and smaller packages.[66]

The purpose of some of these transport events is nuclear waste disposal—among the last steps in the nuclear fuel chain. Permanently securing wastes from commercial plants remains nuclear power's most vexing, lingering, and unresolved problem. Whereas low-level, long-lived military wastes are being entombed at the Waste Isolation Pilot Plant in New Mexico, no approved site yet exists for high-level civilian waste storage anywhere in the world. Billions of dollars were expended to develop one at Yucca Mountain, Nevada, about 90 miles north of Las Vegas, but that effort has been stymied for over a decade by opposition from various groups and the state of Nevada. In contrast, Finland is actively working on establishing a small one there, the Onkalo spent nuclear fuel repository. For on overview of the consternating matter of nuclear waste, I recommend *The Legacy of Nuclear Power,* an excellent and readable summary of the disposal issue in several countries by the British geographer Andrew T. Blowers. (For specific details on the Finnish efforts, I recommend Vincent Ialenti's 2020 book *Deep Time Reckoning.*)

Meanwhile, spent fuel accumulates with nowhere to go. As a stopgap protocol, the current approach in the United States is to store spent fuel temporarily in dry casks at the sites of the nation's nuclear power plant fleet (Figure 12.31). This

Figure 12.31 Dry casks for nuclear waste storage. After residing in cooling ponds for several years, spent nuclear fuel is removed and stored, usually on site, above ground in dry casks. *Source:* https://en.wikipedia.org/wiki/Dry_cask_storage.

approach sustains the risks to the public from human error, leakage, sabotage, or terrorism. To be fair, the risks to the public are measurably lower than from the documented daily hazards of operating fossil fuel power plants. But the potential consequences of nuclear accidents are more widespread, longer-lasting, and less than fully understood.

From the viewpoint of public safety, protection from the hazards of waste transport and disposal take similar formats, with each using a multilayered approach. Again, while the risks of public exposure may appear to be slight, they are not inconsequential, especially when viewed from a political perspective. Moreover, the vast monetary commitment already made to study and mitigate risk from these activities suggests that the public remains worried. Indeed, transportation and especially the disposal of radioactive wastes have played a significant role in slowing the expanded use of nuclear power in many countries.

Transmission of Electricity

The intended product of most power plants is electricity, the most visible thread of energy. Maintaining these lines is hazardous. According to the Electrical Safety Foundation International, the construction industry had the highest rate of fatal electrical injuries (0.7/100,000) followed by utility (0.4/100,000) in

2019.[67] The four main types of electrical injuries are electrocution (fatal), electric shock (non-fatal), burns, and falls resulting from contact with electrical energy.

Health risks also exist for the public, such as coming into contact with lines downed by weather events or accidents. The public generally avoids such hazardous situations, but they may not be aware of other hazards of electrical transmission. The principal question is whether it is dangerous to just be in physical proximity to such power lines. The fact that millions of people live near transmission lines of various voltages suggests that the perceived risk is low. On the other hand, public protests against proposed power line projects are often vociferous, usually related to aesthetic intrusion.[68]

The other concern is electromagnetic fields (EMF), physical fields produced by electrically charged objects. All transmission lines produce EMFs. If you have ever been listening to AM radio while driving under such lines, you have experienced EMF as a crackling sound or buzz from the radio. Moreover, if you stand under them in an otherwise quiet area, you can hear their hum. If it is windy, the noise is even louder.

You can demonstrate the existence of EMF to yourself by simply holding a fluorescent tube in your hand while standing under such lines at night; the tube lights up even though it is not connected to any power source. Richard Box, once an artist-in-residence at the University of Bristol in England, incorporated this phenomenon into his art when he installed 1,301 fluorescent tubes in a regular pattern beneath high-voltage lines. The electric field set up inside the tubes excited atoms of mercury gas, making them emit ultraviolet light. This invisible light strikes the phosphor coating on the glass tube, making it glow (Figure 12.32).[69]

While dramatic and memorable, does this demonstration prove that EMF poses a public hazard? It might not matter. A 2015 survey found that 85 percent of respondents were concerned about the health effects of transmission lines.[70]

Some jurisdictions are taking no chances, at least with their young people. Because of possible associations between magnetic field exposure and childhood leukemia, for example, the California EMF Program contracted with Enertech Consultants in 1994 to assess the electric and magnetic field exposure from power-line and non-power-line sources for California public school environments. In response to the published findings of this study, the California Department of Education enacted regulations requiring new schools to be separated from the edge of transmission line rights-of-way (the area immediately surrounding the power line). Specifically, the regulations require a setback of 100 feet for 50–133 kV lines, 150 feet for 220–230 kV lines, and 350 feet for 500–550 kV lines.[71]

In 2014, the California State Department of Education reaffirmed its guarded position on the safety of transmission lines: "Research continues on the effects of electromagnetic fields (EMF) on human beings. However, school districts should

Figure 12.32 Richard Box's *The Field*, demonstrating the invisible effects of electromagnetic radiation that emanates from the overhead transmission lines and stimulates the gases within the fluorescent tubes. *Source:* http://www.richardbox.com/archive.htm.

be cautious about the health and safety aspects relating to overhead transmission lines. School districts should take a conservative approach when reviewing sites situated near easements for power transmission lines." However, such caution is not evident everywhere (Figure 12.33).[72]

In November 2018, a different hazard of transmission lines attracted international public attention in tragic and deadly form. Improperly maintained electrical transmission lines, it is suspected, sparked a conflagration in northern California that destroyed more than 6,700 buildings, almost all of them homes, and killed about ninety people in the small retirement community of Paradise. So massive was the damage and so high was the legal liability to the power line operator, Pacific Gas and Electric Company, that the company admitted liability and declared bankruptcy.

Summing Up

Despite the risks and hazards of energy, it is doubtful that you or anyone you know will be abandoning the use of energy resources anytime soon. Nevertheless,

Figure 12.33 Is this such a good idea? Volleyball court directly beneath multiple transmission lines, which are close to houses, subjecting people to the possible hazards of electromagnetic fields. Gilbert, Arizona, 2011.

each step in the fuel chains brings both costs and benefits. You must attempt to minimize the former, just as you strive to maximize the latter.

What should you conclude about the hazards and risks of energy production and use? Here are a few ideas to consider: (1) Be aware of both sides of the cost-benefit equation and take full responsibility for the risks and hazards energy poses throughout the entire cycle of its development and use. Do this both for the well-being of yourself and your family, but also for those you do not know who live elsewhere in the world. (2) Always keep the risks and hazards of your energy use in mind. To put it simply, use your head. (3) Minimize your risks by adhering to the principle of "prudent avoidance"—there is no need to take more risks than is necessary. (4) Incorporate the personal and community risks and hazards of energy when you decide which resources you favor. In other words, don't let price and convenience alone dominate your decisions.

Energy resources are valuable and helpful to all of us. No one, regardless of their economic station, can live securely without affordable, available energy. As we have just learned, however, the benefits of their use are tightly tied up with their risks. If the risk gets too high, we may curtail or abandon the use of a particular resource. These factors are already evident in our reduced use of coal, our reticence to expand the use of nuclear power, and the increasing popularity of renewables. In this way, the relative risks and hazards of our energy choices are influencing the transition to a more sustainable energy future.

13

Business

Jobs and the Workspace

The Business of Energy

Energy is the hub of the life you can live. When you have enough, your life can feel secure and comfortable. When you do not, your life can be fraught with worry and suffering. At every twist and turn of your life, you need reliable energy supplies to satisfy your never-ending energy demands. You understand these relationships because you are actively and directly involved in operating your personal energy system every time you turn on your lights, start your car, or pay your energy bill. What might not be as apparent to you is that energy is also the critical ingredient supporting whatever business you are in, whether you make concrete, manage a corner grocery store, or sit in a corporate office tower. I label these relationships as "energy in business." This is the demand side of the equation.

Knowing that businesses of every type create some degree of energy demand, we might ask where that energy—in whatever form—originates. In the future, we may produce some of our energy we need on-site, but that is not true yet. For now, just about all the energy that every business requires comes through the efforts of millions of people who work within a large and interwoven assemblage of operations that I call "the business of energy." This is the supply side of the equation.

Imagine your life without energy. You leave your cold, dark house intending to drive to work, only to find you cannot start your car because of the cold outside. You reenter your home to try to do some work only to find that without power, your computer has become little more than an expensive paperweight. You cannot watch TV, surf the internet, play video games, or charge your cellphone.

While you might gasp at this scenario, don't worry; it is not likely to happen anytime soon, at least to most of you. Why? The answer is that you likely are living in the Global North. If you are, thousands of people and the companies they work for toil ceaselessly to keep the energy coming to your residence every hour of every day without interruption. That is your reality. Enjoy it. You are a lucky person.

The Thread of Energy. Martin J. Pasqualetti, Oxford University Press. © Oxford University Press 2021.
DOI: 10.1093/oso/9780199394807.003.0013

Even though you probably can satisfy all your energy needs by flipping a switch or by visiting a fueling station, it is unlikely that you give much thought to the vast network that is in place to keep supplies coming. What you probably *do* realize is that the energy you use is not free. You have to pay for all of it. Do you wonder where the collected revenues go? The answer is that the revenues gradually thread their way from one company to another and another, on and on, until everyone gets paid for the role they play in the world's largest enterprise. They are all a part of the business of energy.

While everyone needs energy, not everyone has access to the commercial variety that you use. Indeed, not even half the population of the world gets all the energy it needs. One billion people get no electricity at all for one reason or another. Sometimes the energy is simply not available at any price. Other times it is available but unaffordable. For these people, the notion of a commercial energy system is, at best, an unreachable mirage.

Even for those who *do* have full access to all the energy they need, the many bits and pieces that make it available can be akin to a black box. Let's take a look inside.

Energy Is Big Business

Commercial energy is big business. Collectively, the elements of the energy chain—exploration, extraction, transportation, processing, generation, distribution, and waste disposal—merge into the world's most valuable commercial enterprise. Think of oil as bread. One slice is exploration. Another slice is production. A third slice is transportation. A fourth slice is processing. A fifth slice is distribution. A sixth is waste disposal. The entire loaf feeds energy-starved countries and job-hungry people everywhere from Austria to Zambia.

The business of energy is also varied. It is technical, political, economic, critical, and vastly profitable. The Chinese oil company Sinopec had 2018 revenues of over $400 billion; Saudi Aramco's were about $360 billion; the largest of the U.S. oil companies, Exxon, had about $280 billion in revenue. In December 2019, Saudi Aramco stock briefly reached the $2 trillion mark, making it half again as valuable as Apple, more than twice the value of Amazon, and four times the value of Facebook (Figure 13.1).[1]

The profit accumulation motive of these companies compels them to grow to a large size, with vast investments in human and financial capital. In turn, their size can underpin industrialized societies and provide millions of jobs. In the United States alone, the oil and gas industries support over 7 million jobs; one in twenty jobs is in the energy sector. This sector provides the energy we need, where we need it, when we need it, in the amount we need it, and it does so with

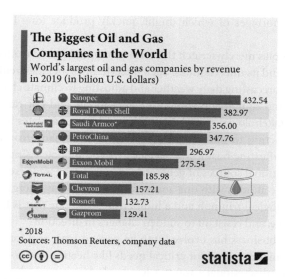

Figure 13.1 The biggest oil and gas companies in the world, January 2020. *Source:* Katharina Buchholz, "The Biggest Oil and Gas Companies in the World," January 10, 2020, https://www.statista.com/chart/17930/the-biggest-oil-and-gas-companies-in-the-world/.

a degree of reliability that belies its size, intricacy, and the internal shenanigans that seem inescapable whenever there is a prodigious amount of money involved and a demand that never ends.

Part of the challenge of being in the energy business is trying to match energy supply with energy demand in real time. Another is to be able to match supply with demand over weeks and months. When there is more supply than demand, prices usually go down, as they did for oil in 2020. In the first quarter of that year, demand was much lower than supply, and the price of oil was actually less than zero for a time. By contrast, when there is more demand than supply, prices rise, sometimes so high that some people have to choose between heating their home and buying food.

Making energy resources available at the time and in the form and place they are needed involves several specific steps, each one attuned to the whims of users and the peculiarities of the energy resource. The role of the coal companies, for example, is to separate coal from the rock that enfolds it. Other companies, such as railroads, then move that coal from mines to power plants. Still other companies receive, crush, and burn the coal to make electricity. The smooth operation of each step is dependent upon the equally smooth operation of every additional step in meeting the demand that exists at that moment (in the case of electricity) or gasoline (shortages of which might quickly produce long lines at fueling stations).

These examples merely scratch the surface of fundamental realities in the energy business and its uncertainties. Such uncertainties include predicting supply and demand; forecasting daily weather and adjusting to climate change; building and maintaining equipment; computing appropriate and fair rates; communicating with stakeholders; complying with legislative mandates and public expectations; reacting to unforeseen acts of nature, human error, and political vagaries; and adjusting business strategies to accommodate rising contributions of non-conventional sources of generation. Considering all these factors, many of which can be at odds with one another, the present-day energy business is at once a technological triumph and a lumbering beast, one that is hard to corral, difficult to tame, and resistant to sudden changes in direction.

The energy business has evolved into its present scale and form in response to growing demand for fuel for critical needs like heating and cooking, but also in response to discretionary wants such as electric carving knives and dune buggies. Whether for essentials or playthings, the business of energy is that sector of the economy dedicated to supporting the lives we live and the lives we aspire to achieve.

The Massive Energy Sector

We begin our look at the energy sector by gazing into the future, a future trending toward worldwide trade, travel, communication, responsibility, and the prospects of a full-fledged global economy of interdependence and mutual long-term goals. As is the case now, in the future energy will be the thread that holds it all together.

We see the marks of energy wherever we look. It is energy that in just hours moves people and products over distances that took weeks or longer just a century ago. It is energy that allows Yankee sluggers to pick up a 90-mile-per-hour fastball coming at them after sunset and construction workers to repair freeways after everyone else is asleep. It is energy that powers our mobile phones and the internet. It is energy that cools our homes in summer and warms them in winter, that lifts us to the top of the tallest building without walking up a single step, that carries us across the city in cars and across the country in airplanes.

So routinely available is energy that we have little awareness of the complexities of its supply. So intuitively valuable is energy that we willingly compensate others to make sure it is available wherever and whenever we want it, and in any form and amount we desire. Where would we be without energy?

Without modern supplies of energy, our lives would better resemble the time of Abraham Lincoln. These would be lives with few if any appliances, no lightbulbs, no motor vehicles, no high-rise buildings, no movies, no air-conditioning, no airplanes, no mobile phones, no televisions or radios, no internet. How would you feel? Could you cope? You would probably feel not only deprived but confused. (It should be pointed out that millions of people on the planet still live that way.) Because few of us wish to give up the modern convenience that we now enjoy, energy demand keeps spiraling upward. Moreover, because energy demand keeps rising, so too does the number of people needed within the energy workforce to meet this demand.

Those involved in the business of energy face a daunting challenge. Consider the following numbers. Consumers in the United States use about 100 quadrillion Btu (100 quads) of energy each year. Each Btu is equivalent to the amount of energy released by burning a wooden match. To give you a better appreciation of the scale of this energy demand, consider laying 100 quadrillion matches end to end. It would exceed the diameter of our solar system! Now imagine the heat released by 100 thousand trillion matches going up in flames every year. With such images in mind, you perhaps can appreciate that the total energy demand of the United States is large indeed. Moreover, the United States represents just 24 percent of the world's demand for commercial energy.[2] (Also worth noting is that the United States has just 5 percent of the world's population—but uses 24 percent of the energy available on the worldwide marketplace.)[3]

Energy development is big business. Just for starters, it is the most capital-intensive industry in the United States.[4] It has created an infrastructure of breathtaking reach, intricacy, and capital. Plus, its growth continues accumulating like a snowball rolling downhill. Just for 2016, the U.S. industry's capital investments exceeded $135 billion. That amounts to more than twice what it was a decade earlier. In total, the industry's economic impact is $880 billion annually (approximately 5 percent of the nation's total GDP).[5]

To take another example of the business of energy, let us look just at oil. Even at the relatively low mid-2019 price of $50 for a barrel of oil, it multiplies out to $4.5 billion per day worldwide or almost $2 trillion per year. To reiterate, this is just for oil, *not* all the energy we need. If we consider all energy, the number is much larger. The global energy consumption sector grew to a value of about $9.1 trillion in 2015. The Asia Pacific region is emerging as the largest regional market, accounting for more than 50 percent of the global market's value. (In 2017 the worldwide GDP was about $80 trillion.) The $9.1 trillion value for energy roughly equals the combined GDP of Japan, Germany, and the United Kingdom. China and the United States form the two largest national energy consumption markets. They are valued at $2.7 trillion and $1.2 trillion, respectively. Most experts expect China to continue growing at a fast pace once the global worries of the Covid-19 pandemic subside.

The business of energy has many moving parts. Imagine combining every element of every energy resource with every aspect of every possible type of energy-related business. Then, once you have completed that task, mix every-thing thoroughly. What would result? Thousands and thousands of com-binations, interactions, dependencies, and symbioses. There are so many combinations that it is impossible to expose or explain all of them here. However, what we *can* do is provide a realistic impression by considering, one at a time, the several stages of the energy fuel chain by melding each phase—exploration, extraction, processing, transportation, combustion, transmission, and waste disposal—with each energy resource. Considered together, the fuel chain creates business opportunities, challenges, risks, and rewards across a broad spectrum of activities. And millions of jobs.

For a broad view of energy as big business, let us consider the job market as outlined in the 2020 "U.S. Energy and Employment Report" (USEER). The four sectors of the USEER report are (1) electric power generation and fuels, (2) transmission, wholesale distribution, and storage, (3) energy efficiency, and (4) motor vehicles. Together these four sectors account for almost all the U.S. en-ergy production and distribution systems, plus roughly 70 percent of U.S. energy consumption. Here's a summary of some findings:[6]

- Approximately 6.8 million Americans work in traditional energy and en-ergy efficiency industries. These industries added over 120,000 net new jobs in 2019, accounting for 14 percent of the nation's job growth.
- Oil and natural gas employers added the newest jobs—more than 18,000—employing 615,500 and 276,000, respectively.
- Electric power generation jobs gained about 21,000 jobs in 2019.
- In 2018 and 2019, coal generation employment declined by over 14 percent, losing over 13,000 jobs.
- Jobs in transmission, distribution, and storage (our energy infrastructure) totaled about 1.4 million in 2019.

Should you wish to learn about an individual state, those data are also available for the first time in the 2020 USEER. State reports highlight the growth and loss of jobs in specific energy sectors across the country, thereby providing a useful tool for state energy offices and economic development agencies. Such profiles demonstrate the unevenness of growth in new energy technologies. If, for ex-ample, you live in Arizona, you should know that Arizona has a low concentra-tion of energy employment, with 46,951 traditional energy workers statewide (representing 1.4 percent of all U.S. traditional energy jobs). Of these traditional energy workers, 24,080 are in electric power generation, 2,095 are in fuels, and 20,776 are in transmission, distribution, and storage. The traditional energy sector in Arizona is 1.7 percent of total state employment (compared to 2.3 per-cent of national employment). Arizona has an additional 44,782 jobs in energy efficiency (1.9 percent of all U.S. energy efficiency jobs). Solar and wind energy produce about 10,000 jobs. Energy Star and efficient lighting plus high-efficiency and renewable heating and cooling produce another 21,000 jobs.

The neighboring state of California has an especially robust energy sector, one that results not just from its greater population but also because of the much stronger state energy policies, especially those that favor renewables and efficiency. It is a good place to look for energy jobs. California has an average concentration of energy employment, with 411,811 traditional energy workers statewide (representing 12 percent of all U.S. traditional energy jobs). Of these traditional energy workers, 182,559 are in electric power generation, 77,049 are in fuels, and 152,204 are in transmission, distribution, and storage. The traditional energy sector in California is 2.3 percent of total state employment (compared to 2.3 percent of national employment). California has an additional 323,529 jobs in energy efficiency (13.6 percent of all U.S. energy efficiency jobs). About 125,000 jobs are in solar energy.[7]

Not surprisingly, given the concentration of major oil and gas ventures that are located there, the leading state for energy jobs is Texas. There are 607,626 traditional energy workers statewide (representing 17.8 percent of all U.S. traditional energy jobs).[8] Of these Traditional Energy workers, 58,476 are in electric power generation, 344,256 are in fuels, and 204,894 are in transmission, distribution, and storage. The traditional energy sector in Texas is 4.8 percent of total state employment (compared to 2.3 percent of national employment). Texas has an additional 169,398 jobs in energy efficiency (7.1 percent of all U.S. energy efficiency jobs). Wind makes up the largest segment of employment related to electric power generation, with 25,507 jobs, a result of a major wind resource in West Texas and an aggressive policy to develop it.

Exploration and Extraction

Exploration and extraction are two early steps in the energy fuel chain. Once we identify the location and extent of energy resources, two groups are called to action. One group focuses on accessing and recovering oil and gas by drilling wells that bring the product to the surface of the land. The other group focuses on the extraction of the solid fuels of coal and uranium by using underground and open-pit mining procedures.

Changes in Techniques

Recovery techniques have evolved and improved over the years for each group, although they may seem fundamentally the same to outsiders. Oil and gas recovery still rely on wells, with hydraulic fracturing (fracking) increasing the productivity of such wells in recent years. Coal and uranium extraction have shifted away from the age-old methods of underground mining. Today, the majority of coal mining in the United States uses the safer and more profitable strip (surface) mining methods that became popular once large excavating machines became available, as in Wyoming.[9] Uranium extraction has evolved from open pits and underground mining to the use of in-situ leaching (ISL). Aside from ISL techniques, the most visible shift in resource recovery is the now-popular brute-force practice of mountaintop removal. Literally, it uses explosives to blow off mountain tops and expose coal hidden beneath, an effective if extremely damaging approach.

With each improvements in mining, safety and productivity, per-unit costs have gone down, and job opportunities have shrunk. Idled miners retired or, if feasible, retrained in adjusting to these changes. As of 2017, there were about 756,000 total mining and extraction jobs in the United States. Of these, 62 percent involved oil, gas, coal, and nuclear fuel stocks, producing about 468,000 jobs.[10] Surprisingly, this is approximately equal in number to those employed in farming, fishing, and forestry occupations.[11]

Thanks to fracking, the extent of extraction in the United States became an impressively robust and widespread enterprise. For example, based on data from 2014–2015, three out of four states had drilling activity, accounting for approximately 1.2 million facilities across the country. These activities include well drilling, active production wells, natural gas compressor stations, and processing plants.[12] There were about 18,500 wells in the United States in 2017, a number that is almost half of the wells drilled *worldwide* in the same period. But, as the price per barrel of crude oil fell from $66 per barrel in December 2017 to a bit under $20 per barrel in April 2020, drilling slowed as well.[13] Many hundreds of wells were closed owing to weak market demand, again illustrating the volatility of employment opportunities in much of the energy sector.

Before this recent downturn, the improvements in drilling technology had led to increased employment, as long as the price of fuel held above $50 a barrel. All totaled, the industry was supporting 9.8 million jobs or 5.6 percent of total U.S. employment, according to the accounting firm of PricewaterhouseCoopers. In fact, due in large part to the oil and natural gas industries, the Texas state comptroller estimated in 2017 that the state had recovered 100 percent of the jobs lost during the Great Recession that started in 2008. It added 597,000 jobs across the entire state economy above the previous peak in August 2008, with 178,000 people employed just in oil and gas extraction as of August 2017.[14] Price wars within OPEC, oversupply of product, and the Covid-19 pandemic reduced job opportunities, and thousands of workers were laid off. Nonetheless, oil and gas production remains a major employer in the United States.

Coal

In contrast to the impressive statistics for enhanced oil recovery, coal production is in decline in many parts of the world. In the United States, coal production fell from 1,00,049 tons in 2014 to 534,302 tons in 2020.[15] Lower production reflects a declining demand for coal in the domestic power sector and lower demand for U.S. exports (Figure 13.2).[16]

The declining fortunes for coal result from the increasing rate of retirement of coal-burning power plants. More than 16,000 MW of coal-fired capacity in the United States were retired through November 2018. An additional 17,000 MW in retirements is expected by 2025, according to the EIA. For a reference to the scale of this reduction, there was 340,000 MW of capacity in 2011 and 249,000 MW in 2019.[17] (China has been going in the opposite direction, however, with 1 million MW of capacity in 2019, up from 700,000 MW in 2011.)

As a direct result of the decline in the demand for coal, coal mining jobs have also been on a steep downward trajectory. From 860,000 employed in the U.S. coal mining industry in 1923, the total dropped over 90 percent to about 103,000 in 2015 (Figure 13.3). By 2017, the number employed in the coal mining industry had dropped even further, to about 50,000, or less than the student population of many university campuses.

When we follow the coal to the power plants, we also see a follow-on decline in employment there as well. Coal-fired electric power generation employed a total of 79,711 workers across the nation in 2019. This represents a nearly 8 percent decrease in jobs from 2018. Electric utilities held almost half (48 percent) of coal jobs in 2019, having lost approximately 7,700 workers in the previous year.

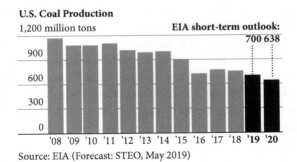

Figure 13.2 The precipitous decline of coal production from almost 1,200 million tons in 2008 to 638 million tons in 2020. *Source:* EIA Short-Term Energy Outlook, May 2019.

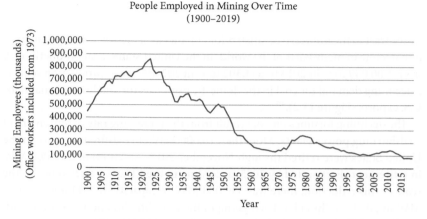

Figure 13.3 In 1923, 862,536 coal miners were working in the United States. By 2015, there were only 102,804, a drop-off of almost 90 percent. The decline continues. *Source:* http://www.energytrendtracker.org/2017/01/century-long-decline-for-u-s-coal-jobs/.

In contrast to what we have been witnessing in the burgeoning renewable energy sector, employment in coal power plants is withering.

As alarming as those declining numbers are for the coal mining industry, direct job losses are only the most visible manifestation of the dilemma. The lost mining jobs radiate their influence outward to the loss of jobs in the supply chain and supporting businesses. For example, in an increasingly rare bid for a (small) new coal mine in the Midwest, the owners calculated that hiring 300 direct employees would create 3.88 times that number of jobs in indirect employment. In other words, 300 mining jobs created 1,464 total jobs. If we extrapolate that to the 50,000 jobs lost in the three years prior, it means over 200,000 total lost jobs and close to 1 million people seriously impacted when we include family members. In a desperate search for any possible new means of economic development, coal mining areas have been willing to try just about anything. In southeastern Ohio, for example, the substitute source of income has been marijuana.[18] Undoubtedly, this new emphasis will alter the social and cultural fabric of life in this and similar regions in the coming years. Whether the change will be for the good is another matter.

As underground coal mining in Appalachia continues to disappear, you might ask whether there is anywhere in the coal-rich areas of the United States where coal mining has been less affected. has been holding steady. The answer, at least for now, the strip mines in Wyoming, a state that leads all others in coal production, are continuing to be productive, although even its coal production dropped 23 percent between 2012 and 2018.

Coal mining in Wyoming holds many advantages over coal mining in other states. The coal is low in sulfur, close to the surface, much cheaper to mine, and is in uncommonly thick beds. For all these reasons, strip mining years ago supplanted sub-surface mining in national importance. Currently, Wyoming produces six times more coal than West Virginia at one-fifth of the cost (Figure 13.4). Lower costs of production allow long-distance transport to market; Wyoming supplies consumers in thirty states. Wyoming coal is so low in sulfur and ash and so inexpensive to mine, it can even be transported economically for use by customers in Asia.[19]

Natural Gas

Even given the advantages of Wyoming coal, the use of coal is on the wane because natural gas is cheaper and cleaner-burning (Figure 13.5). Hundreds of thousands of natural gas and gas condensate wells dot the landscape. Recently, each well has been producing more product, and this increase has allowed a drop in rig count of 80 percent in recent years, while output has climbed 35 percent.

Figure 13.4 Open-cast lignite mine on the outskirts of Most, Czech Republic. Lignite is a low-grade dirty coal found in massive deposits in many parts of Europe.

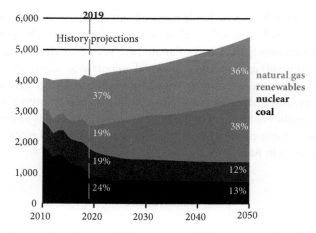

Figure 13.5 U.S. electricity generation by fuel type through 2050, showing the expected continued drop in nuclear and coal and the substantial rise in natural gas and renewables. *Source:* Adapted from Annual Energy Outlook 2020, https://www. eia.gov/outlooks/aeo/.

This increase is all part of the "shale revolution," and it is reflected in the increased job opportunities within this sector. Natural gas electric power generation employed a total of 121,812 workers across the nation in 2019; of these, 75,661 jobs, or 62 percent, are in the category of advanced/low-emissions natural gas generation. Over the preceding year, natural gas electric power generation added more than 9,100 jobs—an 8 percent increase, exceeding the predicted growth of 5 percent.[20] The development of America's vast shale gas reserves could add more than 1 million U.S. manufacturing jobs by 2025, according to PricewaterhouseCoopers. Employment in the oil and gas extraction industry, however, can be volatile, as demand fluctuates (Figure 13.6).

Uranium

Enthusiasm for nuclear power continues to decline in several countries. Nuclear power plant construction in the United States has been almost at a standstill for decades, as reflected in employment within the nuclear industry. Nuclear electric power generation employed a total of 60,916 workers across the nation in 2019. Over the previous year, more than 2,000 jobs were lost from nuclear generation—a decrease in employment of more than 3 percent. And it does not look as though a reversal in this trend is likely.

People Employed in Oil & Gas Extraction Over Time
(2010–2020)

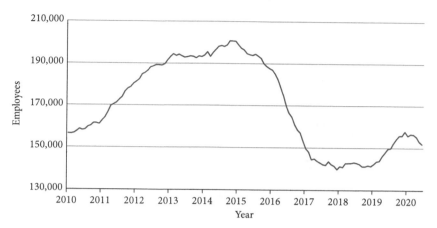

Figure 13.6 All employees working in oil and gas extraction (in thousands), seasonally adjusted. *Source:* U.S. Bureau of Labor Statistics, https://data.bls.gov/timeseries/CES1021100001?data_tool=XGtable.

Despite great sums dedicated to a hoped-for nuclear power renaissance, Vogtle units 3 and 4 in Georgia are the only power plants under construction, albeit with massive cost overruns. Germany and France have signaled their desire to reduce reliance on nuclear power, and Japan has idled most of its power plants in reaction to the meltdown and explosions at Fukushima in 2011. New Zealand and Australia have banned nuclear power entirely. This downward trend is not found everywhere. According to the International Atomic Energy Agency, more than four dozen reactors are under construction—fourteen of them in China and six in India.[21]

Uranium mining also continues. Although the source of the uranium fluctuates a bit, more than two-thirds of the global production comes from Kazakhstan (39 percent in 2016), Canada (22 percent), and Australia (10 percent). If we broaden our search, we find that more than half of the world's uranium production comes from just ten mines. Canada, the second-biggest uranium producer in the world, hosts the world's biggest mine—the McArthur River uranium mine. Canada also has the highest grade of uranium, at almost 18 percent U_3O_8 (the most-desired form). The mine with the next-highest grade, Rabbit Lake, produces uranium with about 1 percent U_3O_8 (Figure 13.7). Kazakhstan, the largest uranium-producing country, operates three of the ten top-producing mines. The uranium industry supplies the 440 commercial reactors in the world

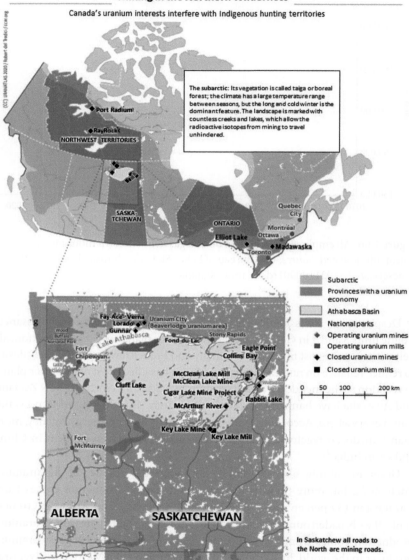

Mining in the Northern Wilderness

Canada's uranium interests interfere with Indigenous hunting territories

The subarctic: Its vegetation is called taiga or boreal forest; the climate has a large temperature range between seasons, but the long and cold winter is the dominant feature. The landscape is marked with countless creeks and lakes, which allow the radioactive isotopes from mining to travel unhindered.

Port Radium

RayRock

NORTHWEST TERRITORIES

SASKA-TCHEWAN

QUEBEC

Quebec City

ONTARIO

Montréal

Elliot Lake

Ottawa

Madawaska

Toronto

Subarctic

Provinces with a uranium economy

Athabasca Basin

National parks

◆ Operating uranium mines

■ Operating uranium mills

◆ Closed uranium mines

■ Closed uranium mills

Fay-Ace-Verna

Lorado

Gunnar

Uranium City (Beaverlodge uranium area)

Wood Buffalo National Park

Stony Rapids

Fort Chipewyan

Lake Athabasca

Fond-du-Lac

Eagle Point

Collins Bay

McClean Lake Mill

McClean Lake Mine

Cluff Lake

Cigar Lake Mine Project

Rabbit Lake

McArthur River

Key Lake Mine

Key Lake Mill

Fort McMurray

0   50   100   200 km

ALBERTA   **SASKATCHEWAN**

In Saskatchew all roads to the North are mining roads.

Figure 13.7 Uranium mining in Canada. *Source:* Uranium Atlas.

as of March 2020, making up a significant sector of the business of energy. So much uranium is available, however, that an oversupply continues to suppress the yielded price. As of June 2021, the price was $32.30 per pound.[22]

During the 1990s, takeovers, mergers, and closures consolidated the global uranium production industry. This trend toward consolidation then switched direction, and the industry diversified again with Kazakhstan's multinational ownership structure. Over half of uranium mining production is from state-owned mining companies in Kazakhstan, some of which prioritize secure supply over market considerations. In 2015, eleven companies marketed 89 percent of the world's uranium mined production.[23] KazAtomProm, the world's largest uranium mining company, is headquartered in Kazakhstan, and it is a profitable business, as attested in its financial statements.[24]

Cameco is the world's largest publicly traded uranium company, with over 3,000 employees internationally in 2012 and revenues of $2.4 billion Canadian in 2013. The present low price of uranium is, however, causing the company to shrink operations. By the end of 2018, Cameco laid off 845 workers (temporarily, according to the company). The affected employees were from the company's McArthur River mine and Key Lake milling operations in northern Saskatchewan.[25]

Three methods of uranium mining prevail at present: (1) traditional underground mining, (2) surface mining, and (3) in-situ leaching. Each method produces a different surface manifestation, and each has its own employment requirements. The trend has been moving toward ISL, both because it affords greater safety and because it is less labor-intensive. ISL is now responsible for 48 percent of uranium production, although this technology can be used only for uranium deposits located in an aquifer in permeable rock that is itself confined in non-permeable rock.[26] Today virtually all Kazakh—and most U.S. uranium production—is from ISL mining.

Taken together, the business of uranium mining fluctuates with the present and potential contribution of nuclear power on the global scene. Again, such closures result in decreased demand for uranium mining, although these activities may increase once the power plants being constructed in China go online.

Processing

Many native forms of energy are subjected to extensive processing after original extraction to meet specific demand specifications best and to increase their market value. Oil and uranium offer the best examples.

Oil

Designing, siting, constructing, and safely operating oil refineries are parts of an expensive and complicated series of steps that—even though sometimes dangerous—support thousands of jobs, generate large profits, and produce countless consumer products from TNT to perfume.

The most common product of oil refining is motor fuel, and it is expensive to produce. A refinery of average complexity processing 100,000 barrels of crude oil per day may cost a billion dollars (or more) to build. Economic viability depends on minimizing operating costs. Profits from refining are usually expressed as the gross refining margin (GRM). We can derive the GRM of a refining company by subtracting the cost of the crude oil it consumes from the total market value of refined products it produces. Refining margins are thus dependent on the cost of input crude, the types of products produced, and the prices of those refined products.

The refining margin is an indicator of the overall profitability of a company's refining operations.[27] The margins (or profitability) are calculated on a per-barrel basis. A barrel of crude can produce an entire range of so-called fractionates such as gasoline, diesel, LPG (liquefied petroleum gas) and furnace oil. The price of each of these is different, and the GRM for a refinery is higher if it produces more high-value products.

Refining oil is a business proposition, meaning that costs should be low and sale price high. Keeping refinery operating costs as low as feasible within the margins of safety and competition starts with understanding the wide variety of component costs that keep refineries in the black. Such operating expenses fall into roughly nine categories, each of which supports a secondary set of businesses. The categories are:

- *Personnel cost.* Includes salaries and wages of regular employees, employee benefits, contract maintenance labor, and other contracted services.
- *Maintenance cost.* Includes maintenance materials, contract maintenance labor, and equipment rental.
- *Insurance.* Needed for the fixed assets of the refinery and its hydrocarbon inventory.
- *Depreciation.* Assessed on refinery assets: plant machinery, storage tanks, marine terminal, and the like.
- *General and administrative costs.* Includes all office and other administrative expenses.
- *Chemicals and additives.* The compounds used in processing petroleum and final blendings, such as antioxidants, anti-static additives and anti-icing

agents, pour point depressants, anti-corrosion agents, dyes, water treatment chemicals, and so forth.

• *Catalysts.* Proprietary catalysts used in various process units.
• *Royalties.* Paid either in a lump sum or as running royalties for purchased know-how.
• *Purchased utilities.* May include electric power, steam, water, and so on.
• *Purchased refinery fuel.* May include natural gas purchased by the refinery for use as a refinery fuel and feedstock for hydrogen production.

These operating costs can vary substantially among the 135 operating oil refineries in the United States. For example, for the massive Exxon Baytown refinery in Texas, which employs 3,000 people in the refinery alone and twice that number nearby, and handles 584,000 barrels per day (bpd), the cost is $3–4 per barrel. At smaller operations, such as the 155,000-bpd refinery in Torrance, California, the price is $10–12 per barrel.

After calculating the refining costs, one can identify the profit margin, which has been declining recently but is still attractive. The average worldwide gross refining margin for Phillips 66 stood at $7.30 per barrel in the third quarter of 2016, down by $5 per barrel from the first quarter of 2015. This means Phillips was earning $7.30 per barrel from the sale of its refined products after deducting the costs of input and processing.[28] Such a profit margin signals the attraction the refining business holds for investors.

Natural Gas

Natural gas is almost entirely usable-quality methane when brought to the surface. This means that it requires less processing than oil, mostly to remove impurities including water, to meet the specifications for marketable natural gas. The byproducts of this processing include ethane, propane, butanes, pentanes, higher-molecular-weight hydrocarbons, hydrogen sulfide (convertible into pure sulfur), carbon dioxide, water vapor, and sometimes helium and nitrogen.[29] As with petroleum, natural gas has substantial value as a feedstock. For this reason, we can increase the value of natural gas by using it to make many useful products, from nitrogen fertilizer and antifreeze to plastics, pharmaceuticals, fabrics, and hundreds of other products.

Natural gas has been spiking in popularity as a fuel for power plants. This has resulted from its rising abundance from fracking, its resulting drop in price, and carbon emissions that are much lower than for coal. For these reasons, utility companies are substituting natural gas power plants for coal power plants. It is now the dominant source of electricity generation in the United States.[30] Indeed, hundreds of coal plants are being decommissioned, demonstrating the vagaries of the fuel market and the risks of putting all your chips on one fuel. All these factors have driven up employment and business opportunities in the natural gas sector of the energy economy. How long this love affair with natural gas will last is an open question.

Uranium

Designing, building, and operating the various types of equipment necessary to process uranium ore into useful fuel involves a multibillion-dollar array of businesses. Once uranium ore is extracted and separated from the other materials that accompany it (such as copper, silver, gold, and soil), the remaining product is mostly U-238 and U-235. These isotopes are chemically identical, each having 92 protons. U-235 is the desired isotope because it is fissile (i.e., it splits) with thermal neutrons, whereas U-238 does not.[31] For it to be useful in light-water reactors that make up all commercial operations in the United States, the concentration of U-235 in fuel must be raised from 0.7 percent to about 3 percent. Given their identical-twin chemical characteristics, the question of how to accomplish this bit of alchemy perplexed scientists working to develop the first atomic bomb.

The answer to the problem of separation lay in the three-neutron difference between the two isotopes. U-238 is ever so slightly heavier than U-235. After milling, the resulting "yellowcake" is converted into a gas—uranium hexafluoride—to allow for the enrichment operations that gradually increase the concentration of U-235. The earliest successful way to accomplish such enrichment is through a process called gaseous diffusion (Figure 13.8). This gaseous diffusion enrichment technique sucked up a lot of electricity, a requirement that influenced the siting of the three now-closed gaseous enrichment facilities within easy reach of cheap hydropower from the Tennessee Valley Authority.

In recent years, the critical need for cheap hydropower to enrich uranium has been largely set aside. Improved enrichment technology is responsible for this change, with high-speed centrifuges preferred today. Instead of using the energy-intense method of repeatedly forcing uranium hexafluoride gas through extremely fine filters, centrifuges now accomplish the separation through centrifugal force with comparatively little energy. Such centrifuges require less than 2 percent of the energy of gaseous diffusion.[32] The sole enrichment facility presently in the United States is near Eunice, New Mexico. It supplies 25 percent of all the fuel needed to run the country's civilian nuclear power plants. In a few years, this percentage is expected to grow to 50 percent.

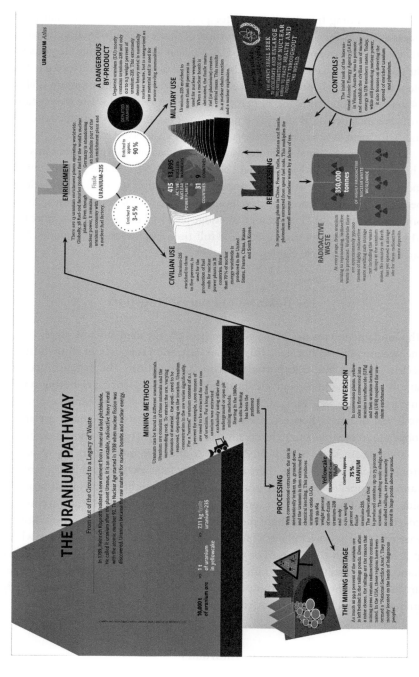

Figure 13.8 The uranium pathway. *Source:* Adapted from the Uranium Atlas.

Even including the energy needs for processing, the fuel for nuclear power plants accounts for a much smaller portion of the costs of production than electricity derived from fossil fuels. The U.S. Nuclear Energy Institute suggests that the cost of fuel for a coal-fired plant is 78 percent of total costs, and for a gas-fired plant the figure is 87 percent, but for nuclear power plants the uranium is about 14 percent of total costs (or 34 percent, if all front-end and waste management costs are included).[33] It should be noted, however, that costs can fluctuate substantially, even over a relatively short period, for a variety of reasons.

Many of the national laboratories have long had an energy-related mission. This is true of Oak Ridge National Laboratory (ORNL), where gaseous enrichment was first developed. ORNL and the sixteen other national laboratories all became engines of economic development in their local communities. As just one example, the 2017 budget for Oak Ridge National Laboratory was more than $1 billion. This budget supported 4,368 full-time employees, 520 students, 3,115 facility users, and 2,280 visiting scientists.[34]

The Business of Energy Transport

The fundamental feature of the geography of energy demand and supply is that the two are rarely collocated. This spatial separation supports an entire business sector—namely, the transportation of fuels. Sometimes the distance between supply and demand is relatively short, as at the Four Corners generating station in northwestern New Mexico, which is adjacent to a coal mine. More commonly, however, distances between areas of supply and areas of demand are much longer, thereby supporting the creation of an entire energy transport industry.

Many businesses are dedicated largely to the movement of energy resources from one place to another. There are primarily four ways to do this job: tankers, railroads, pipelines, and transmission lines. Each resource lends itself best to a single type of transport. For example, coal moves principally by unit trains that cover hundreds and even thousands of miles. This coal sometimes moves directly to users (such as power plants), but sometimes it arrives at a transfer point where it is loaded onto river barges or ocean-going freighters. Tankers, on the other hand, are best suited to the transport of oil and liquefied natural gas. Pipelines are the dominant means of transporting these products on land and occasionally under water, such as between Africa and Europe.

Railroads

Transporting coal has long been a staple business for railroads, but this role is on the wane. The Association of American Railroads reports that rail coal volume has declined in recent years. Nonetheless, coal remains a crucial commodity for U.S. freight railroads, in 2016 accounting for 31.6 percent of originated tonnage for U.S. Class I railroads, far more than any other commodity (Figure 13.9). Coal also accounted for 13.9 percent of rail revenue in 2016, behind only intermodal and chemicals. Power plants consume most of the coal mined in the United States, with railroads delivering approximately 70 percent of that coal.

Coal's share of U.S. electricity generation dropped 50 percent between 2007 and 2019, to less than 1 billion MWh. Compare this to the 1.6 billion MWh generated from natural gas in 2019.[35] This has led to sharp declines in rail carloads as well. According to the EIA, U.S. coal production in 2016 was 728.2 million tons, down 18.8 percent from 2015's 896.9 million tons, down 37.9 percent from the all-time high of 1.17 billion tons set in 2008, and the lowest annual total since 1978. Wyoming accounted for 41 percent of U.S. coal production in 2016, followed by West Virginia (11 percent) and Kentucky (6 percent). Seven of the nation's top ten highest-producing U.S. coal mines are in Wyoming. All these downward trends have meant railroad revenues are declining as well.

As demand for coal declines, railroad companies that rely on coal transport are hurting. For example, the Norfolk Southern Company owns the huge Conway Terminal, 22 miles west of Pittsburgh. Furloughs that began in January 2016

served as a wake-up call for Heath Vezza, vice president of the workers union representing about 325 track maintenance employees in the central operating

Figure 13.9 Coal unit train, Richmond, Virginia.

division in Pennsylvania. Mr. Vezza said 12 of his union workers—younger employees hired from 2012 to 2015—were furloughed. There have been handfuls of layoffs at other divisions through January, he added, and it is hard to know how many jobs in total disappeared. CSX Corp., which owns a competing rail line across the river from Conway Yard and follows a similar path through Pittsburgh, plans to continue its years-long effort to reshape its business after coal revenue dived 19 percent in 2015.[36]

The fortunes of coal mining and coal trains are closely tied to the reduced demand by coal-burning power plants. Reduced demand from power plants means reduced business for railroad and their employees.[37] For this reason, the coal train business is not a secure place for employment, again demonstrating the vagaries of the business of energy. Those seeking employment in the energy sector must remain agile and accustomed to change.

There is a more positive note to report, however. Plans are afoot to modernize terminal facilities for the use of intermodal transportation. This shift favors the merging of different types of transportation—usually trains and trucks—without interfering with the product itself. For example, CSX is making a significant investment at the Pittsburgh and Lake Erie rail yards in Stowe and McKees Rocks. The new facilities will allow for stacking containers four high. This project will provide approximately 150 construction jobs, 30 to 40 permanent operating jobs, and around 150 indirect jobs, including drayage, once operational.[38]

With markets shrinking in the United States, coal and railroad companies are working to bolster overseas markets. In 2018, the United States exported about 116 million tons of coal—equal to about 15 percent of U.S. coal production—to at least fifty-two countries. Almost all of this coal leaves from East Coast ports such as Baltimore, Maryland, and Newport News, Virginia, bound in large part heading for Europe. However, attention is now turning toward Asia.

Currently, less than 25 percent of exported U.S. coal reaches Asian markets. Still, there is an increasing lobbying effort to support building port facilities—creating hundreds of jobs—on the Pacific coast of Oregon and Washington. The goal is to sell coal from the massive reserves in the Powder River Basin to China, Japan, and South Korea, among other countries. Such plans have run aground, at least temporarily, with growing public resistance to the location of the planned docking facilities in either state.

Tankers

Once products such as oil get to the coasts, oil tankers take over. They are the most common form of shipping on the planet and, at about 2–3 cents per gallon, the cheapest way to move oil. Of the 17,000 bulk tankers in the world, about 7,000 of them are oil tankers. By 2019, the capacity of seaborne tankers was 568 million deadweight tons. The larger tankers can carry more than $100 million worth of product. Such valuable cargo, of course, makes a tempting target for pirates, as depicted in the Tom Hanks movie *Captain Philips*, a 2013 thriller inspired by the true story of the 2009 *Maersk Alabama* hijacking.[39]

Crew size also can influence the efficiency of movement. Large tankers need crew numbers of twenty to twenty-five depending on the size of the ship. (By contrast, coal trains—which, of course, carry less product—usually need one to three people.) Despite the temptations of transport efficiency, oil tankers are responsible for a substantial amount of oil lost at sea between loading and unloading. Of the approximately 700 million gallons of waste oil deposited in the ocean each year, offshore drilling operations contribute about 2.1 percent, but transportation accidents account for an additional 5.2 percent.[40]

Pipelines hold several advantages over other means of transportation. For example, they are more efficient than railroads and are not prone to derailments, are much less likely to spill massive amounts of product into bodies of water, require few people per unit of energy delivered to keep them operating, and are usually located underground, where the vagaries of the weather have less impact on reliability. While burying them has aesthetic advantages, the practice can produce life-threatening mishaps, as when a large-diameter gas line ruptured without warning beneath a residential neighborhood in San Bruno, California, in 2010, killing several people and devastating a neighborhood.

The Business of Generating Electricity

The magic of converting fuels into electricity is arguably the most significant invention of all time. Nothing compares to electricity in importance to modern-day lifestyles, and power plants are the most recognized link in the energy supply chain. This status is guaranteed by their ubiquity as well as their large and recognizable landscape imprint, one that is at the convergence of highways, pipelines, railroads, and transmission lines. They can be very obvious, especially in the treeless regions of the American West. When you spot one, think about the diverse team of specialists who keep it running. Their collective mission is to operate each power plant safely, efficiently, reliably, profitably, yet with the smallest possible environmental impact. It is no simple task. Electricity is the result.

Power generation is not only the most important single step within the overall business of energy, it is a huge undertaking. The total U.S. installed electricity generation summer capacity at the end of 2020 was 1,117,475 MW—or about 1.12 billion kilowatts (kW)—of total utility-scale electricity generating capacity.[41]

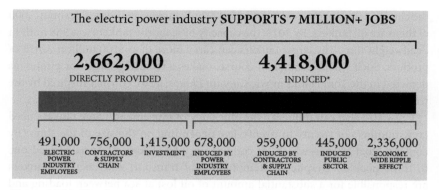

Figure 13.10 The electric power sector creates 7 million jobs in America both directly and indirectly. *Source:* M. J. Bradley and Associates, "Powering America: The Economic and Workforce Contributions of the U.S. Electric Power Industry," 2017.

Running and managing about 7,000 power plants of 1 MW or more in generating capacity in the United States requires more than 7 million people (Figure 13.10).[42] Millions more are needed to support the total worldwide generating capacity of 6,581 GW. As one example, the 4 GW Palo Verde Nuclear Generating Station near Phoenix employs 2,000 workers, with an annual payroll of $255 million, and it annually pays over $54 million in state and local taxes.[43] If we focus on the nuclear industry, it directly employs nearly 100,000 people in long-term jobs. This number climbs to 475,000 when you include secondary jobs.[44]

Power plant operations vary by technology and fuel, with more than 90 percent of electricity in the United States coming from power plants fueled by natural gas, coal, uranium, or falling water. The first three types are thermal plants, creating steam that turns a turbo-generator. (Hydropower requires no steam, just the kinetic energy of falling water striking the turbine blades.) The actual operation of all thermal power plants is complicated, and the training necessary within the workforce covers several types of engineering disciplines. Taken together, siting, designing, building, operating, and managing power plants is a multibillion-dollar enterprise, involving thousands of companies and millions of people working in dozens of trades, all of whom have just one job: meeting the demand for electricity (Table 13.1).

With so many people working around concentrated forms of energy such as high-pressure steam lines, blisteringly hot boilers, and miles of transmission wires, dangers are always present. Employees receive never-ending reminders and mandatory safety training. Nonetheless, even the most routine operations can become dangerous in short order. As just one example, sixteen people were severely burned and six died in 1985 when a massive reheat pipe broke at the

Table 13.1 Power Plant Job Listings by Title (Partial List)[45]

Accountant | Analyst | Area Manager | AutoCAD Drafter | Auxiliary Operator | Boiler Engineer | Boiler Technician | Boiler Tuner | Chemist | Civil Engineer | Commissioning Engineer | Component Engineer | Construction Manager | Construction Superintendent | Consultant | Controls Engineer | Designer | Director | Distribution Designer | Electrical Design Engineer | Electrical Designer | Electrical Engineer | Electrical Technician | Electrician | Electronic Engineer | Electronics Technician | Engineering Manager | Environmental Engineer | Estimator | Field Service Engineer |Field Service Technician | Financial Analyst | HVAC Engineer | Instrument Technician | Instrumentation Technician | Maintenance Manager | Maintenance Supervisor | Maintenance Technician | Manager | Mapping Drafter | Mechanic | Mechanical Engineer | Mechanical Field Engineer | Mechanical Technician | Metallurgical Engineer | Mine Development Manager | Mining Engineer | Nuclear Engineer | Operations Manager | Operator | Performance Engineer | Planning Engineer | Plant Engineer | Plant Manager | PLC Engineer | Programmer | Project Controls Manager | Project Director | Project Engineer | Project Manager | Purchasing Manager | Recruiter | Regulatory Affairs | Relay Engineer | Relay Technician | Reliability Engineer | Reservoir Engineer | Rotating Equipment Engineer | Safety Engineer | Safety Manager | Sales Engineer |Sales Manager | Salesmen | SCADA Engineer | Service Manager | Shift Supervisor | Site Manager | Structural Engineer | Substation Engineer | Superintendent | Technical Specialist | Technician | Telecommunications Engineer | Thermal Engineer | Trainer | Transmission Engineer | Transmission Planning Engineer | Transmission Technician

---

Mojave Generating Station in Laughlin, Nevada.[46] Employees are aware of these hazards; they train to minimize the risks they face, and they then get back to the work of providing electricity to the rest of us with the high level of reliability we take for granted.

The Business of Waste Disposal

Every stage in the chain of activities between energy exploration and energy use produces waste products in one form or another. As we discussed in Chapter 5, on the environment, these wastes pose hazards that cannot be responsibly ignored. Safe disposal of such wastes has become a very big business, amounting to at least half a trillion dollars of expense in the United States alone. That level of financial commitment signals the importance we all attach to waste treatment, and it supports thousands of jobs. Such jobs are within those companies able and willing to overcome higher barriers to entry such as high capital expense and a volatile array of waste management practices, acceptable techniques, and governing policies.

Each year the United States spends an estimated $30 billion just on nuclear waste disposal. This is but a fraction of the estimated $300 billion needed to clean up the legacy wastes from the Manhattan Project of the 1940s. Moreover, that expense is incurred just for waste from defense-related projects.[47] The estimated cost of cleaning up America's nuclear waste has jumped more than $100 billion in only one year, according to a DOE report—and a watchdog warns the cost may climb still higher. The Energy Department's projected cost for waste cleanup projects continues to rise. For example, it expanded from $383.78 billion in 2017 to $493.96 billion just in one year.[48] Globally, the amount is at least a trillion dollars. If you cannot quite imagine the scale of that number, consider there are more than one trillion seconds in about 32,000 years

The disposal of low-level radioactive waste presents its own particular challenges. Over the years, many organizations have tried their hand at establishing low-level waste disposal sites in the United States. Presently, only four are in operation: Barnwell, South Carolina; Richland, Washington; Clive, Utah; Andrews, Texas. One near Andrews is run by Waste Control Specialists (WCS), which received a license to operate in 2009 (Figure 13.11).[49] WCS collaborated with AREVA and NAC International. AREVA—now called Orano—had 18,604 employees in 2019, and its salaries ranged from an average of $53,081 to $139,939 a year in 2019. Perhaps such salaries would tempt you, assuming you did not consider handling such waste personally risky.

A few miles to the northwest of the WCS facility lies another nuclear waste storage operation, but one with a much different mission. Known as the Waste Isolation Pilot Plant (WIPP), the facility accepts defense-generated transuranic (TRU) waste from DOE disposal sites around the country. TRU waste consists of clothing, tools, rags, residues, debris, soil, and other items contaminated with small amounts of plutonium and other human-made radioactive elements. The waste received at WIPP since 1999 is permanently entombed 2,150 feet underground in bedded salt. The entire process creates hundreds of different types of jobs, including driving the transport trucks, manufacturing the transport casks, the on-site work of unloading and preparing for entombment, and excavation of the salt beds to receive the wastes (Figure 13.12).

Whereas WIPP continues receiving TRU defense waste, disposal of high-level waste from commercial power plant operations is in a precarious state of suspended animation and is the source of mounting concern. High-level radioactive wastes are the highly radioactive materials that are byproducts of the reactions that occur inside nuclear reactors. High-level wastes take one of two forms: spent (used) reactor fuel intended for disposal and the waste materials remaining after spent fuel is reprocessed. Although the spent fuel from commercial power plants is much smaller in volume than defense waste—some 80,000 metric tons sit at seventy-five sites in thirty-five states—the total amount of radioactivity is roughly 20 to 30 times greater than defense waste. Because the

Figure 13.11 Part of the 1,338-acre WCS Texas Compact Waste Facility. WCS provides for the treatment, storage, and disposal of Class A, B and C low-level radioactive waste, hazardous waste, and byproduct materials. *Source:* Photo credit, http://www.wcstexas.com/. Used with permission.

Figure 13.12 Mining in the salt bed in preparation for receipt of TRU waste 2,150 feet underground at the WIPP site. *Source:* WIPP Facebook site, https://www.facebook.com/pg/WIPPNews/photos/?tab=album&album_id=927355184080513&ref=page_internal.

hazards of radioactive waste are invisible and often long-lasting, safe disposal presents a vexing array of complicated challenges, which can also be seen as business opportunities.

At present, no permanent waste storage facility for civilian high-level radioactive waste exists in the United States or anywhere else in the world. (However, some are in the initial stages of preparation in Sweden and Finland.) Currently, operators of all civilian nuclear power plants in the United States store their spent fuel on-site—first in pools and then in dry casks—as the operators await the availability of an approved final resting place. Eventually, all spent nuclear fuel might be interred under Yucca Mountain, about 90 miles northwest of Las Vegas, Nevada. In 2008, the life cycle costs for the repository at Yucca Mountain were revised upward from $58 billion to $96 billion to cover 150 years of research, construction, and operation of the geologic repository.[50] To date, the cost of developing the repository at Yucca Mountain amounts to approximately $13.5 billion. However, despite incurring such expenses preparing just this single site for commercial waste disposal, its eventual full activation and licensing remain uncertain.

Regardless of the specific location and availability of various types of nuclear waste storage, both short- and long-term, a range of businesses associated with waste disposal will continue operating—and continue receiving funding—indefinitely. Funds committed for the Nuclear Waste Fund in 2017 totaled $46.7 billion, an amount accumulated from a 1/10th of a cent per kWh surcharge for electricity generated at nuclear power plants, plus interest, since 1983.[51]

Meanwhile, the decommissioning of commercial nuclear power plants is beginning. The approaches to accomplish such decommissioning range from complete dismantlement (after which the site can be released for unrestricted use) to permanent entombment.[52] Dismantlement requires available sites for disposal. One unique example of complete dismantlement was at the 60 MW Shippingport, PA, reactor. In 1988 the 956-ton reactor pressure vessel/neutron shield tank assembly was lifted out of the containment building.[53] To get to its final burial site, it was barged down the Ohio and Mississippi Rivers to the Gulf of Mexico, through the Panama Canal, and then northward along the west coast to the mouth of the Columbia River. From there, it was pushed eastward up the Columbia River to the Hanford Site in central Washington, where it rests today.

More recently, the decommissioned San Onofre reactor vessel was transported by rail to the Apex Industrial Park, northeast of Las Vegas, and then by highway to Clive, Utah, 75 miles west of Salt Lake City, where low-grade (Class A) nuclear waste is buried. It was a big job. At more than 1.5 million pounds, it will be the

Figure 13.13 Part of a nuclear reactor being transported from San Onofre, California, to Clive, Utah. Here it is about 100 miles northeast of Las Vegas in Alamo, Nevada, in July 2020.

largest and heaviest object ever moved on a Nevada road (Figure 13.13). Such removal, transport, and burial at sites mainly in the western states will become increasingly common as more nuclear power plants are shut down. Waste disposal from nuclear power, as well as the handling of these wastes for disposal or recycling—produces business opportunities for tens of thousands of workers.

The Energy Business and You

The energy sector underpins the quality of the life you experience. This function applies to everyone on the planet. However, everyone does not have equal access to all the energy they want or the form in which they prefer it. Those for whom supplies are plentiful, affordable, and reliable tend to live in relative comfort. Those for whom supplies are scarce, expensive, and unreliable often live in discomfort and peril. Sustaining the lives of the first group, or improving the lives of the second group, will depend on which energy future transpires. Will we be conventional and work to maintain the existing patterns of supply and demand, or will we be bold and seek the benefits of a more sustainable energy future? Whichever path we follow will have a determinant influence on the quality of the lives of young people alive today as they approach adulthood and of all those people yet unborn.

Energy transitions are afoot, and we will have to accept their inevitability if we are to survive with the lifestyles we wish to have. Already, we are witnessing some of the early stages of change, including the waning of the coal industry. A similarly dim future is visible for the nuclear industry, particularly in Germany, the United States, and Japan, despite the advantage of carbon-free power generation.[54] Natural gas would seem likely to have a lasting presence, but with the demise of coal, diesel, and nuclear generation, energy efficiency and alternative energy resources will be increasingly called upon to fill the need. As traditional energy jobs disappear, we must consider alternative job opportunities, identify where and in what quantity such new jobs will emerge, and how we might be able to transfer old skills and experiences to the new energy economy:[55]

- Of the 1.9 million workers in electric power generation and fuels, 800,000 employees work in electric power generation, including renewables, plus about 97,000 in full-time or temporary solar jobs.
- Roughly 32 percent of the 6.5 million employees in the U.S. construction industry work on energy or building energy efficiency projects.
- Solar industry employment jumped by 5,700 jobs to reach 248,000 in 2019, up 2.3 percent over 2018.
- Solar and wind technologies rank first and second in the construction industry, with 127,000 and 38,000 jobs, respectively.
- Wind industry employment added 3,600 new jobs to total 114,800 jobs in 2019, up 3.2 percent over 2018.

These statistics summarize the data you can use to make decisions about employment opportunities that may be open to you in the near future. It might also help you avoid mistakes in terms of the training and education you pursue. For example, you would want to avoid devoting time and money to training for a job in the coal industry. The same advice might well hold for the nuclear industry, given its sensitivity to the political winds of change, especially in the United States. Instead, you might wish to acquire skills in energy efficiency, alternative energy, electric vehicles, and even the associated energy policies. It is easy to see that jobs are already increasing in those sectors (Figure 13.14; Table 13.2).

Development and incorporation of renewable energy resources are growing globally with unanticipated speed, as illustrated by the work of the International Renewable Energy Agency (IRENA). IRENA reports that the renewable energy industry created more than 500,000 new jobs around the world in 2017, with the total number of people employed in renewables (including large hydropower) surpassing 10 million for the first time.[56]

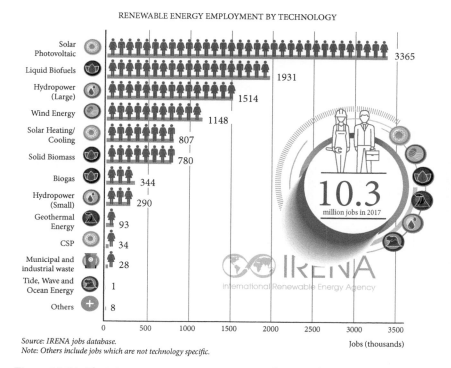

RENEWABLE ENERGY EMPLOYMENT BY TECHNOLOGY

Source: IRENA jobs database.
Note: Others include jobs which are not technology specific.

Jobs (thousands)

Figure 13.14 Electric power generation sector, employment by detailed technology application, 2018–2019. *Source:* 2020 U.S. Energy and Employment Report; International Renewable Energy Agency. © IRENA. Used with permission.

Among the various technologies based on renewables, the solar photovoltaic (PV) industry supports the most jobs. PV jobs increased by almost 9 percent to reach 3.4 million around the world in 2017, reflecting the year's record 94 GW of PV installation (Figure 13.15). IRENA has found that renewable energy employed 9.8 million people around the world in 2016. China, Brazil, the United States, India, Japan, and Germany accounted for most of the renewable energy jobs through training in the appropriate scientific and technical/engineering skills. The shift to Asia continues, with 62 percent of the global total located in that part of the world.[57]

When you are ready to start seeking jobs in the energy sector, you could begin by consulting the many sites available on the internet to increase your understanding of energy employment trends and needs. You might also wish to consult updated versions of the annual USEER and yearly updates from IRENA.[58]

Table 13.2 Employment in Electric Power Generation Sector, by Detailed Technology Application and Industry, Q2 2019

	Total	Utilities	Construction	Manufacturing	Wholesale Trade	Professional Services	Other Services
Majority-Time Solar Employment	248,034	3,682	126,979	34,243	23,913	37,479	21,738
Wind	114,774	6,360	37,910	26,408	12,305	28,873	2,918
Geothermal	8,794	1,095	5,184	295	361	1,830	29
Biofuels	13,178	1,897	5,809	1,133	576	3,317	446
Low Hydro (<50MWe)	12,304		1,913	3,440	2,593	4,283	75
Trad Hydro	55,468	17,464	8,934	14,458	6,075	8,301	236
Adv Nat Gas	75,661	47,224	9,638	2,791	4,983	10,118	907
Nuclear	60,916	44,366	2,217	1,901	2,639	9,705	89
Coal	79,711	38,158	8,847	1,083	6,104	24,508	1,011
Natural Gas	46,151	19,276	10,551	3,635	3,180	8,371	1,139
Combined Heat and Power	30,342	1,608	4,361	2,100	3,944	18,133	196
Other	41,417	1,961	20,119	3,788	3,083	11,777	690
TOTAL	799,742	183,565	242,462	101,065	74,906	182,688	43,134

Source: 2020 U.S. Energy & Employment Report, https://www.usenergyjobs.org.

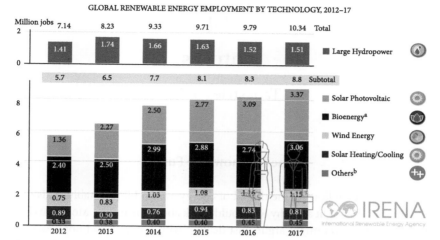

Source: IRENA jobs database.
Note: The numbers shown in this Figure reflect those reported in past editions of the Annual Review.

a Includes liquid biofuels, solid biomass and biogas
b Other technologies include geothermal energy, hydropower (small), concentrated solar power (CSP), heat pumps (ground-based), municipal and industrial waste, and ocean energy.

Figure 13.15 Global renewable energy employment by technology, 2012–2017. *Source:* International Renewable Energy Agency, *Renewable Energy and Jobs— Annual Review 2018.* © IRENA 2018. Used with permission.

A Takeaway Thought

I hope this brief look inside the business of energy has conveyed some sense of its complexity, significance, and opportunity. No enterprise has more bearing on the well-being of society. Those promoting and facilitating the transition toward a more sustainable energy future will be lifting a heavier load than they might realize as they continue the arduous task of exchanging the energy portfolio of the present for what we will need if we are to prosper in the future. If you want to endorse that transition, the first step is to understand the energy business that is now in place. The goal of this chapter was to provide you with some insights into such an understanding. The rest will be up to you.

14

Futures

Tapestries of Life

The Unknowable Future

As we continue following the thread of energy, we find it inevitably leads us to the future. You might ask why we should think about the future. Most of us have our hands full, you might argue, just dealing with the here and now. As some wag once proposed, "What have the future generations ever done for us?" Setting cynicism aside, why we should think about the future is a serious question. And, the blunt answer is, we cannot help it. We are hard-wired to think about the future, sometimes quickly and emotionally, and sometimes more slowly and in a more reasoned fashion.[1] We all hold somewhere inside our bodies a natural thread that connects our present status and what is to come, and energy is one of its most important strands.

While most of us have little confidence that we can accurately predict what our future will look like, we are nevertheless quite confident that it will be different from the present. We can speculate on a host of possible futures, and we can set policies to help us reach them, but there are far too many variables to consider for us to say with certainty where we will end up. All we know is that the future will be different than the present, and that change will arrive faster than we anticipate.

This reality leaves us to speculate and to pose questions about what is on the horizon. Will global use of nuclear power expand, as it is doing in China, or will it contract, as is happening in Germany? Will per capita use of fossil fuels—and the emission rate of greenhouse gases—spike south of the Sahara, or will this area make the leap to low-carbon renewable generation and give the rest of the world more time to address global warming? Will coal mining remain a viable enterprise in the United States in support of increased demands from abroad, or will it die a natural death because reduced domestic demand and increased environmental controls collude to make coal mining here uneconomic? Will deployment of renewable energy change the quality of life for the billions of people who don't have electricity? Will we be transporting ourselves in hybrid cars, all-electric cars, hydrogen-fueled cars, bullet trains, hyper loops, supersonic jets,

The Thread of Energy. Martin J. Pasqualetti, Oxford University Press. © Oxford University Press 2021. DOI: 10.1093/oso/9780199394807.003.0014

rockets, flying pods? As the old saying goes, the problem with predicting the future is that it is in the future.

Part of the problem each of us faces is that as the speed of change accelerates, we will find it increasingly difficult to keep pace, to stay current, to remain adaptable, and to be appropriately trained. The unhappy reality of living in a time of rapid change, at least on the personal level, comes down to this: if you cannot match it stride for stride, you may soon be fraught trying to support yourself and your family using skills that are outdated, redundant, unwanted, or undervalued. You may feel confident that you are agile enough to overtake the rate of change, but coming out ahead in that race requires that you stay curious, remain flexible, and acquire aptitudes and skills that allow you to adapt to whatever future you might encounter. If you are reading this, you are already likely to be on that path.

Because the thread of energy is everywhere you look and in everything you do, all changes in energy supply and energy demand will influence you in ways you might be able to imagine and in ways you probably cannot. These changes will affect where you live, how you travel and communicate, what you eat, whether you are comfortable and secure, how the global economy develops, and which countries dominate world politics. The most resilient among us—from individuals to nations—will realize that trying to influence our energy future will bring challenges, but they will also bring opportunities. The cold reality is this: we will not be in a position to benefit from such opportunities if we maintain a stance of disinterest or just throw up our hands in defeat.

While we do not know exactly which future will develop or which future we might be able to influence, a first step is to make some informed guesses at what might lie ahead. Let's start.

Energy Portfolios

Envisioning energy futures prompts several fundamental questions: What will be the scale of our energy demands? How will we meet energy demands in developed and developing countries? What resources will dominate? What will be the proportions of foreign and domestic supplies? How will geopolitics affect energy supply, demand, and policies? Will we "run out," as with "peak-oil" predictions? Can we conserve resources enough to make a difference in supply requirements? How will environmental concerns affect energy development and availability? Will we master meaningful energy storage? How close can we get to a sustainable energy scenario? These questions do not have sure answers, only guesses based on predictive models that combine a host of variables to arrive at a prediction. The primary challenge we face is how to predict the future, how to identify the degree to which we can trust our predictions, and how to determine what actions are needed to adjust accordingly.

Predicting the future is, of course, tricky. It usually involves developing a series of models based on combinations of assumptions. It also depends on what future you would like to appear and whose perspective dominates. There are lobbying efforts for just about every possible scenario. Will it be heavy on nuclear, or solar, or efficiency, or hydrogen, or fusion? Will future fracking be prohibited? Will wilderness areas be opened for exploration? Will electricity storage technology have made breakthroughs? Will we take measures to increase domestic production or continue to trade for what we prefer?

Bruce Tonn and his colleagues have recently offered us a thoughtful glimpse of these several such perspectives, formulated as a series of questions, based on whose views dominate the conversation.[2] Here are a few to consider. For example, will the future portfolio resemble the base-case forecast produced by the U.S. Energy Information Administration, one that includes a large amount of imported energy and domestic coal? Or should we think about the future in terms of which group of like-minded people dominates?

- Will it be dominated by the Technophiles, who would push nuclear power (fission and fusion)?
- Will it be the Environmentalists, who encourage higher amounts of renewables and no domestic coal production?
- Will it be the Bottom-Liners, who would place even more reliance on low-cost coal?
- Will it be the Individualists and America-Firsters, who advocate policies to increase U.S. production of liquid fuels to reduce the cost of driving and to achieve energy independence, respectively?

Whichever scenario you think will dominate, none of them can operate entirely independently of some or all the others. Such is the nature of predicting the future, especially in a democracy, where protracted debate can slow decisions.

As Tonn and his colleagues make clear, while coal-based electricity production might decline as alternative energies grow, it might also continue nevertheless as a fallback option. Similarly, "the America-Firsters portfolio is similar to the Technophiles portfolio in that all energy components are represented, but with less aggressive growth targets for nuclear and renewable energies. The Bottom-Liner portfolio is closest to the current status quo (and the EIA base case) and is the least successful in reducing imports" (Figure 14.1).[3]

You can glean quite a lot from a detailed study of Figure 14.1, and I encourage you to take on that assignment. One of the most important messages is how each scenario results in a different outcome for CO_2 emissions.[4] The

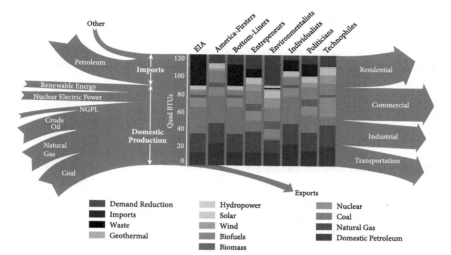

Figure 14.1 Summary of the seven perspectives' energy portfolios (year 2030). Coal has continued to fall on hard times since this figure was prepared in 2009. *Source:* Bruce Tonn et al., "Power from Perspective: Potential Future United States Energy Portfolios," *Energy Policy* 37, no. 4 (2009): 1432–1443. See plates for color version.

Environmentalists' approach would produce the most significant reduction. It emphasizes demand reduction and renewable energy development. Direct Air Capture (DAC) should also be included. None of the other scenarios can accomplish enough, fast enough, to sufficiently slow the warming of the planet.

For the increasing number of people who worry about such problems, the common question is this: how do we get off our present rutted path to disaster, and get on a smoother path to a more sustainable future, one less reliant on carbon-based fuels? The sobering reality is that the concentrations of greenhouse gases already in the atmosphere are sufficient to wreak a lot of havoc, even if we stopped using all fossil fuels immediately.

A common way to address this critical concern in capitalistic economic systems is to consider the economic elements of potential changes. Consider, for example, how much it will cost to continue as we are. A recent study by the Rocky Mountain Institute estimates that U.S. electricity generators may be committing their customers and investors to "as much as $1 trillion" in future costs through 2030 as they "rush to build new gas-fired power plants."[5]

Other options are becoming increasingly attractive alternatives to business as usual. Advances in renewable energy have rendered them comparably priced to or cheaper than conventional energy resources in many locations. Several of them produce no emissions and use little to no water for cooling. As we move into the future, we will become increasingly conscious of the characteristics of these resources as compared with conventional resources. Whatever we do (or don't do) might well depend on landscape commitments, patterns of population growth, which form of government prevails, and how clever we are in the design of where most of us are living—namely, cities.

Landscape Change

Everyone knows that the energy path we choose affects the air we breathe, the water we drink, and the land we occupy. You have all heard of the choking smog in China and India, dying coral in the South Pacific, acidic streams in Appalachia, melting glaciers in the Arctic and high mountains, rising sea levels from Miami to the Maldives, and lost visibility just about everywhere. These environmental stresses come to our attention in an endless stream of bad news, often accompanied by dire warnings about the permanent damage to the health of the biosphere we count on to sustain ourselves and our planet. While we read about problems with air and water all the time, damage to another category of natural resources is often overlooked.

Our ignorance is not entirely our fault. While we endlessly need air and water—and we quickly notice when changes occur—most of us are less aware of changes to the landscape. The reason for this oversight is simple: landscapes hold a sense of indestructibility, of permanence. We tend to think of changes to them in geologic time. Yet, landscapes do change, and changes resulting from energy development are among the starkest. Moreover, changed landscapes tend to persist in their damaged condition, especially when they are in remote areas where there might be little competition for the land or public demand for remediation.

Changes to the lands from energy development continue to be blatant, apparent, and noticeable, regardless of which type of resource we use. It has been this way for a long time; for most of two and a half centuries, we have been turning the earth inside out to recover fossil fuels. We have created a legacy of litter and upheaval from these actions. In sites of energy development everywhere, we can see huge open pits, surfaces pockmarked from subsidence and littered with piles of discarded ash, canyons flooded, and mountaintops blown asunder to more cheaply access subterranean treasures. We have left behind hollowed buildings, rusting equipment, denuded forests, and even entire abandoned towns.

In many instances, particularly in spacious countries like the United States, derelict and abandoned energy landscapes are commonly invisible to people living their city lives some distance away. Such landscapes are even more unlikely to be on display when the energy is imported from a foreign country such as Venezuela.

Figure 14.2 The Euro Speedway Laurits is a racetrack located near Letitia in the state of Brandenburg in Eastern Germany, near the borders of Poland and the Czech Republic, on the site of a played-out lignite mine. Nearby, over two dozen closed open-cast mines have been converted into recreational lakes.

But things can be different in densely settled countries where city life and energy extraction are cheek by jowl. Such is the case around the gritty coal fields near Katowice, Poland, and across the border in eastern Germany. In such places, people live among energy development, and the resulting energy landscapes are hard to avoid. In some of these places, however, where there is public support, sufficient funds, and perhaps a profit motive, some energy landscapes are being recycled for tourism, recreation, or cultural purposes, depending upon opportunities, prior use, and land values (Figure 14.2).[6]

One of the most impressive and attractive recycling has been in Essen, Germany. There, the Zollverein Coal Mine Industrial Complex is thriving in a new form instead of disappearing at the hands of demolishing crews. It has been refitted with museums, parks, jogging and bike paths, a swimming pool for summer, and an ice-skating track for winter (Figure 14.3). So impressive is this repurposed energy landscape that it has been designated as a UNESCO World Heritage site.[7]

Opting for reclamation, remediation, and repurposing of energy landscapes in Europe is beginning to pick up momentum elsewhere, including in the United

Figure 14.3 Zollverein legacy coal mining and manufacturing facility in Essen, Germany. Nowadays, it is a major tourist attraction, including museums, a swimming pool, and (in winter) ice skating.

States. Even energy landscapes created before the enactment of mandatory reclamation laws are being revisited and repurposed. Sometimes these are brought back to useful life as farms or ranches, but other times the landscapes are being remediated for increased public safety. The trend toward recycling of energy landscapes seems likely to continue as public tolerance for the accumulation of discarded and hazardous landscapes decreases, and as the value of remediated lands rises. Literally thousands of damaged landscapes remain scattered across the country as if it were a giant junkyard, so there is still much work to do. However, this growing trend toward greater attention to reusing and repurposing the land should be considered a prelude of what can and should be done worldwide in the future.

While conventional energy activities can alter vast tracts of land, sometimes in irretrievable ways, renewable energy developments are producing landscape changes of their own. These new landscape changes, however, differ in appearance, flexibility, and longevity, as with wind power. Especially over the past two decades, wind power installations have created the most noticeable energy landscape changes. Although each turbine takes only a small parcel for foundations of concrete and steel, wind power installations rotate on their axis with changes

Figure 14.4 Wind turbines in fields of corn and soybeans near Dixon, Illinois. This option reduces the land commitment for wind developments while providing additional income for the landowners.

in wind direction and, of course, they spin, thereby attracting attention. Another characteristic of wind power results from their low energy density. This means that they are typically distributed on the landscape in a pattern that is discernable from a commercial aircraft flying at an altitude of 30,000 feet.

Significantly lowering the impact, however, is the third characteristic of wind power: whereas the turbines cannot be sited in a dense pattern, each turbine structure has quite a small footprint. Other activities, such as farming and even solar power, can double up on the land and make it even more valuable (Figures 14.4 and 14.5).

Solar energy is another renewable energy resource quickly growing in both affordability and popularity. It holds significant advantages over conventional generation, including that it is predictable, endless, free, and quiet, and it does not need cooling water. However, it has some obvious limitations. It is unavailable during nighttime hours if there is no means of energy storage, and—like wind power—it is a low-density energy resource and therefore requires a lot of land when deployed at utility scale.

Despite its low-density nature, solar energy is less of a landscape gobbler than you might think after you compare its entire fuel chain with that of conventional energy resources. For example, several steps are absent, including fuel

Figure 14.5 Wind turbines and solar installations collocate at the base of Mt. San Jacinto, near Palm Springs, California. This option reduces the land commitment for both solar and wind developments.

collection, processing, transportation, and waste disposal. If you tallied up the land requirements for every step in the fuel chain for coal, it would be more than the amount of land needed to accommodate solar deployments on a per-kWh basis. Moreover, distributed solar applications do not take up any additional land.

Population Changes

Especially regarding the demand for energy, speaking of the future always leads to a discussion of the impacts of population changes going forward. While some believe that having more people will unleash more creativity and innovations, it is fair to say that more people are concerned about the difficulties of adjusting to the extra 2 billion more people we expect within the next thirty years.[8] Even though the populations in some countries are contracting—Italy and Germany, for example—other countries are bursting with people. Already, many thousands of people are on the move in search of better economic opportunities, others are trying to escape war and violence, and still others are trying to avoid ethnic persecution. Think Sudan, Ethiopia, Honduras, Venezuela, Nigeria, Myanmar, Indonesia, Pakistan, El Salvador, Syria, Lebanon, Kazakhstan, Bosnia-Herzegovina, and many other countries. Population pressure is often a driving force of relocations, and it could continue until something slows or halts the momentum.

What do such population scenarios mean for the future of energy? One thing that seems inevitable is that continuing to rely on "fund" resources such as oil will be the plan of those trying to sustain the status quo. Those with the military and political power to control the development and sale of their national resources will continue to do so. Those who lack these resources will be held back until they can afford them, acquire them through public largesse, or take them by force. What can slow or halt this predictable outcome? Reduced population pressure could help, as could widespread access to freely available and abundant renewable energy resources.

The preceding discussion leads us to the notion of widespread availability and affordability of energy. Certainly, the age of fossil fuels has reached maturity. Coal is on the way out in North America and Europe for a variety of environmental and economic reasons. Natural gas use is on the rise in many parts of the world, partly because it has become less expensive than other fuels and because it produces fewer carbon emissions than coal. We have already consumed about 1 trillion barrels of oil. There are an estimated 1.5 trillion barrels still underground and under water. Fracking—and future technological breakthroughs—may enhance rates of recovery, but environmental costs will rise as well as the retail price we will have to pay. And, inevitably, oil and natural gas are diminishing resources. No one is making more.

In a time when fossil fuels sustain the world's industrial capability, what will their increasing costs and declining availability mean for us in the future, with a more crowded world? The answer is that we do not know. So we turn to two options that we have already had some success with: cutting consumption and developing sustainable supplies. Reducing consumption—at least at a per capita level—can be accomplished through energy efficiency (which is often technical) and conservation (which is often behavioral). There is plenty of room for improvement on both these fronts, and both strategies are good at reducing energy demand, but there is naturally a limit to what can be done. And, of course, as long a population continues growing, per capita improvements will be neutralized by the sheer number of new consumers. And, regardless of how much we improve efficiency, if we keep inventing new sources of demand, eventually enhanced energy efficiency will reach its limits.

This discussion takes us, then, to the next most obvious approach: reducing the rate of population growth or—more controversial—reversing population growth. All such considerations are sensitive. It is an especially thorny problem because much of the anticipated population growth is forecast to be within the borders of developing countries whose challenging socioeconomic conditions would be mitigated with greater energy access.

Democracies Versus Authoritarian Societies

Now, the big question: Is a democratic society such as ours structured to promulgate changes at scale that can move us meaningfully closer to a sustainable energy future before it is too late? The increasing trends toward partisan politics and nationalism suggest that it is not. We built our democracy based on debate, compromise, and consensus. This multi-step process is deliberative, and it is usually slow. Vested interests favoring the status quo are often well-funded, and they play to the natural human instinct to resist change. We are already seeing how this has retarded progress toward reducing greenhouse gases and increasing renewable energy.

On the other hand, one could argue that countries such as China, with its centralized and authoritarian governmental systems, can implement changes more quickly than democratic societies. Yet even China has obstacles to overcome, just like every other country. For example, despite its rapid progress in developing wind and solar power, and despite the government's acknowledgment of the threats from global warming, the construction of new coal-burning power plants and the widespread acceptance of the automobile for personal transportation continue to neutralize progress toward a more sustainable energy future. Pragmatism is outweighing idealism even with a system that represses dissent.

So which system of government should we rely on to lead, drag, or force us all toward a more sustainable future? How will our energy choices influence that determination? On which would you place your bets to move more decisively and meaningfully to save the planet: large centralized economies or large democratic societies? Which will take the necessary steps? These are abiding questions for us all to consider, whether our education is already behind us or continues.

China is moving along several paths simultaneously. They are building a dozen nuclear plants as well as many new coal plants. Still, they are also improving energy efficiency,[9] erecting huge solar arrays,[10] installing solar water heating,[11] building massive wind farms,[12] and improving energy storage capabilities. It is a different story in the United States, which continues stutter-stepping into the future, unsure which path is most likely to lead us to a stable tomorrow. We tend to dawdle by endorsing "all the above," but we are actually traveling along a circular path that leads us back to where we started. We have been subsidizing renewables, but these subsidies are diminishing and will soon vanish altogether. We have run up massive cost overruns at the only two nuclear power plants currently approved for construction. We are opening up wilderness areas to oil and gas exploration in Utah, continue to encourage fracking in North Dakota and West Texas, and support oil sands development in Canada.

Additionally, many of those people in leadership positions—not to mention the efforts of lobbyists and investors—continue to insist that coal and nuclear have a future and that solar and wind do not. Such support for maintaining our present ways has had the effect of creating a public resistance to change that will be hard to overcome. That inertia will prevail regardless of how cost-competitive alternatives become, irrespective of how dramatically sea levels rise, and notwithstanding the form of governmental rule that prevails. Likely, the place where we can make the quickest measurable progress toward overcoming this inertia will be in our cities, where energy demand concentrates.

Energy-Efficient Cities

The tightest weave of the thread of energy is in cities. Cities are insatiable consumers of energy, and even when you are outside a city, there is abundant evidence of the intended destination of energy supply infrastructure. Just follow any transmission line, any pipeline, any railroad, any canal or barge, any ship, any highway, or any airplane route, and you will find that almost all of them head to cities.

The reason is apparent: more than half the people on the planet live in cities, and two-thirds of the world's population will live in cities by 2050. In other words, cities are where energy demand is greatest. Given these facts, cities would seem the most fruitful target for concentrated efforts at overcoming our traditional patterns of urban energy use, while moving toward a more sustainable energy future.

Without changes in the form and function of our cities, the impacts of rising energy demand and the consequences of energy use will become increasingly difficult to manage. It is already challenging to satisfy all city energy demands on time and with a high degree of reliability.

What can we do to improve on the existing patterns? If we cannot or do not make improvements, there will be massive increases in energy infrastructure, including transmission lines, railroads, pipelines, and highways. In other words, cities will be much the same, only bigger and bigger. This solution is not sustainable. You may remember the adage "Insanity is doing the same thing over and over again and expecting different results." It would seem we need to make some adjustments in the future.

One overarching policy change in starting the move toward greater energy sustainability is to maximize energy efficiency in every sector of urban life, including transportation, construction and maintenance, commerce, manufacturing, architecture and design, engineering, landscaping, water supply, and all other operations. The basic idea is to consolidate the breadth of possible actions into a single policy decision to raise energy efficiency and thereby lower energy demands on the complicated and expensive infrastructure we need to supply it.

In large part, the technological and design competencies needed to increase efficiency already exist. The tougher challenges are how to institutionalize such a robust policy, how to put it into practice, and how to enforce it uniformly. Bringing improved energy efficiency to city life involves the allocation of funds, public and private support, education, time, economic competitiveness, and, most important of all, political will. As alluded to earlier, implementing such steps might be easier to envision in centrally controlled, authoritarian societies than in democratic societies. For this reason, authoritarian countries might have some advantages, especially given that many of the largest cities in the world are in China.

If we are realistic, we must acknowledge that driving up energy efficiency in existing cities is difficult, costly, and slow. The other approach, one that has appeal, is to build cities afresh, ones that are designed from inception to meet such higher standards. Masdar City, in Abu Dhabi, is an example of where such a strategy now exists. As initially envisioned in 2006, Masdar City was to occupy 6 km² (2.3 mi²) and cost an estimated $18 billion to $22 billion. Masdar is a sustainable mixed-use development designed to be friendly to pedestrians and cyclists.

Matching ambitious plans against reality often carries inherent difficulties, and Masdar has been no exception. For example, the original master plan envisioned a city functioning on its own grid with full carbon neutrality. Instead, the development was later attached to the public system, and by 2016 its managers determined that the city would never reach net zero carbon levels.[13] Yet many of the concepts developed for Masdar City are sensible and achievable components of any strategy to increase urban sustainability. These steps include the use of super-insulating building materials, maximizing shade, incorporating solar and other renewable energy sources, reusing waste heat, and minimizing the need for private transportation.

Another venture into newly built urban sustainability, also in the UAE, is the Dubai Sustainable City, a 46-hectare (114 acres) property development, that is intended to be the first net zero energy development in the Emirate of Dubai. Phase 1 would include villas, apartments, mixed-use areas, a healthcare facility and a child-care center, plus food and beverage outlets. Phase 2 would expand on the facilities to include a hotel and school. More ambitions future expansions would include greenhouses, organic farms, water recycling, solar production, and passive cooling design. The entire development is intended to be car-free.[14]

A third example of a planned sustainable city is Amaravati, in southeast India. Here, 217 km² (84 mi²) would be transformed from farmland along the Krishna River to an urban utopia. A green-spine would be 5.5-km-long and 1-km-wide (3.4 miles by .62 miles). These design features are the work of the British architectural firm Foster + Partners, which is leading the project. (It is the same company working on Dubai's sustainable city project.) Amaravati, when complete, is to be the new capital of Andhra Pradesh state, and one of the most sustainable cities in the world. The general approach to Amaravati is like that of similar projects elsewhere. The estimated total cost of the project is $6.5 billion, according to the Andhra Pradesh authorities.[15]

On the grandest scale of them all, Saudi Arabia plans to invest $500 billion to build a mega-city called Neom. It will rely, as much as possible, on renewable energy.[16] The project is in the far northwest of Saudi Arabia near Tabuk. It will include marine land located within the borders of Egypt and Jordan. It will extend 460 km (286 mi) along the coast of the Red Sea and is intended to provide many investment opportunities within a total area of 26,500 km^2 (10,200 mi^2). The idea behind the project is to transform Saudi Arabia into a leading global model in various aspects. One of the main objectives of the project is to seek cooperation and investment with an extensive network of international investors and innovators, all to focus on advanced industries and advanced technology. The initiative emerged from Saudi Vision 2030, a plan that seeks to reduce Saudi Arabia's dependence on oil, diversify its economy, and develop public service sectors.[17] The first phase is to be completed by 2025.[18]

These and other such conceptual cities—despite their attractions—do not directly address the problems of existing cities. Instead, they are aspirational and reflect the vision of a future where present-day approaches are no longer viable. They are a welcome recognition that business as usual cannot persist in a world of growing population, existential threats from global warming, and a growing awareness that change in the energy sector must be embraced and promoted.

Such model cities may become realities someday, and they seem novel, innovative, and futuristic. However, they conceptually reflect ideas about future cities that have been expressed for decades. For example, Richard Register first coined the term "ecocity" in his 1987 book *Ecocity Berkeley: Building Cities for a Healthy Future*. Other visionaries of sustainable cities include architect Paul F. Downton, who later founded the company Ecopolis, as well as authors Timothy Beatley and Steffen Lehmann, who have written extensively on the subject. What is new about Masdar City and the other examples discussed here is that such ideas have gained traction—and investment—in recent years.

Coming Full Circle

Humans evolved with sunlight, and in the earliest days of human existence only the sun provided the energy needed for survival. The sun provided warmth, evaporated moisture into the sky that returned as rain, and made possible foods in the form of fruits, vegetables, nuts, and animal flesh.

With a better and more consistent diet rich in protein, early humans' brains got bigger; as brain size increased, people got smarter. Technologies improved, and people lived longer. Once fire was controlled, the burning of wood (a substance holding stored solar power) allowed humans to migrate into colder climes, expand their food supplies, and increase personal protection. Refined tools came into use, and language sophistication increased. Wherever numbers remained low and people remained mobile, and as long as there was fuel to burn, people usually could meet their personal energy needs. As populations grew and as densities increased, wood supplies became insufficient to meet the demand for heat, construction materials, and the making of charcoal (an important key to industrialization). Coal was the first of what became known as the fossil fuels to come into widespread use, to be followed in the twentieth century by oil and natural gas. All three types of fossil fuels are, in essence, stored solar energy. The world would never be the same.

By the middle of the twentieth century, we learned how to harness the power of the atom, converting mass into pure energy. This discovery led to the development of weapons of mass destruction and, ten years later, nuclear generating stations. Nuclear power was to be the energy source of the future, its fuel holding such a high energy density that fuel replenishments could be months if not years apart. It was an ideal solution for naval vehicles because it removed the need to surface to charge batteries or to return to harbor. So promising was nuclear energy in its earliest days that it was promoted as an energy source that could be used to generate electricity at such a low cost that it would be too cheap to meter.

Yet despite its promise, over seventy-five years after it was first developed, nuclear power still generates only about 5 percent of the total energy used in industrialized countries. It supplies about 20 percent of electricity in the United States and about 72 percent in France; it supplied about 30 percent in Japan before Fukushima. It was embraced especially by countries with limited fossil fuel reserves, and it was employed in other countries such as India and Pakistan largely for reasons of national pride and weapons development. As we have seen, despite the advantages of nuclear power, it also holds several drawbacks. Nonetheless, it is still considered a viable source of electrical generation. It might have a future—at least in some countries—especially if improvements in safety continue and concerns about greenhouse gas emissions do not abate.

Meanwhile, dozens of countries are leaping on the renewable energy bandwagon. Renewable energy sources are now easier to deploy and are more widely distributed; they permit a reduced reliance on foreign energy supplies, and offer a low-carbon footprint. Now that renewables have gained momentum, places that have never had much in the way of electricity can get it. In other parts of the world where conventional energy resources are the norm, renewables have become less expensive than other options.

We seem, despite resistance, to be at the beginning of a solar renaissance. If we can just loosen the reins on renewables and tighten the reins on fossil fuels and nuclear—at least enough to level the playing field a bit more—we might leave a lot of the angst of our modern age behind. We still eat solar energy in the form of food. We have added photovoltaics and solar concentrators; solar water heating, pumping, and desalination; wind turbines; and biofuels to the mix. Solar energy holds the promise to become our dominant energy resource once again. When it does, we will have returned to where we started. We will have come full circle.

What Awaits Us?

Successfully adapting to whatever future may roll over the horizon starts with identifying and softening the worst energy practices of the past. Can we stop adding new greenhouse gases to the atmosphere while also learning to remove the problematic GHGs already there? Can we restore the quality of rivers, lakes, and oceans that have been degraded from energy exploration and development, and then move on to rely more heavily on energy resources that do not hold such damaging potential? Can we reverse, remediate, or at least minimize damage to the landscapes we have scarred by the quest for the energy we need? Can we maintain a balance between the standard of living people desire and the energy needs of a growing population?

Looking ahead, a profound question continues to overshadow all others: is there any hope for the future, or are we already doomed—by population increase, by climate change, by the limits of fossil fuel resources, by tendencies toward nationalism, by the reluctance of world leaders to take decisive action in the face of obvious threats? Plenty of thoughtful and knowledgeable people have been painting a pessimistic picture of the future, partly because they are themselves alarmed—and possibly because they wish to shake people awake to the realities that are crowding in on us everywhere.

But, you may ask, even if we humans survive, will our future be like the dystopian one detailed in the Cormac McCarthy book and movie *The Road*?[19] Or will we be able to cobble together some version of sustainability, some form of equilibrium between the needs of the population and the capacities of the natural systems?

Pessimists believe that, on balance, the odds of prosperous, safe, and happy survival into the future are small, outweighed by a tilt toward some version of growing calamity. Such a dismal future is not inevitable, but it seems more in fashion these days to project an inevitability of doom than a dream of paradise (especially when the goal is to sell books and movie tickets!).

Nonetheless, while for a very long time we have conveniently overlooked the warning signs, we are gradually accepting that multiple threats to humanity exist.

We have gotten better at predicting them and measuring them. We have come to accumulate more knowledge about how to make corrective adjustments that will move us toward some version of a steady-state lifestyle.

We have, most important of all, come to accept that the ability of the earth's natural systems is limited to support us. These include limits of habitable space, soil fertility, clean water, landscape permanence, and rates of population growth. We have come to understand that we are reaching the limits of natural systems to cleanse themselves and repair human damage to flora, fauna, water, air, and the land itself. We have created a host of problems by disrupting these balances and exceeding nature's buffering capabilities. But at least we are coming to a communal state of awareness. Our next move is to take the steps needed to slow the degradation of such systems and turn the corner toward global sustainability.

Fortunately, humans have proved themselves a resilient species, and so there is still room for optimism. And it is not just optimism based on faith, but confidence based on history. We survived the Black Plague. We survived the influenza pandemic of the early twentieth century. We survived genocides in Europe, Africa, and Asia. We survived two catastrophic world wars and many disastrous regional ones. Hopefully, the same will happen when we face future pandemics.

One should caution that counting on our innate resiliency does not mean that we should simply step back from activism and hope for the best. It means that in addition to curtailing our most harmful practices, we urgently need to develop counterbalancing actions. It is not sufficient just to reduce greenhouse gas emissions (CO_2, methane, nitrous oxide, chlorofluorocarbons), for a simple reason; there is already too much in the atmosphere. Thus, we face not one challenge but two. We must stop making our problems worse, *and* we must introduce measures to reverse the damage we have already created. At present, we are not yet even holding steady, and we cannot hold back the flood forever.

If we simple-mindedly continue what we are doing, we are doomed. Following the thread of energy shows us all the ways energy weaves in and out of everything we do, everything we value, and everything we hope for the future. It ties together the accumulating problems of environmental quality, geopolitics, human welfare and dignity, justice, the law, personal comfort, the humanities, and the form and function of cities. While the challenges are daunting, following the thread of energy can also lead us to the solutions we seek.

Following the thread of energy will increase our awareness of the place of energy in our lives. Acknowledging and acting upon this accumulated awareness is the first step in illuminating the path to the solutions we must achieve to survive. When we do so, we will have accepted that *energy is a social issue with a technical component, rather than the other way around.*

Notes

Chapter 1

1. Benjamin Sovacool, Sarah Ryan, Paul C. Stern, Katy Janda, Gene Rochlin, Daniel Spreng, Martin J. Pasqualetti, Harold Wilhite, and Loren Lutzenhiser, "Integrating Social Science in Energy Research," *Energy Research & Social Science* 6 (March 2015): 95–99.

2. *Energy, Sustainability and Society*, https://energsustainsoc.biomedcentral.com/; *Sustainable Cities and Society*, https://www.journals.elsevier.com/sustainable-cities-and-society/; *Energy and Buildings*, https://www.journals.elsevier.com/energy-and-buildings/; *Renewable and Sustainable Energy Review*, https://www.scimagojr.com/journalsearch.php?q=27567&tip=sid&clean=0; *Energy Economics*, https://www.scimagojr.com/journalsearch.php?q=29374&tip=sid&clean=0; *Food and Energy Security*, https://www.scimagojr.com/journalsearch.php?q=21100775662&tip=sid&clean=0; *Energy Policy*, https://www.scimagojr.com/journalsearch.php?q=29403&tip=sid&clean=0.

3. On Scimago journal rankings, see https://www.scimagojr.com/journalrank.php. For details on *Energy Research & Social Science*, see https://www.scimagojr.com/journalsearch.php?q=21100325067&tip=sid&clean=0.

4. Fred Cottrell, *Energy and Society*, revised ed. (Bloomington, IN: AuthorHouse, 2009 [originally published in 1955 as *The Relation Between Energy, Social Change, and Economic Development*]); Harold Schobert, *Energy and Society: An Introduction*, 2nd ed. (Boca Raton, FL: CRC Press, 2014 [originally published 2002]).

5. Kolya Abramsky, ed., *Sparking a Worldwide Energy Revolution: Social Struggles in the Transition to a Post-Petrol World* (Oakland, CA: AK Press, 2010).

6. Gavin Bridge, Stewart Barr, Stefan Bouzarovski, Michael Bradshaw, Ed Brown, Harriet Bulkeley, and Gordon Walker, *Energy and Society: A Critical Perspective* (London: Routledge, 2018).

7. One Btu is approximately equivalent to the energy released by burning a wooden match.

8. Martin J. Pasqualetti and Marilyn A. Brown, "Ancient Discipline, Modern Concern: Geographers in the Field of Energy and Society," *Energy Research & Social Science* 1 (March 2014): 122–133. Also of relevance are the following: Martin J. Pasqualetti, "The Geography of Energy and the Wealth of the World," *Annals of the American Association of Geographers* 101, no. 4 (2011): 971–980. ; Kirby Calvert, "From 'Energy Geography' to 'Energy Geographies': Perspectives on a Fertile Academic Borderland," *Progress in Human Geography* 40, no. 1 (2015): 105–125; Barry D. Solomon and Kirby E. Calvert, eds., *Handbook on the Geographies of Energy*

(Cheltenham, UK: Edward Elgar, 2017); Stefan Bouzarovski, Martin J. Pasqualetti, and Vanesa Castán Broto, eds., *The Routledge Research Companion to Energy Geographies* (London: Routledge, 2017).

Chapter 2

1. Michael T. Klare, *The Race for What's Left: The Global Scramble for the World's Last Resources* (New York: Picador, 2012)

2. For an excellent treatment of energy transitions, see Vaclav Smil, *Energy Transitions: History, Requirements, Prospects* (Santa Barbara, CA: Praeger, 2010).

3. See Vaclav Smil, *Energy in World History* (Boulder, CO: Westview Press, 1994).

4. Rosita Fitch, "The History of the Noria," Noria Corporation, n.d., http://www.machinerylubrication.com/Read/1294/noria-history.

5. Gavin Menzies, *1421: The Year China Discovered America* (New York: Harper Perennial, 2002).

6. Robert Wilde, "The Development of Canals in the Industrial Revolution," ThoughtCo, 2019, https://www.thoughtco.com/development-of-canals-the-industrial-revolution-1221646.

7. For a series of maps of world pipelines, see http://www.theodora.com/pipelines/russia_former_soviet_union_pipelines.html.

Chapter 3

1. United Nations University and World Health Organization, *Human Energy Requirements: Report of a Joint FAO/WHO/UNU Expert Consultation: Rome, 17–24 October 2001*, vol. 1 (Rome: FAO, 2004).

2. You may calculate your own MDER here: https://www.vitfit.com.au/how-to-calculate-your-daily-energy-requirement/.

3. World Health Organization *The State of Food Security and Nutrition in the World 2018: Building Climate Resilience for Food Security and Nutrition* (Rome: FAO, 2018).

4. As of 2020, the Organization for Economic Cooperation and Development (OECD) has thirty-seven member countries that span the globe; these are the more highly developed economies. The BRIC countries (Brazil, Russia, India, and China) are considered non-OECD.

5. Tom McCann and Ron Lunt, "CAP Energy Needs," Central Arizona Project, 2013, http://www.cap-az.com/documents/meetings/10-17-2013/CAP_energy_needs_Board_retreat_10-16-13.pdf.

6. Often you will see the term "calorie" (lowercase *c*) used to denote a heat unit commonly used by physicists, and "Calorie" (capital *C*) to measure the heat content of food. One Calorie equates to 1,000 calories. Food system energy use data are from Patrick Canning et al., *Energy Use in the US Food System*, Economic Research Report

94 (Washington, DC: USDA, 2010) and from the USDA Food Availability (Per Capita) Data System.

7. Canning et al., *Energy Use in the US Food System*; Eric Garza, "The Energy Cost of Food," July 22, 2013, http://www.resilience.org/stories/2013-07-22/the-energy-cost-of-food; https://ericgarza.info/.

8. Garza, "The Energy Cost of Food."

9. Michael Bluejay, "Vegetarianism and the Environment: Why Going Meatless Is Important," http://michaelbluejay.com/veg/environment.html.

10. John Robbins, *The Food Revolution: How Your Diet Can Help Save Your Life and Our World* (New York: Red Wheel/Weiser, 2010).

11. Earl Ferguson Cook, *Man, Energy, Society* (San Francisco: Freeman, 1976).

12. Water Footprint Networks, "How to Calculate the Water Footprint of Any Food," http://waterfootprint.org/en/resources/interactive-tools/product-gallery/.

13. Robbins, *The Food Revolution*.

14. Christopher Matthews, "Livestock a Major Threat to Environment," FAO Newsroom, 2006, http://www.fao.org/newsroom/en/news/2006/1000448/index.html.

15. NRDC, "CO_2 Smackdown, Step 6: Trimming Out Beef and Pork," 2011, http://www.nrdc.org/living/energy/co2-smackdown-step-6-trimming-out-beef-and-pork.asp.

16. Bryan Walsh, "Meat: Making Global Warming Worse," *Time*, September 10, 2008.

17. Natalie Browne, Ross Kingwell, Ralph Behrendt, and Richard Eckard, "Comparing the Profitability of Sheep, Beef, Dairy and Grain Farms in Southwest Victoria Under Different Rainfall Scenarios," paper prepared for the 56th annual AARES conference, Fremantle, Western Australia, February 7–10, 2012, http://ageconsearch.umn.edu/bitstream/124249/2/2012AC%20Browne,%20Natalie%20CP.pdf.

18. Jonathan Safran Foer, "The End of Meat Is Here," *New York Times*, May 21, 2020.

19. Jeff Tollefson, "*US electrical grid on the edge of failure.*" *Nature*, August 25, 2013.

20. "9 of the Worst Power Outages in United States History," Electric Choice, https://www.electricchoice.com/blog/worst-power-outages-in-united-states-history/.

21. "Micro-Grids and Distributed Power Generation," *Trends*, December 14, 2013, https://trends-magazine.com/micro-grids-and-distributed-power-generation-2.

22. Kristina Hamachi LaCommare and Joseph H. Eto, "Understanding the Cost of Power Interruptions to US Electricity Consumers," Lawrence Berkeley National Lab, LBNL-55718, 2004.

23. U.S. Department of Energy, "Demand Response" https://www.energy.gov/oe/activities/technology-development/grid-modernization-and-smart-grid/demand-response.

24. American Society of Heating, Refrigerating, and Air-Conditioning Engineers, https://www.ashrae.org/.

25. U.S. Department of Energy, "Thermostats," https://www.energy.gov/energysaver/thermostats.

26. "20 Ways to Cut Your Utility Bill Without Sacrificing Comfort," *Consumer Reports*, August 26, 2015.

27. California Public Utilities Commission, "What Are TOU Rates?," http://www.cpuc.ca.gov/General.aspx?id=12194.

28. https://www.energy.gov/energysaver/save-electricity-and-fuel/lighting-choices-save-you-money/when-turn-your-lights.

29. U.S. Department of Energy, "Energy Efficient Computers, Home Office Equipment, and Electronics," http://energy.gov/energysaver/articles/energy-efficient-computer-use.

30. For example, just for matters of climate: Michael C. Baechler et al., "Guide to Determining Climate Regions by County," Building America Best Practices Series vol. 7.3, U.S. Department of Energy, August 2015, https://www.energy.gov/sites/prod/files/2015/10/f27/ba_climate_region_guide_7.3.pdf.

31. U.S. Department of Energy, "Resolve to Save Energy This Year," http://energy.gov/articles/resolve-save-energy-year.

32. For more information on energy codes, see International Code Council, http://www.iccsafe.org.

33. J. W. Gunnison, *A History of the Mormons* (Philadelphia: Lippincott, 1852), 151, republished at https://mmarkmiller.wordpress.com/2011/09/24/a-tale-jim-bridgers-descriptions-of-yellowstone-wonders-gunnison-1852/.

34. Martin V. Melosi, "The Automobile Shapes the City," Automobile in American Life and Society Project, 2004–2010, http://autolife.umd.umich.edu/Environment/E_Casestudy/E_casestudy2.htm.

35. https://blog.midwestind.com/how-much-road-in-the-us-in-miles/.

36. U.S. Department of Transportation, "Traffic Volume Trends," 2012, http://www.fhwa.dot.gov/policyinformation/travel_monitoring/12dectvt/12dectvt.pdf.

37. For the latest data, consult Oak Ridge National Laboratory, *Transportation Energy Data Book*, 39th ed. 2021. https://tedb.ornl.gov/data/.

38. "How Many Light Bulbs Are There in the World?," Answers.com, March 22, 2009, http://www.answers.com/Q/How_many_light_bulbs_are_there_in_the_world.

39. "China National Highway 110 Traffic Jam," Wikipedia, accessed May 22, 2021, http://en.wikipedia.org/wiki/China_National_Highway_110_traffic_jam.

40. "Fuel Efficiency: Modes of Transportation Ranked by MPG," True Cost (blog), May 27, 2010, http://truecostblog.com/2010/05/27/fuel-efficiency-modes-of-transportation-ranked-by-mpg/. Note: conversion factors used in these calculations are found on the referenced web page.

41. Gordon, "Concorde Technical Specs," http://www.concordesst.com/powerplant.html; "Boeing 747-8," Boeing, http://www.boeing.com/commercial/747/.

42. https://www.nrcan.gc.ca/energy/efficiency/communities-infrastructure/transportation/idling/4463.

43. https://www.anl.gov/es/idle-reduction-research.

44. "Save Fuel, Save Money," Trucker to Trucker, December 23, 2011, http://www.truckertotrucker.com/blog/save-fuel-save-money/.

45. Transportation expenditures per person, 2012, from U.S. Department of Energy, "How Much Do You Spend?," July 2, 2014, http://energy.gov/articles/how-much-do-you-spend.

Chapter 4

1. Derek Hodgson, "What's the Point of Art?," *The Conversation*, 2017, https://theconversation.com/whats-the-point-of-art-77118.

2. Alan S. Cowen and Dacher Keltner, "Self-Report Captures 27 Distinct Categories of Emotion Bridged by Continuous Gradients," *Proceedings of the National Academy of Sciences* 114, no. 38 (2017): E7900–E7909.

3. Danielle De Wolfe, "15 Things You (Probably) Didn't Know About *There Will Be Blood*," ShortList, 2014, https://www.shortlist.com/news/15-things-you-probably-didnt-know-about-there-will-be-blood.

4. Paul Roberts, *The End of Oil: On the Edge of a Perilous New World* (New York: Houghton Mifflin Harcourt, 2005).

5. Tamir Kalifa and Clifford Krauss, "This Feels Very Different," *New York Times*, May 1, 2020.

6. Gary Weimberg, "The China Syndrome: Film + Reality = Awareness," *Jump Cut: A Review of Contemporary Media* 24–25 (March 1981): 57–58.

7. John Wills, "Celluloid Chain Reactions: The China Syndrome and Three Mile Island," *European Journal of American Culture* 25, no. 2 (2006): 109–122.

8. Tony Shaw, "Rotten to the Core: Exposing America's Energy-Media Complex in *The China Syndrome*," *Cinema Journal* 52, no. 2 (2013): 93–113.

9. Stephen J. Dubner and Steven D. Levitt, "The Jane Fonda Effect," *New York Times Magazine*, September 16, 2007.

10. "Energy at the Movies," energyatthemovies.com.

11. The gun battle at Matewan, along with events at the so-called "Ludlow Massacre" in Colorado six years earlier over similar issues, marked an important turning point in the battle for miners' rights.

12. U.S. Department of Energy, "2017 US Energy and Employment Report," January 13, 2017, https://www.energy.gov/downloads/2017-us-energy-and-employment-report.

13. William Feaver, *Pitmen Painters: The Ashington Group 1934–1984* (Ashington, UK: Mid Northumberland Arts Group, 1993).

14. "Visions of the Valleys: Kim Howells on Art from South Wales," *Gerry in Art* (blog), April 1, 2015, https://gerryco23.wordpress.com/2015/04/01/visions-of-the-valleys-kim-howells-on-art-from-south-wales/.

15. "Visions of the Valleys."

16. Valerie Ganz, "Breakfast at the Coalface," http://www.valerieganz.co.uk/product/breakfast-at-the-coalface/.

17. Vitaly Savelyev, "Kogalym, a Town at the End of the Earth," Oil of Russia, 2009.

18. One can observe on his website many additional photographs of the human costs of energy, whether in Siberia or Chernobyl. See his images at http://www.gerdludwig.com/.

19. Coal: http://fineartamerica.com/art/photographs/coal?page=2). Oil: http://fineartamerica.com/art/photographs/oil. Nuclear: http://fineartamerica.com/art/photographs/nuclear.

20. For dozens of other passive designs, see https://www.pinterest.com/explore/passive-solar/.

21. Paul Poplawski and John Worthen, *D. H. Lawrence: A Reference Companion* (Westport, CT: Greenwood Publishing Group, 1996).

22. As stated on https://www.amazon.com/King-Coal-Gem-Putanto-Novel/dp/9811111952. Lawrence came of age in the coal district of north England, and the conditions above and below ground informed his writing. Refer especially to Lawrence's *Son and Lovers* and *Odour of Chrysanthemums*.

23. William G. Williams, *The Coal King's Slaves: A Coal Miner's Story* (Shippensburg, PA: Burd Street Press, 2002). Quotation is from the back cover.

24. Many books have taken on theme of coal mining. For a further sampling, consult https://www.goodreads.com/shelf/show/coal-mining.

25. Tom Frazier, "Coal Mining, Literature, and the Naturalistic Motif: An Overview," in *The Caverns of Night: Coal Mines in Art, Literature, and Film*, edited by William B. Thesing (Columbia: University of South Carolina Press, 2000), 201.

26. Anthony Bimba, *The Molly Maguires*, Intl Pub; third printing edition, June 1, 1970.

27. Karl Grossman, "Chernobyl Death Toll: 985,000, Mostly from Cancer," Global Research, 2013, http://www.globalresearch.ca/new-book-concludes-chernobyl-death-toll-985-000-mostly-from-cancer/20908.

28. Svetlana Alexievich, *Voices from Chernobyl: The Oral History of a Nuclear Disaster* (New York: Macmillan, 2006).

29. Julia Voznesenskaya, *The Star Chernobyl* (London: Quartet Books, 1987).

30. Vladimir Gubaryev, *Sarcophagus* (Harmondsworth, UK: Penguin Books, 1987).

31. Trevor Rahman, *Fukushima Nuclear Catastrophe* (n.p.: Organic eBooks, 2012).

32. One becquerel is defined as the activity of a quantity of radioactive material in which one nucleus decays per second; http://www.world-nuclear.org/information-library/safety-and-security/safety-of-plants/chernobyl-accident.aspx.

33. "TEPCO Wary of Fukushima Radiation Leak Exceeding Chernobyl," Reuters, April 12, 2011.

34. Hideo Furukawa, *Horses, Horses, in the End the Light Remains Pure: A Tale That Begins with Fukushima* (New York: Columbia University Press, 2011); Kenichi Hasegawa, *Fukushima's Stolen Lives: A Dairy Farmer's Story*, translated by Amy C. Franks (n.p.: Reportage Laboratory, 2016); Gretel Ehrlich, *Facing the Wave: A Journey in the Wake of the Tsunami* (New York: Vintage, 2014).

35. The entire poem is available here: http://www.thebeckoning.com/poetry/auden/auden1.html.

36. Marsha Bryant, "W. H. Auden and the Homoerotics of the 1930s Documentary," in *The Caverns of Night: Coal Mines, in Art, Literature, and Film*, edited by William B. Thesing (Columbia: University of South Carolina Press, 2000), 111.

37. Chris Green, ed., *Coal: A Poetry Anthology* (Frankfort, KY: Blair Mountain Press, 2007).

38. Gumey Norman, quote on the back cover of *Coal: A Poetry Anthology*, edited by Chris Green (Frankfort, KY: Blair Mountain Press, 2007).

39. Jason Frye, "Buffalo Creek," in *Coal: A Poetry Anthology*, edited by Chris Green (Frankfort, KY: Blair Mountain Press, 2007).

40. Tsutomu Sakai, "Pollution of Our Ancestral Land," in *Reverberations from Fukushima*, edited by Leah Stenson and Asao Sarukawa Aroldi (Portland, OR: Inkwater Press, 2014), 63.

41. Thomas G. Andrews, *Killing for Coal* (Cambridge, MA: Harvard University Press, 2010). The event was also the theme of several other non-fiction books, including one by 1972 US presidential candidate George McGovern and Leonard F. Guttridge, *The Great Coalfield War* (1972; reprint, Boulder: University Press of Colorado, 1996), and novels, including *Fire in the Hole*, by Sybil Downing (Boulder: University Press of Colorado, 1998), and *Horrors of History: Massacre of the Miners: A Novel*, by T. Neill Anderson (New York: Penguin Random House, 2015).

42. John Blair, "American Lung Association Report Highlights Toxic Health Threat of Coal-Fired Power Plants, Calls for EPA to Reduce Emissions and Save Lives," ValleyWatch, March 8, 2011, http://valleywatch.net/?p=1891.

43. Tennessee Ernie Ford: https://www.youtube.com/watch?v=jIfu2A0ezq0. LeAnn Rimes: https://www.youtube.com/watch?v=gwwIMWTZhhw. Red Army Choir: https://www.youtube.com/watch?v=dI9KBLb_8ro. ZZ Top: https://www.youtube.com/watch?v=J2aqvKY6zLc.

44. Frank J. Oteri, "Julia Wolfe Wins 2015 Pulitzer Prize in Music," New Music Box, April 21, 2015, https://nmbx.newmusicusa.org/2015-pulitzer-prize-in-music/.

45. Deborah Vankin, "Julia Wolfe's 'Anthracite Fields' Wins 2015 Pulitzer Prize in Music," *Los Angeles Times*, April 20, 2015.

46. "The Fracking Song," YouTube, posted by RT America, May 12, 2011, https://www.youtube.com/watch?v=oHQu3SeUwUI&index=6&list=PL04DA8F3AF9BE26AE.

47. Noriko Manabe, *The Revolution Will Not Be Televised: Protest Music After Fukushima* (New York: Oxford University Press, 2015).

48. "Cultural Impact of the Chernobyl Disaster," Wikipedia, accessed June 1, 2020, https://en.wikipedia.org/wiki/Cultural_impact_of_the_Chernobyl_disaster. There are also many songs about the Fukushima disaster, some of which are mentioned here: Dan Grunebaum, "Japan's New Wave of Protest Songs," *New York Times*, July 1, 2011.

49. "Chernobyl Song + Lyrics | Song About Chernobyl," YouTube, posted by Tomas Novak, February 2, 2013, https://www.youtube.com/watch?v=vsgf1dM9ACY.

Chapter 5

1. Michael S. Asante, *Deforestation in Ghana: Explaining the Chronic Failure of Forest Preservation Policies in a Developing Country* (Lanham, MD: University Press of America, 2005).

2. Christopher Helman, "America's 20 Dirtiest Cities," *Forbes*, December 10, 2012; Poulomi Banerjee, "Gone to Waste: How India Is Drowning in Garbage," *Hindustan Times*, February 9, 2016.

3. "Environmentalism in 1306," Scribol, 2018, http://scribol.com/art-and-design/green-design/environmentalism-in-1306/.

4. "Houses of Parliament (Monet Series)," Wikipedia, accessed June 1, 2020, https://en.wikipedia.org/wiki/Houses_of_Parliament_(Monet_series).

5. Christine L. Corton, *London Fog: The Biography* (Cambridge, MA: Harvard University Press, 2015).

6. Joanna Bourke, "*London Fog: The Biography* by Christine Corton, Review: 'Brilliant,'" *The Telegraph*, November 18, 2015.

7. Tom Kington, "First Scrub in 2,000 Years for Rome's Colosseum," *The Scotsman*, May 17, 2014.

8. Anthony Kleven, "Will China and India Choke to Death?," *The Diplomat*, February 26, 2016.

9. Alan Yuhas, "Scientists: Air Pollution Led to More than 5.5 Million Premature Deaths in 2013," *The Guardian*, February 12, 2016.

10. John Vidal, "Indian Coal Power Plants Kill 120,000 People a Year, Says Greenpeace." *The Guardian*, March 10, 2013.

11. For example, arsenic is a known poison and causes various cancers, including lymphoma; lead affects almost every body system; mercury causes irreversible damage to the nervous system; cadmium causes high blood pressure and kidney damage; vanadium causes lung damage and birth defects; selenium causes liver failure and fluid in the lungs. See https://vault.sierraclub.org/sierra/costofcoal/default.aspx.

12. "Summary of the Clean Air Act," U.S. Environmental Protection Agency, http://www.epa.gov/laws-regulations/summary-clean-air-act.

13. Search results for "acid rain" at lyrics.net, accessed June 1, 2020, http://www.lyrics.net/lyrics/acid%20rain.

14. Wikipedia, "Clean Air Act of 1963," accessed June 1, 2020, https://en.wikipedia.org/wiki/Clean_Air_Act_(United_States).

15. Richard Martin, *Coal Wars: The Future of Energy and the Fate of the Planet* (New York: St. Martin's Press, 2015).

16. Richard Martin, "For $20 Million, a Coal Utility Bought an Ohio Town and a Clear Conscience," *The Atlantic*, October 2014.

17. Rachel Price, "15 Abandoned Places in West Virginia That'll Give You Goosebumps," Only in Your State, April 1, 2015, http://www.onlyinyourstate.com/west-virginia/wv-abandoned-places/.

18. "The Energy Cost of Pumping Water from a Well," Code Green Prep, 2012, http://codegreenprep.com/2012/05/the-energy-cost-of-pumping-water-from-a-well/.

19. Bill Peacock, "Energy and Cost Required to Lift or Pressurize Water," University of California Cooperative Extension, Tulare County, 1998, https://www.hort360.com.au/wordpress/wp-content/uploads/2019/11/Energy-and-Cost-Required-to-Lift-or-Pressurise-Water.pdf.

20. And that is just to deliver the water to a local distribution network. Even more energy is needed to purify the water, pressurize the water, and treat the wastewater that occupants produce each year from showers, bathtubs, dishwashers, clothes washers, and toilets.

21. "Fracking Across the United States," Earthjustice, accessed June 1, 2020, http://earthjustice.org/features/campaigns/fracking-across-the-united-states#.

22. Eric Ng, "China Turns Increasingly to Unconventional Energy Production Such as Shale Gas," *South China Morning Post*, November 22, 2013.

23. Suzanne Goldenberg, "A Texan Tragedy: Ample Oil, No Water," *The Guardian*, August 11, 2013.

24. Christopher Bateman, "A Colossal Fracking Mess," *Vanity Fair*, June 21, 2010.

25. Scott K. Johnson, "Magnitude 5.6 Earthquake in Oklahoma Ties Biggest Area Has Seen," ArsTechnica, September 8, 2016, http://arstechnica.com/science/2016/09/magnitude-5-6-earthquake-in-oklahoma-ties-biggest-area-has-seen/.

26. One of the best sources of information and data about spills is held by the Centre of Documentation, Research and Experimentation on Accidental Water Pollution: http://wwz.cedre.fr/en/.

27. "A Deadly Toll," Center for Biological Diversity, April 2011, https://www.biologicaldiversity.org/programs/public_lands/energy/dirty_energy_development/oil_and_gas/gulf_oil_spill/a_deadly_toll.html.

28. "A Deadly Toll."

29. Some of the pollution comes from operational discharges, such as flushing out the bunkers, but this practice has been discouraged and reduced substantially in recent years, at least in U.S. territorial waters.

30. "What Is MARPOL Convention? IMO Convention for the Prevention of Pollution from Ships," EduMaritime, February 11, 2020, https://www.edumaritime.net/imo/marpol-convention.

31. Brian Westenhaus, "Trucks, Trains, or Pipelines—The Best Way to Transport Petroleum." OilPrice.com, August 13, 2013, https://oilprice.com/Energy/Energy-General/Trucks-Trains-or-Pipelines-The-Best-Way-to-Transport-Petroleum.html.

32. "Gulf War Oil Disaster," Counterspill, accessed June 1, 2020, http://www.counterspill.org/disaster/gulf-war-oil-disaster#photos-20. The on-land spillage from the sabotage of the well fields, by contrast, spilled 42 *billion* gallons. The Lakeview blowout (March 14, 1910–September 1911) spilled 9 million barrels, creating a river of oil. Less than half that oil was saved; the rest seeped into the soil or evaporated. "Lakeview Gusher," Wikipedia, accessed June 1, 2020, https://en.wikipedia.org/wiki/Lakeview_Gusher.

33. On territorial disputes: Alastair Leithead, "The 'Water War' Brewing over the New River Nile Dam," BBC News, February 24, 2018. On negative environmental and human effects: In the case of anadromous fish, such as salmon, the installation of large dams can bring about financial calamity, huge cultural changes to fish-dependent indigenous peoples, and even extinction of the fish themselves, as has already been documented for the Columbia River, among many other places.

34. Eliot Porter, *The Place No One Knew: Glen Canyon on the Colorado* (San Francisco: Sierra Club, 1966).

35. April R. Summit, *Contested Waters: An Environmental History of the Colorado River* (Boulder: University Press of Colorado, 2012).

36. "About," Glen Canyon Institute, http://www.glencanyon.org/about.

37. Jonathan Watts, "Three Gorges Dam May Force Relocation of a Further 300,000 People," *The Guardian*, January 22, 2010.

38. "List of Dam-Threatened World Heritage Sites," International Rivers, accessed June 1, 2020, https://www.internationalrivers.org/list-of-dam-threatened-world-heritage-sites.

39. For a summary of these themes, see Julian Simon, *The Ultimate Resource* (Princeton: Princeton University Press, 1981), and the work of Donella Meadows (e.g., Donella Meadows et al., "The Limits of Growth: A Report to the Club of Rome," 1972, http://web.ics.purdue.edu/~wggray/Teaching/His300/Illustrations/Limits-to-Growth.pdf) and Paul Ehrlich ("*The Population Bomb*," Wikipedia, accessed June 1, 2020, https://en.wikipedia.org/wiki/The_Population_Bomb).

Chapter 6

1. For comprehensive studies on climate change, refer to the website of the Intergovernmental Panel on Climate Change: https://www.ipcc.ch/.

2. "The Intergovernmental Panel on Climate Change (IPCC) is an intergovernmental body of the United Nations that is dedicated to providing the world with objective, scientific information relevant to understanding the scientific basis of the risk of human-induced climate change, its natural, political, and economic impacts and risks, and possible response options. . . . The IPCC produces reports that contribute to the work of the United Nations Framework Convention on Climate Change (UNFCCC), the main international treaty on climate change. The objective of the UNFCCC is to "stabilize greenhouse gas concentrations in the atmosphere at a level that would prevent dangerous anthropogenic (human-induced) interference with the climate system." "Intergovernmental Panel on Climate Change," Wikipedia, accessed May 22, 2021, https://en.wikipedia.org/wiki/Intergovernmental_Panel_on_Climate_Change.

3. IPCC, "Global Warming of 1.5°C: An IPCC Special Report," World Meteorological Organization, 2018, https://www.ipcc.ch/sr15.

4. IPCC, "Global Warming of 1.5°C."

5. "Special Report on Global Warming of 1.5°C," Wikipedia, accessed May 22, 2021, https://en.wikipedia.org/wiki/Special_Report_on_Global_Warming_of_1.5_%C2%B0C#cite_note-SR15_Headlines-5.

6. World Meteorological Organization, "WMO Statement on the State of the Global Climate in 2018," 2019, https://library.wmo.int/index.php?lvl=notice_display&id=20799#.XoeZVYipGMq.

7. Rebecca Lindsey, "Climate Change: Global Sea Level," Climate.gov, January 25, 2021, https://www.climate.gov/news-features/understanding-climate/climate-change-global-sea-level.

8. A series of interactive maps from Climate Central vividly illustrate some of the alarming annual flooding projections for several areas in India and Southeast Asia. Jessica Aldred, "These Maps Triple Number of People Threatened by Rising Seas," China Dialogue Ocean, November 20, 2019, https://chinadialogueocean.net/11740-sea-level-rise-map-interactive-asia/.

9. CO_2 emissions from fossil fuel combustion were almost zero prior to 1750. The United Kingdom was the world's first industrialized nation—and first fossil-fuel CO_2 emitter. In 1751 its (and global) emissions were less than 10 million metric tons—3,600 times less than global emissions today. We can conclude that emissions prior to 1750 were very low (and inconsequential compared to the numbers today).

10. "Coal Explained: How Much Coal Is Left?," U.S. Energy Information Administration, last updated October 9, 2020, https://www.eia.gov/energyexplained/coal/how-much-coal-is-left.php

11. "Global Warming and Hurricanes: An Overview of Current Research Results," NOAA, last revised February 5, 2020, https://www.gfdl.noaa.gov/global-warming-and-hurricanes/.

12. "Climate Misinformation by Source: Fred Singer," Skeptical Science, accessed May 22, 2021, https://skepticalscience.com/skeptic_Fred_Singer.htm.

13. "Data Snapshots: Reusable Climate Maps," Climate.gov, accessed May 22, 2021, https://www.climate.gov/maps-data.

14. T. F. Stocker et al., "Summary for Policymakers," in *Climate Change 2013: The Physical Science Basis, Contribution of Working Group I to the Fifth Assessment Report of the Intergovernmental Panel on Climate Change* (Cambridge: Cambridge University Press, 2013).

15. "The Effects of Climate Change," NOAA, accessed May 22, 2021, https://climate.nasa.gov/effects/.

16. D. Ehhalt et al., "Atmospheric Chemistry and Greenhouse Gases," chapter 4 in *IPCC Third Assessment Report Climate Change 2001: The Scientific Basis* (Cambridge: Cambridge University Press, 2001), https://www.ipcc.ch/site/assets/uploads/2018/03/TAR-04.pdf.

17. Hannah Ritchie, "The Carbon Footprint of Foods: Are Differences Explained by the Impacts of Methane?," Our World in Data, 2020, https://ourworldindata.org/carbon-footprint-food-methane.

18. Yadav Bajagai, "Role of Ruminant Animals in Global Climate Change," Food and Environment, December 2012, https://www.foodandenvironment.com/2012/12/role-of-ruminant-animals-in-global.html.

19. "Overview of Greenhouse Gases," EPA, accessed May 22, 2021, https://www.epa.gov/ghgemissions/overview-greenhouse-gases.

20. The terms "CO_2" and "CO_2e" seem similar but have different definitions. However, they are related to each other because they include the major greenhouse gas components. CO_2 is carbon dioxide gas. It is a colorless gas. The term "CO_{2e}" stands for carbon dioxide equivalents. It is a measure of how much global warming is produced by a greenhouse gas as a function of the amount or concentration of CO_2 gas. The key difference between CO_2 and CO_2e is that CO_2 is a gaseous compound whereas

CO_{2e} is a measure of the greenhouse effect. For a fuller discussion, see: https://www. differencebetween.com/difference-between-co2-and-vs-co2e/.

21. https://www.iea.org/data-and-statistics/data-product/co2-emissions-from-fuel-combustion.

22. Brian Wang, "China's Next Five Year Energy Plan Will Shape the Global Energy Future," Next Big Future, January 23, 2020, https://www.nextbigfuture.com/2020/01/chinas-next-five-year-energy-plan.html.

23. "Car Emissions and Global Warming," Union of Concerned Scientists, July 18, 2014, https://www.ucsusa.org/resources/car-emissions-global-warming.

24. "Top 20 Electric Vehicles in Norway (Jan–Dec 2019)," Clean Technica, accessed May 22, 2021, https://cleantechnica.com/2020/01/13/100-electric-vehicles-collected-42-of-norways-vehicle-sales-in-2019/.

25. Hristina Nikolovska, "Electric Car Statistics to Know if You're Going Green in 2020," CarSurance, February 2, 2021, https://carsurance.net/blog/electric-cars-statistics/.

26. "How Much Carbon Dioxide Is Produced from U.S. Gasoline and Diesel Fuel Consumption?," U.S. Energy Information Administration, last updated May 3, 2021, https://www.eia.gov/tools/faqs/faq.php?id=307&t=10.

27. Hannah Ritchie and Max Roser, "CO_2 and Greenhouse Gas Emissions," Our World in Data, last revised August 2020, https://ourworldindata.org/co2-and-other-greenhouse-gas-emissions#annual-co2-emissions.

28. For an addictive interactive graphic on different emissions rates from many of the world's counties, go to this web site: https://www.wri.org/blog/2017/04/interactive-chart-explains-worlds-top-10-emitters-and-how-theyve-changed.

29. Solar Power to Help Eliminate Poverty in Developing Countries. https://borgenproject.org/solar-power-to-help-eliminate-poverty/.

30. Carbon Emissions Outsourced to Developing Countries. https://borgenproject.org/carbon-emissions-outsourced-developing-countries/.

31. https://www.theguardian.com/environment/2015/nov/23/paris-climate-talks-developed-countries-must-do-more-than-reduce-emissions.

32. Carbon Capture and Storage (CCS). *Innova Magazine* (21 March 2016). https://russelsmithgroup.com/think-green/carbon-capture-storage-ccs/.

33. https://www.eia.gov/todayinenergy/detail.php?id=41433.

34. Within the U.S. Department of Energy, there is an office for this called EERE—the Office of Energy Efficiency and Renewable Energy. https://www.energy.gov/eere/about-office-energy-efficiency-and-renewable-energy.

35. Joseph E. Fargione. "Natural Climate Solutions for the United States," *Science Advances* 4, no. 11 (14 Nov 2018), eaat1869. DOI: 10.1126/sciadv.aat1869.

36. https://www.wri.org/blog/2020/01/2020-could-be-year-carbon-removal-takes.

37. Jean-Francois Bastin et al. "The Global Tree Restoration Potential," *Science* 365, no. 6448 (5 Jul 2019): 76–79. DOI: 10.1126/science.aax0848.

38. David Roberts, "Sucking Carbon Out of the Air Won't Solve Climate Change," *Vox,* July 16, 2018, https://www.vox.com/energy-and-environment/2018/6/14/17445622/direct-air-capture-air-to-fuels-carbon-dioxide-engineering. For more on Carbon Engineering's approach, see "Our Technology" at the Carbon Engineering website,

https://carbonengineering.com/our-technology/. For a sample of university-based approaches, see the website for the Center for Negative Carbon Emissions at Arizona State University, https://cnce.engineering.asu.edu/, and view the short Inside Science video presentation "What's White, Shaggy and Could Help Reduce Carbon Dioxide by 80%?" (posted September 28, 2016), at https://www.youtube.com/watch?v=HBatnQZvN64.

39. An updated list of announcements and commitments can be found at Kelly Levin et al., "What Does 'Net-Zero Emissions' Mean? 8 Common Questions, Answered," World Resources Institute, last updated May 2021, https://www.wri.org/blog/2019/09/what-does-net-zero-emissions-mean-6-common-questions-answered.

40. Seth Wynes and Kimberly A. Nicholas. "The Climate Mitigation Gap: Education and Government Recommendations Miss the Most Effective Individual Actions." Environ. Res. Lett. 12 (2017), 074024. https://iopscience.iop.org/article/10.1088/1748-9326/aa7541/pdf. The EPA has identified several dozen ways you can do your part as well: https://www.epa.gov/ghgemissions/overview-greenhouse-gases#carbon-dioxide.

41. You can begin by calculating your personal carbon footprint at the EPA's "Household Carbon Footprint Calculator," https://www.epa.gov/ghgemissions/household-carbon-footprint-calculator.

Chapter 7

1. Brad Plumer, "Trump Orders a Lifeline for Struggling Coal and Nuclear Plants," *New York Times*, June 1, 2018.
2. "President Obama Announces National Fuel Efficiency Policy," White House, May 19, 2009.
3. Todd Spangler and Eric D. Lawrence, "Trump's EPA Seeks to Cut Obama Rules on Gas Mileage in Cars, Trucks," *USA Today*, April 2, 2018.
4. U.S. Bureau of Land Management, "Methane and Waste Prevention Rule," 2018, https://www.blm.gov/programs/energy-and-minerals/oil-and-gas/operations-and-production/methane-and-waste-prevention-rule.
5. When we start listing questions of energy policy, we soon grasp the complexity of choosing one resource over another, or even deciding whether to establish any policies at all. We also begin to understand why we have often been relatively unsuccessful in setting national policy in the past. To provide some examples of the range and intricacies of energy policy, take a look at the 100 episodes of the podcast *Global Energy Policy*, produced by the Center for Global Energy Policy at Columbia University. In Apple Podcasts it can be found at https://itunes.apple.com/us/podcast/center-on-global-energy-policy/id920531348?mt=2.
6. Charles Homans, "Energy Independence: A Short History," *Foreign Policy*, January 3, 2012.

7. Other policy initiatives have also been proposed, especially regarding controlling the emissions affecting climate change. For example, there were eight different carbon pricing bills introduced in the 116th Congress. A good summary is available from Resources for the Future's "Carbon Pricing" webpage: https://www.rff.org/topics/carbon-pricing/.

8. Approximately every four years, the policies of individual member countries are reviewed in depth by a team of peers led by the IEA. For more country information, please consult the IEA country pages: https://www.iea.org/publications/countryreviews/.

9. Plummer, "Trump Orders a Lifeline for Struggling Coal and Nuclear Plants," *The New York Times*, 2018, https://www.nytimes.com/2018/06/01/climate/trump-coal-nuclear-power.html.
   Kelsey Tamborrino, "The Story of the Hurricanes," *Politico*, May 29, 2018.

10. Robert J. Brulle, "The Climate Lobby: A Sectoral Analysis of Lobbying Spending on Climate Change in the USA, 2000 to 2016," *Climatic Change* 149, nos. 3–4 (2018): 289–303.

11. "Energy Policy," Wikipedia, accessed June 16, 2020, https://en.wikipedia.org/wiki/Energy_policy.

12. Michael S. Hamilton, *Energy Policy Analysis: A Conceptual Framework* (Armonk, NY: M. E. Sharpe, 2015).

13. Jordan Wirfs-Brock, "Data: Explore 15 Years of Power Outages," Inside Energy, August 18, 2014, http://insideenergy.org/2014/08/18/data-explore-15-years-of-power-outages/.

14. Try clicking on www1.udel.edu/DRC/emforum/recordings/20111012.pdf and compare what you find to what your community has in place for outages of energy supplies.

15. "Speed Kills MPG," Shane Labs, accessed June 16, 2020, http://www.mpgforspeed.com.

16. https://www.rff.org/publications/reports/toward-a-new-national-energy-policy-assessing-the-options/.

17. "Energy Policy Act of 1992," Wikipedia, accessed June 16, 2020, https://en.wikipedia.org/wiki/Energy_Policy_Act_of_1992.

18. "Energy Policy Act of 2005," Wikipedia, accessed June 16, 2020, https://en.wikipedia.org/wiki/Energy_Policy_Act_of_2005.

19. John Quiggin, "Reviving Nuclear Power Debates Is a Distraction. We Need to Use Less Energy," *The Guardian*, November 8, 2013.

20. "Why Do Americans Fear Nuclear Power," *Frontline* (PBS) #1511, April 22, 1997, https://www.pbs.org/wgbh/pages/frontline/shows/reaction/etc/script.html.

21. "The Yom Kippur War: Background & Overview (October 1973)," Jewish Virtual Library, accessed June 16, 2020, https://www.jewishvirtuallibrary.org/background-and-overview-yom-kippur-war.

22. The chart is from "Nuclear Power in France," Wikipedia, accessed June 16, 2020, https://en.wikipedia.org/wiki/Nuclear_power_in_France.   Details   are   from

"Electricité de France History," Funding Universe, http://www.fundinguniverse.com/company-histories/electricit%C3%A9-de-france-history/.

23. Broomby, Rob Broomby, "France Struggles to Cut Down on Nuclear Power," BBC News, January 11, 2014, https://www.bbc.com/news/magazine-25674581.

24. International Energy Agency, "Japan," https://www.iea.org/countries/membercountries/japan/.

25. Hardy Graupner, "What Exactly Is Germany's 'Energiewende'?," Deutsche Welle, 2013, https://www.dw.com/en/what-exactly-is-germanys-energiewende/a-16540762.

26. Austin Davis, "Tiny Wolfhagen, Germany Leads the Country's Green Energy Transition," *The World*, May 25, 2018, https://www.pri.org/stories/2018-05-25/tiny-wolfhagen-germany-leads-countrys-green-energy-transition.

27. Volker Quaschning, "Statistics: Renewable Electricity Generation in Germany," https://www.volker-quaschning.de/datserv/ren-Strom-D/index_e.php.

28. Luigi Grossi, Sven Heim, and Michael Waterson, "The Impact of the German Response to the Fukushima Earthquake," *Energy Economics* 66 (2017): 450–465.

29. "Nuclear Energy in Denmark," World Nuclear Association, last updated January 2021, http://www.world-nuclear.org/information-library/country-profiles/countries-a-f/denmark.aspx. Denmark does, however, rely on nuclear power plants in other countries for 10 percent of its electricity.

30. "Nuclear Power in Italy," Wikipedia, accessed June 17, 2020, https://en.wikipedia.org/wiki/Nuclear_power_in_Italy.

31. "Nuclear Power Phase Out," Wikipedia, accessed June 17, 2020, https://en.wikipedia.org/wiki/Nuclear_power_phase-out.

32. Fred Pearce, "Industry Meltdown: Is the Era of Nuclear Power Coming to an End?," *Yale Environment 360*, May 15, 2017, https://e360.yale.edu/features/industry-meltdown-is-era-of-nuclear-power-coming-to-an-end.

33. "Plans for New Reactors Worldwide," World Nuclear Association, 2021, https://www.world-nuclear.org/information-library/current-and-future-generation/plans-for-new-reactors-worldwide.aspx.

34. "Emerging Nuclear Energy Countries," World Nuclear Association, 2021, https://www.world-nuclear.org/information-library/country-profiles/others/emerging-nuclear-energy-countries.aspx.

35. Andrew Blowers, *The Legacy of Nuclear Power* (New York: Routledge, 2016).

36. "EDF Energy Expects 20% Cost Saving for Sizewell C," World Nuclear News, January 18, 2018, http://www.world-nuclear-news.org/C-EDF-Energy-expects-20-cost-saving-for-Sizewell-C-18011801.html, "Nuclear Power in the United Kingdom," World Nuclear Association, 2017, http://www.world-nuclear.org/information-library/country-profiles/countries-t-z/united-kingdom.aspx.

37. "List of Cancelled Nuclear Reactors in the United States," Wikipedia, accessed June 17, 2020, https://en.wikipedia.org/wiki/List_of_cancelled_nuclear_reactors_in_the_United_States.

38. James Conca, "How Far Do You Have to Run After a Small Modular Nuclear Meltdown?" *Forbes*, August 29, 2018.

39. "Nuclear Reprocessing," Wikipedia, accessed June 17, 2020, https://en.wikipedia.org/wiki/Nuclear_reprocessing.

40. "Nuclear Reprocessing," Wikipedia.

41. A graphic reconstruction of the dispersal of the radioactive cloud from the Chernobyl accident can be found at "In Depth: Chernobyl's Accident Path and Extension of the Radioactive Cloud," Ratical Earth Journal, https://ratical.org/radiation/Chernobyl/IRSN14dayPlume.html. And NOAA's HYSPLIT model shows the dispersal pattern for Fukushima: "Fukushima Radioactive Aerosol Dispersion," NOAA SOS, posted May 1, 2012, https://www.youtube.com/watch?v=HCzuPm4T4qo.

42. "China: Energy Development Strategy Action Plan (2014–2020)," Asian and Pacific Energy Forum, accessed May 22, 2021, https://policy.asiapacificenergy.org/node/138.

43. Database of State Incentives for Renewables and Efficiency, NC Clean Energy Technology Center, North Carolina State University, http://www.dsireusa.org.

44. Bob Christie, "SolarCity Sues SRP over New Fee for Rooftop Solar Customers," *San Diego Union*-Tribune, March 3, 2015.

45. "Public Utility Regulatory Policies Act," Wikipedia, accessed May 22, 2021, https://en.wikipedia.org/wiki/Public_Utility_Regulatory_Policies_Act.

46. Nichola Groom, "Decades-Old Green Power Law Is a Fresh Nuisance to U.S. Utilities," Reuters, March 29, 2017.

47. Database of State Incentives for Renewables and Efficiency, NC Clean Energy Technology Center, North Carolina State University, http://programs.dsireusa.org/system/program/detail/658.

48. "Solar Market Insight Report 2021 Q2," Solar Energy Industries Association, https://www.seia.org/research-resources/solar-market-insight-report-2021-q2; "Solar Investment Tax Credit (ITC)," Solar Energy Industries Association, https://seia.org/initiatives/solar-investment-tax-credit-itc.

49. "Solar Investment Tax Credit (ITC)," June 2018, https://seia.org/sites/default/files/inline-files/SEIA-ITC-101-Factsheet-2018-June.pdf.

50. SCF News, "The ITC Cliff: Will Solar Be Economically Viable Without the ITC?," November 22, 2017, https://scf.com/blog/itc-cliff-will-solar-economically-viable-without-itc/.

51. "Solar Investment Tax Credit (ITC)," https://seia.org/initiatives/solar-investment-tax-credit-itc.

52. Policies of IEA Countries: United Kingdom. 2019 Review, https://www.iea.org/reports/energy-policies-of-iea-countries-united-kingdom-2019-review.

53. Christoph Böhringer et al., "The Impact of the German Feed-in Tariff Scheme on Innovation: Evidence Based on Patent Filings in Renewable Energy Technologies." *Energy Economics* 67 (2017): 545–553.

54. Chris Baynes, "Germany Produces Enough Renewable Energy in Six Months to Power Country's Households for an Entire Year," *Independent*, July 2, 2018

55. "Accomplishments," California Energy Commission, accessed May 22, 2021, http://www.energy.ca.gov/commission/accomplishments/.

56. Other common names for the same concept include Renewable Electricity Standard (RES) at the U.S. federal level and Renewables Obligation in the United Kingdom.

57. National Renewable Energy Laboratory, https://www.nrel.gov/state-local-tribal/basics-portfolio-standards.html.

58. For more details, see the website of the Intergovernmental Panel on Climate Change, http://www.ipcc.ch.

59. Union of Concerned Scientists, "The IPCC: Who Are They and Why Do Their Climate Reports Matter?," last updated October 11, 2018, https://www.ucsusa.org/global-warming/science-and-impacts/science/ipcc-backgrounder.html#bf-toc-1.

60. Oliver Milman, "James Hansen, Father of Climate Change Awareness, Calls Paris Talks 'a Fraud,'" *The Guardian*, December 12, 2015.

61. "Corporate Average Fuel Economy," NHTSA, https://www.nhtsa.gov/laws-regulations/corporate-average-fuel-economy.

62. "Energy Efficiency," California Energy Commission, https://www.energy.ca.gov/programs-and-topics/topics/energy-efficiency.

63. Weihua Zhao, University of Louisville, "Do HOV Lanes Save Energy? Evidence from a General Equilibrium Model of the City," 2018, Corpus ID: 46991225, Semantic Scholar, https://www.semanticscholar.org/paper/Do-HOV-Lanes-Save-Energy-Evidence-from-a-General-of-Zhao/51a791d6b3e84e8f7a2874627616f87468f66073.

64. United Kingdom. Transport for London. "Congestion Charges," Transport for London, https://tfl.gov.uk/modes/driving/congestion-charge.

65. Nicole Badstuber, "London Congestion Charge: Why It's Time to Reconsider One of the City's Great Successes," *The Conversation*, last updated April 11, 2019, http://theconversation.com/london-congestion-charge-why-its-time-to-reconsider-one-of-the-citys-great-successes-92478.

66. Arthur Van Benthem, "The Unintended Consequences of Ambitious Fuel-economy Standards," Wharton School, University of Pennsylvania, February 3, 2015, https://knowledge.wharton.upenn.edu/article/unintended-consequences-ambitious-fuel-economy-standards/.

67. "More U.S. Coal-Fired Power Plants Are Decommissioning as Retirements Continue," U.S. Energy Information Administration, July 26, 2019, https://www.eia.gov/todayinenergy/detail.php?id=40212.

68. Benjamin Storrow, "And Now the Really Big Coal Plants Begin to Close," *Scientific American*, August 16, 2019, https://www.scientificamerican.com/article/and-now-the-really-big-coal-plants-begin-to-close/.

69. "Regulating Power Sector Carbon Emissions," Center for Climate and Energy Solutions, Arlington, VA, accessed May 22, 2021, https://www.c2es.org/content/regulating-power-sector-carbon-emissions/.

70. U.S. EPA, "Electric Utility Generating Units: Repealing the Clean Power Plan," https://www.epa.gov/stationary-sources-air-pollution/electric-utility-generating-units-repealing-clean-power-plan.

71. Simon Evans and Rosamud Pearce, "Global Coal Power," Carbon Brief, March 26, 2020, https://www.carbonbrief.org/mapped-worlds-coal-power-plants.

72. Russ Baker, "What They Don't Tell You About Oil Industry Tax Breaks," *Business Insider*, May 23, 2011.

73. Baker, "What They Don't Tell You About Oil Industry Tax Breaks."

74. Justin Gillis and Hal Harvey, "Why a Big Utility Is Embracing Wind and Solar," *New York Times*, February 6, 2018.

75. Production Tax Credit for Renewable Energy, https://www.ucsusa.org/resources/production-tax-credit-renewable-energy.

76. Benedict Clements and Vitor Gaspar, "Act Local, Solve Global: The $5.3 Trillion Energy Subsidy Problem," *IMFBlog*, May 18, 2015, https://blogs.imf.org/2015/05/18/act-local-solve-global-the-5-3-trillion-energy-subsidy-problem/.

Chapter 8

1. "Hitler Focuses East, Sends Troops to Romania," History.com, last updated September 9, 2020, http://www.history.com/this-day-in-history/hitler-focuses-east-sends-troops-to-romania.

2. This is being challenged in court.

3. U.S. Department of Energy, Office of Fossil Energy, "Office of Petroleum Reserves," accessed June 19, 2020, http://energy.gov/fe/services/petroleum-reserves.

4. U.S. Department of Energy, "Strategic Petroleum Reserve Inventory," accessed June 12, 2020, http://www.spr.doe.gov/dir/dir.html.

5. Jared Diamond, *Guns, Germs, and Steel: The Fates of Human Societies* (New York: W. W. Norton, 1999).

6. U.S. Energy Information Administration, "The Strait of Hormuz Is the World's Most Important Oil Transit Chokepoint," January 4, 2012, http://www.eia.gov/todayinenergy/detail.cfm?id=4430.

7. Dion Nissenbaum and Julian E. Barnes, "U.S. Sends Ship, Planes as Iranians Seize Commercial Ship," *Wall Street Journal*, April 29, 2015.

8. Laleh Khalili, "How the (Closure of the) Suez Canal Changed the World," *The Gamming* (blog), August 31, 2014, https://thegamming.org/2014/08/31/how-the-closure-of-the-suez-canal-changed-the-world/.

9. Samantha Brletich, "The Crimea Model: Will Russia Annex the Northern Region of Kazakhstan?," *Geopolitics, History, and International Relations* 7, no. 1 (2015): 11–29.

10. Svetlana Burmistrova and Alessandra Prentice, "Conflict-Hit Coal Sector Casts Shadow on Ukraine Electricity Market," Reuters, August 14, 2014.

11. https://qz.com/1329732/what-is-the-eu-ukraine-russia-gas-pipeline-germany-hosts-meeting/.

12. Martin J. Pasqualetti, "The Alberta Oil Sands from Both Sides of the Border," *Geographical Review* 99, no. 2 (2009): 248–267.

13. Carl von Clausewitz, *On War*, trans. J. J. Graham (London: Kegan Paul, Trench, Trubner, 1918). This is a variant translation. More completely, the quotation is: "We see, therefore, that war is not merely an act of policy but a true political instrument, a

continuation of political intercourse carried on with other means. What remains peculiar to war is simply the peculiar nature of its means." The book *On War* was written mostly after the Napoleonic Wars and published posthumously in 1832.

14. Michael T. Klare, *Rising Powers, Shrinking Planet: The New Geopolitics of Energy* (New York: Macmillan, 2009).

15. Javier Blas, "The Biggest Saudi Oil Field Is Fading Faster than Anyone Guessed," Bloomberg, April 2, 2019.

16. "Gulf War Oil Disaster," Counterspill, January 22, 1991 http://www.counterspill.org/disaster/gulf-war-oil-disaster.

17. Antonia Juhasz, "Why the War in Iraq Was Fought for Big Oil," CNN, March 19, 2013.

18. Georgi Kantchev and Margarita Papchenkova, "Outlook for Oil Use Dims amid Supply Glut," *Wall Street Journal*, October 22, 2015: "The International Energy Agency, an energy watchdog, forecasts global oil demand growth falling from 1.8 million barrels a day this year to 1.2 million next year."

19. "How Is Saudi Arabia Reacting to Low Oil Prices?," World Bank, 2016, http://www.worldbank.org/en/country/gcc/publication/economic-brief-july-saudi-arabia-2016.

20. Peter R. Odell, *Oil and World Power* (London: Routledge, 2013).

21. "Saudi Arabia Economy Profile," Index Mundi, last updated November 27, 2020, http://www.indexmundi.com/saudi_arabia/economy_profile.html.

22. Nathaniel Parish Flanner, "More Bad News for Mexico's Economy," *Forbes*, July 30, 2015.

23. "Masdar City," Wikipedia, accessed June 19, 2020, https://en.wikipedia.org/wiki/Masdar_City.

24. Elisabeth Malkin, "With Oil Revenue Dropping, Mexico Announces Budget Cuts," New York Times, January 31, 2015.

25. Tsvetana Paraskova, "President of African Oil Producer Sudan Toppled in Military Coup," Oil Price.com, April 11, 2019, https://oilprice.com/Geopolitics/Africa/President-Of-African-Oil-Producer-Sudan-Toppled-In-Military-Coup.html.

26. For more maps about South Sudan, consult http://southsudaninfo.net/maps/.

27. Klare, *Rising Powers, Shrinking Planet*; Lutz Kleveman, *The New Great Game: Blood and Oil in Central Asia* (New York: Grove Press, 2003); Toyin Falola and Ann Genova, *The Politics of the Global Oil Industry: An Introduction* (Westport, CT: Greenwood, 2005); Klaus Dodds, Merje Kuus, and Joanne Sharp, *The Ashgate Research Companion to Critical Geopolitics* (Farnham, UK: Ashgate, 2013); Michael J. Bradshaw, "Global Energy Dilemmas: A Geographical Perspective," *Geographical Journal* 176, no. 4 (2010): 275–290; Philippe Le Billon, *The Geopolitics of Resource Wars* (London: Routledge, 2017).

28. For more on this project, see C. Alvares and R. Billorey, *Damming the Narmada. India's Greatest Planned Environmental Disaster* (Malaysia: Third World Network/APPN, 1988).

29. "Narmada River," Wikipedia, accessed June 19, 2020, https://en.wikipedia.org/wiki/Narmada_River.

30. "Joint Comprehensive Plan of Action," Wikipedia, accessed June 19, 2020, https://en.wikipedia.org/wiki/Joint_Comprehensive_Plan_of_Action.

31. There are many works on the subject of energy and geopolitics. For one that will open the door to others, see Harvard Kennedy School's Geopolitics of Energy Project, https://www.belfercenter.org/project/geopolitics-energy-project.

32. The Solutions Project, http://thesolutionsproject.org/; Amory Lovins, *Reinventing Fire* (White River Junction, VT: Chelsea Green, 2013).

33. From a press release issued by the International Renewable Energy Agency, http://irenanewsroom.org/2015/07/15/retrack-at-cop21. See also "COP21 Renewable Energy Track," http://irenanewsroom.org/2015/07/15/retrack-at-cop21/.

Chapter 9

1. United Nations Department of Economic and Social Affairs, "World's Population Increasingly Urban with More than Half Living in Urban Areas," July 10, 2014, https://www.un.org/en/development/desa/news/population/world-urbanization-prospects-2014.html.

2. "Arizona Economy at a Glance," Arizona Bureau of Labor Statistics, https://www.bls.gov/eag/eag.az.htm.

3. Andrew Ross, *Bird on Fire: Lessons from the World's Least Sustainable City* (Oxford: Oxford University Press, 2011).

4. You can get some idea of these lines of convergence in the U.S. Energy Information Administration's U.S. Energy Mapping System, https://www.eia.gov/state/maps.php.

5. United Nations Department of Economic and Social Affairs, Population Division, *World Urbanization Prospects: The 2018 Revision* (New York: United Nations, 2019).

6. Joyce Klein Rosenthal, "Evaluating the Impact of the Urban Heat Island on Public Health: Spatial and Social Determinants of Heat-Related Mortality in New York City," Ph.D. dissertation, Columbia University, 2010.

7. "Hot and Getting Hotter: Heat Islands Cooking U.S. Cities," Climate Central, August 20, 2014, http://www.climatecentral.org/news/urban-heat-islands-threaten-us-health-17919

8. Ross, *Bird on Fire.*

9. M. Glaskin, *Cycling Science: How Rider and Machine Work Together* (Chicago: University of Chicago Press, 2013).

10. U.S. Department of Energy, "Where the Energy Goes: Gasoline Vehicles," accessed January 13, 2017, https://www.fueleconomy.gov/feg/atv.shtml.

11. EU SME Centre and the China-Britain Business Council, https://www.eusmecentre.org.cn/partner/china-britain-business-council.

12. U.S. Energy Information Administration, "Country Analysis Executive Summary: China," 2020, https://www.eia.gov/international/content/analysis/countries_long/China/china.pdf.

13. IQ Air "World's Most Polluted Cities, 2020. https://www.iqair.com/world-most-polluted-cities.

14. American Council for an Energy-Efficient Economy, "2019 State Energy Efficiency Scorecard," https://www.aceee.org/research-report/u1908.

15. "Soft Energy Path," Wikipedia, accessed June 24, 2020, https://en.wikipedia.org/wiki/Soft_energy_path.

16. U.S. Energy Information Administration, "State Energy Consumption Estimates, 1960 Through 2014," DOE/EIA-0214(2014), June 2016, Table C13: "Energy Consumption per Capita by End-Use Sector, Ranked by State, 2014."

17. California Energy Commission, "Building Energy Efficiency Standard—Title 24," https://www.energy.ca.gov/programs-and-topics/programs/building-energy-efficiency-standards; and "Appliances Efficiency Regulation—Title 20," http://www.energy.ca.gov/appliances/.

18. Learn more about BREEAM at www.breeam.com, about BREEAM USA at www.breeam.com/usa/, and about LEED at www.usgbc.org/leed.

19. Renewable portfolio standards are regulations that require the increased production of energy from renewable energy sources, such as wind, solar, biomass, and geothermal. Other common names for the same concept include renewable electricity standard at the U.S. federal level and renewables obligation in the United Kingdom.

20. Hawaii State Energy Office, "Hawaii's Emerging Future: State of Hawaii Energy Resources Coordinator's Annual Report 2016," Honolulu.

21. Chris Megerian, "California Senate Leader Puts 100% Renewable Energy on the Table in New Legislation," *Los Angeles Times,* February 21, 2017.

22. "California Solar," Solar Energy Industries Association, https://seia.org/state-solar-policy/california-solar.; Net Summer Capacity of Utility Scale Units. Source: https://www.eia.gov/electricity/monthly/epm_table_grapher.php?t=table_6_02_b

23. Dean Apostol et al., *The Renewable Energy Landscape* (Abingdon, UK: Routledge, 2017).

24. U.S. Energy Information Administration, "Frequently Asked Questions: How Much Carbon Dioxide Is Produced by Burning Gasoline and Diesel Fuel?," last updated May 3, 2021, http://www.eia.gov/tools/faqs/faq.cfm?id=307&t=11.

25. For more about Arcosanti design principles, concepts, and history as well as plans for the future of the community, refer to https://arcosanti.org/.

26. Based on the U.S. Supreme Court decision in *Village of Euclid vs. Ambler Realty* in 1926.

27. J. Brian Phillips, "Houston, We Have a (Zoning) Problem," *The Objective Standard,* Spring 2009, https://www.theobjectivestandard.com/issues/2009-spring/houston-zoning-problem/.

Chapter 10

1. Local Initiatives Support Corporation, "Opportunity Atlas Shows the Effect of Childhood Zip Codes on Adult Success," accessed June 22, 2020, https://www.lisc.org/our-resources/resource/opportunity-atlas-shows-effect-childhood-zip-codes-adult-success.

2. Benjamin K. Sovacool and Michael H. Dworkin, *Global Energy Justice* (Cambridge: Cambridge University Press, 2014).

3. "Lago Agrio Oil Field," Wikipedia, accessed June 22, 2020, https://en.wikipedia.org/wiki/Lago_Agrio_oil_field; Michael I. Krauss, "*Chevron v. Donziger*: The Epic Battle for the Rule of Law Hits the Second Circuit." *Forbes*, April 21, 2015.

4. "40 Years in, Manitoba Apologizes to First Nations for Hydropower Dam Flooding," *Indian Country Today*, last updated September 13, 2018.

5. Kyle Ashmead, "Saskatchewan Uranium Mines Create Toxic Legacy," *Digital Journal*, 2010.

6. Martin J. Pasqualetti, "The Alberta Oil Sands from Both Sides of the Border," *Geographical Review* 99, no. 2 (2009): 248–267.

7. Martin J. Pasqualetti, "Opposing Wind Energy Landscapes: A Search for Common Cause," *Annals of the Association of American Geographers* 101, no. 4 (2011): 907–917.

8. See https://ejatlas.org/.

9. James Wilt, "'Projects of Death': Impact of Hydro Dams on Environment, Indigenous Communities Highlighted at Winnipeg Conference," The Narwhal, November 20, 2019, https://thenarwhal.ca/projects-of-death-impact-of-hydro-dams-on-environment-indigenous-communities-highlighted-at-winnipeg-conference/.

10. World Commission on Dams, *Dams and Development: A New Framework for Decision-Making: The Report of the World Commission On Dams* (London: Earthscan, 2000).

11. World Commission on Dams, *Dams and Development*.

12. Peter Bosshard, "New Independent Review Documents Failure of Narmada Dam," International Rivers, 2008, https://www.huffpost.com/entry/twelve-dams-that-changed_b_6457184.

13. "China's Global Role in Dam Building," International Rivers, accessed June 22, 2020, https://www.internationalrivers.org/campaigns/china-s-global-role-in-dam-building.

14. The Glen Canyon Institute is dedicated to the restoration of Glen Canyon and the removal, or breaching, of Glen Canyon Dam on Colorado River in Arizona (https://www.glencanyon.org/).

15. Sovacool and Dworkin, *Global Energy Justice*.

16. Peter Eigen, "Fighting Corruption in a Global Economy: Transparency Initiatives in the Oil and Gas Industry," *Houston Journal of International Law* 29 (2006): 327.

17. A video on the Extractive Industries Transparency Initiative can be found at https://eiti.org/eiti-videos.

18. David Blair, "Soaring Prices Push Queen Close to 'Fuel Poverty,'" *Financial Times*, October 21, 2011.

19. See https://www.energypoverty.eu/. See also Maricopa County, "Heat Reports," https://www.maricopa.gov/1858/Heat-Surveillance.

20. Jonathan Teller-Elsberg, Benjamin Sovacool, Taylor Smith, and Emily Laine, "Fuel Poverty, Excess Winter Deaths, and Energy Costs in Vermont: Burdensome for Whom?," *Energy Policy* 90 (2016): 81–91.

21. Dave Watson, "Pre-Payment Meters Masking Fuel Poverty," https://unisondave.blogspot.com/2015/05/pre-payment-meters-masking-fuel-poverty.html.

22. Ian Beesley and Ian McMillan, *The Book of Damp* (n.p.: Darkroom Press, 2014), n.p.

23. Simon Read, "Fuel Poverty Deaths Three Times Higher than Government Estimates," *Independent*, February 28, 2012, http://www.independent.co.uk/news/uk/home-news/fuel-poverty-deaths-three-times-higher-than-government-estimates-7462426.html.

24. Brenda Boardman, *Fixing Fuel Poverty: Challenges and Solutions* (London: Routledge, 2013).

25. Imtiaz Saba and Zia Ur-Rehman, "Death Toll from Heat Wave in Karachi, Pakistan, Hits 1,000," *New York Times*, June 26, 2015.

26. Chris Greig, "Energy, Carbon and Poverty: The Dilemma of Our Times," University of Queensland, 2013, https://energy.uq.edu.au/article/2016/05/energy-carbon-and-poverty-dilemma-our-times.

27. Global Commission to End Energy Poverty, accessed June 22, 2020, https://www.endenergypoverty.org/.

28. Gautam N. Yadama, *Fires, Fuel, and the Fate of 3 Billion: The State of the Energy Impoverished* (London: Oxford University Press, 2013).

29. Greig, "Energy, Carbon and Poverty."

30. World Health Organization, "Household Air Pollution and Health: Key Facts," accessed June 22, 2020, http://www.who.int/mediacentre/factsheets/fs292/en/.

31. "Energy Is a Human Right," Solar Electric Light Fund, accessed June 22, 2020, http://www.energyisahumanright.com/.

32. Erin Patrick, "Sexual Violence and Firewood Collection in Darfur," *Forced Migration Review* 27 (2007): 40–41.

33. Sarah K. Chynoweth and Erin M. Patrick, "Sexual Violence During Firewood Collection: Income-Generation as Protection in Displaced Settings," in *Gender-Based Violence: Perspectives from Africa, the Middle East, and India,* edited by Yanyi K. Djamba and Sitawa R. Kimuna, 43–55 (New York: Springer, 2007).

34. Soma Dutta and Tjarda Muller, "Women, Energy and Economic Empowerment," *Boiling Point* 66 (Spring 2015): 1, https://energia.org/assets/2015/07/BP66-Women-Energy-and-Economic-Empowerment-.compressed.pdf.

35. Soma Dutta, "Unlocking Women's Potential Towards Universal Energy Access," *Boiling Point* 66 (Spring 2015): 2, https://energia.org/assets/2015/07/BP66-Women-Energy-and-Economic-Empowerment-.compressed.pdf.

36. Nicholas Shaxson, *Poisoned Wells: The Dirty Politics of African Oil* (New York: St. Martin's Press, 2007).

37. Martin J. Pasqualetti et al., "A Paradox of Plenty: Renewable Energy on Navajo Nation Lands," *Society & Natural Resources* 29, no. 8 (2016): 885–899.

Chapter 11

1. "Xinjiang: A Region with Large Energy Resources and High Energy Consumption," Recast Urumqi, Task Group Energy, accessed June 27, 2020, http://recasturumqi.azurewebsites.net/en/The-Project/Task-Group-Energy/Research-Area.

2. https://worldpopulationreview.com/country-rankings/energy-consumption-by-country.

3. Agence France-Presse, "Dubai to Build Climate-Controlled 'City,' Largest Mall," July 6, 2014, http://news.yahoo.com/dubai-build-climate-controlled-city-largest-mall-130959387.html.

4. For the ecological footprint ranking of the UAE as of November 18, 2015, see https://www.footprintnetwork.org/2015/11/18/united-arab-emirates/.

5. "Human Development Index (HDI)," Wikipedia, accessed June 27, 2020, http://en.wikipedia.org/wiki/Human_Development_Index.

6. 50 GJ is equal to almost 14,000 kWh, or a bit more than the annual demand of a private residence in Phoenix, Arizona.

7. Ed Caryl, "Boosting per Capita Prosperity and Energy Consumption Is the Only Way to Care for Our Planet," *No Tricks Zone* (blog), June 3, 2012, http://notrickszone.com/2012/06/03/boosting-per-capita-prosperity-and-energy-consumption-is-the-only-way-to-care-for-our-planet/. These charts were developed from figures in the CIA World Factbook.

8. Hans Rosling looks at this a bit differently. He takes this idea a step further and suggests that as the fertility rate declines, so too will the rate of population growth, stabilizing at around 10 billion souls in the world. For an entertaining exposition of his idea, see "Dr. Hans Rosling Keynote | 2013 ARPA-E Energy Innovation Summit," YouTube, posted March 6, 2014, by ARPA-E, https://www.youtube.com/watch?v=OjEa6MUM9xM.

9. Michael T. Klare, *Resource Wars: The New Landscape of Global Conflict* (New York: Henry Holt, 2002).

10. See https://uspcase.asu.edu/.

Chapter 12

1. Raye C. Ringholz, *Uranium Frenzy: Boom and Bust on the Colorado Plateau* (New York: W. W. Norton, 1991).

2. This is the Arcosanti project, https://arcosanti.org/.

3. "Upper Big Branch Mine Disaster," Wikipedia, https://en.wikipedia.org/wiki/Upper_Big_Branch_Mine_disaster.

4. Howard Berkes, "An Epidemic Is Killing Thousands of Coal Miners. Regulators Could Have Stopped It," *All Things Considered,* NPR, December 18, 2018.

5. "Coal Worker's Pneumoconiosis (Black Lung Disease)," American Lung Association, accessed May 30, 2021, https://www.lung.org/lung-health-and-diseases/lung-disease-lookup/pneumoconiosis/.

6. https://oshissues.blogspot.com/2012/01/pneumoconiosis-january.html.

7. "Respirable Dust Rule: A Historic Step Forward in the Effort to End Black Lung Disease," Mine Safety and Health Administration, U.S. Department of Labor, accessed May 30, 2021, http://www.msha.gov/endblacklung/.

8. "Worker Safety: Black Marks for the Mines," *Time,* April 13, 1992.

9. Federal Coal Mine Health and Safety Act of 1969, PL 91-173, https://www.cdc.gov/niosh/programs/resp/risks.html.

10. "Respirable Dust Rule." The entire rule is at "Lowering Miners' Exposure to Respirable Coal Mine Dust, Including Continuous Personal Dust Monitors," 79 FR 24813, 24813–24994, https://www.federalregister.gov/articles/2014/05/01/2014-09084/lowering-miners-exposure-to-respirable-coal-mine-dust-including-continuous-personal-dust-monitors.

11. "Faces of Black Lung II," YouTube, posted by the CDC on January 21, 2020, https://youtu.be/X-agtyN4py4.

12. Richard Valdmanis, "A Tenth of U.S. Veteran Coal Miners Have Black Lung Disease: NIOSH," Reuters, July 19, 2018.

13. Cheng Yingqi, "Mines Seek Justice for Black Lung Disease," China Daily, February 16, 2011, http://www.chinadaily.com.cn/cndy/2011-02/16/content_12022012.htm.

14. GBD 2013 Mortality and Causes of Death Collaborators, "Global, Regional, and National Age-Sex Specific All-Cause and Cause-Specific Mortality for 240 Causes of Death, 1990–2013: A Systematic Analysis for the Global Burden of Disease Study 2013," Lancet 385, no. 9963 (2014): 117–71.

15. John G. Mitchell, "When Mountains Move," National Geographic, March 2006.

16. "Aberfan Disaster," Wikipedia, accessed May 30, 2021, http://en.wikipedia.org/wiki/Aberfan_disaster.

17. Glenn B. Stracher, Anupma Prakash, and Ellina V. Sokol, eds., Coal and Peat Fires: A Global Perspective (Amsterdam: Elsevier, 2015).

18. Kristin Ohlson, "Earth on Fire," Discover, July-August 2011.

19. Ohlson, "Earth on Fire."

20. Robert B Finkelman, "Potential Health Impacts of Burning Coal Beds and Waste Banks," International Journal of Coal Geology 59, nos. 1–3 (2004): 19–24.

21. For a Centralia mine fire chronology, see "A Brief History of the Centralia Mine Fire," Offroaders, accessed May 30, 2021, http://www.offroaders.com/album/centralia/chronology.htm.

22. "Abandoned Centralia," November 27, 2014, http://www.centraliapa.org/abandoned-centralia/.

23. For a summary of the history and fate of Centralia, consult Deryl B. Johnson, Centralia (Charleston, SC: Arcadia, 2004). For a theatrical treatment, watch the award-wining film The Town That Was. Many other books provide details and perspective on specific aspects of the fire.

24. G. B. Stracher, "Coal Fires: The Rising Global Interest," Earth 55, no. 9 (2010): 46–55.

25. There is a good summary of the conditions at Jharia, including the fire and relocation program, at "Jharia," Wikipedia, accessed May 30, 2021, http://en.wikipedia.org/wiki/Jharia#cite_note-JhariaTimes-5.

26. India State-Level Disease Burden Initiative Air Pollution Collaborators, "The Impact of Air Pollution on Deaths, Disease Burden, and Life Expectancy Across the States of India: The Global Burden of Disease Study 2017," Lancet Planetary Health 3, no. 1 (December 5, 2018): E26–E39.

27. "Coal and Air Pollution," Union of Concerned Scientists, last updated December 19, 2017, https://www.ucsusa.org/clean-energy/coal-and-other-fossil-fuels/coal-air-pollution#bf-toc-0.

28. "2014 National Emissions Inventory (NEI) Data," U.S. EPA, accessed May 30, 2021,https://www.epa.gov/air-emissions-inventories/2014-national-emissions-inventory-nei-data.

29. "America, Let's Move Beyond Coal and Gas," Sierra Club, accessed May 30, 2021, https://coal.sierraclub.org. For a current interactive map of the status of coal-burning power plants, go to "Coal Pollution in America," Sierra Club, https://coal.sierraclub.org/coal-plant-map.

30. "Oil 2018," IEA, March 2018, https://www.iea.org/oil2018/#section-3-1.

31. "Oil and Gas Extraction," OSHA, U.S. Department of Labor, accessed May 30, 2021, https://www.osha.gov/SLTC/oilgaswelldrilling/. For specifics regarding oil and gas drilling, see OSHA's "Oil and Gas Well Drilling and Servicing eTool," https://www.osha.gov/SLTC/etools/oilandgas/sitemap.html. See also "Fatal Injuries in Offshore Oil and Gas Operations—United States, 2003–2010," *Morbidity and Mortality Weekly Report* 62, no. 16 (April 26, 2013): 301–304.

32. Cayce Peterson, "Offshore Workers and Hurricanes in the Gulf of Mexico," November 7, 2017, https://thelambertfirm.com/?s=hurricanes.

33. "The World's Worst Offshore Oil Rig Disasters," Offshore Technology, last updated December 10, 2020, http://www.offshore-technology.com/features/feature-the-worlds-deadliest-offshore-oil-rig-disasters-4149812/.

34. Eric Nalde, "Oil Refineries a Risky Business," *San Francisco Chronicle,* last updated March 27, 2013.

35. "In Amenas Hostage Crisis," Wikipedia, accessed May 30, 2021, https://en.wikipedia.org/wiki/In_Amenas_hostage_crisis.

36. "Crude Oil by Rail," Association of American Railroads, accessed May 30, 2021, https://www.aar.org/article/crude-oil-by-rail.

37. Jim Wilson, "Accidents Surge as Oil Industry Takes the Train," *New York Times,* January 26, 2014.

38. Jad Mouawad, "Oil Tank Car Fire Forces Evacuation of North Dakota Town," *New York Times,* May 6, 2015.

39. Transportation Safety Board of Canada, "Railway Investigation Report R13D0054," 2013, http://www.tsb.gc.ca/eng/rapports-reports/rail/2013/r13d0054/r13d0054.asp.

40. "A Danger on the Rails | Op-Docs | The New York Times," YouTube, posted by the *New York Times,* April 22, 2015, https://www.youtube.com/watch?v=5JbvvE3x6hY.

41. For an interactive map that includes the various routes of the Keystone pipeline, see http://maps.fractracker.org/latest/?webmap=9b99a65e04fd4229959cc59fa6dd4eea.

42. https://www.phmsa.dot.gov.

43. Customarily, a chemical called mercaptan is added to the gas to give it a unique and identifiable odor. Mercaptan smells like rotten eggs, so it is easily noticed.

44. "List of Pipeline Accidents," Wikipedia, accessed May 30, 2021, https://en.wikipedia.org/wiki/List_of_pipeline_accidents.

45. Vanessa Schipani, "The Facts on Fracking Chemical Disclosure," FactCheck, April 7, 2017, https://www.factcheck.org/2017/04/facts-fracking-chemical-disclosure/.

46. Qingmin Meng and Steve Ashby, "Distance: A Critical Aspect for Environmental Impact Assessment of Hydraulic Fracking," *Extractive Industries and Society* 1 (2014): 124–126.

47. Benjamin K. Sovacool, "Cornucopia or Curse? Reviewing the Costs and Benefits of Shale Gas Hydraulic Fracturing (Fracking)," *Renewable and Sustainable Energy Reviews* 37 (2014): 249–264.

48. http://www.world-nuclear.org/info/Current-and-Future-Generation/Nuclear-Power-in-the-World-Today/.

49. "Nuclear Power in the World Today," World Nuclear Association, last updated March 2021, https://world-nuclear.org/information-library/current-and-future-generation/plans-for-new-reactors-worldwide.aspx.

50. Potential Human Health Effects of Uranium Mining, Processing, and Reclamation, https://www.ncbi.nlm.nih.gov/books/NBK201047/.

51. "Uranium: Known Facts and Hidden Dangers," address by Dr. Gordon Edwards at the World Uranium Hearings, 1992, http://www.ccnr.org/salzburg.html#he.

52. "The Legacy of Abandoned Uranium Mines in the Grants Mineral Belt, New Mexico," U.S. EPA, 2011, http://www.epa.gov/region6/6sf/newmexico/grants/uranium-mine-brochure.pdf.

53. Ringholz, *Uranium Frenzy*.

54. "Moab Utah: Mountain Biking Capital of the World," Durango Roadtripping, May 27, 2009; Christopher Reynolds, "Moab Is Mountain Biking Capital," *Deseret News,* May 22, 1994.

55. "Moab Uranium Mill Tailings Pile," Wikipedia, accessed May 30, 2021, http://en.wikipedia.org/wiki/Moab_uranium_mill_tailings_pile.

56. Anthony Ripley, "City in Colorado Awakens to Scope of Radioactive Waste Problem," *New York Times*, October 4, 1971.

57. Elisa J. Grammer, "The Uranium Mill Tailings Radiation Control Act of 1978 and NRC's Agreement State Program," *Natural Resources Lawyer* 13, no. 3 (1981): 469–522.

58. Sharon Sullivan, "Radioactive Mill Tailings Still an Issue," *Post Independent/Citizen Telegram,* December 3, 2009.

59. "The Nuclear Chain," NuclearRisks.org, accessed May 30, 2021, http://www.nuclear-risks.org/en/hibakusha-worldwide/black-hillspaha-sapa.html; J. Stone et al., "Final Report: North Cave Hills Abandoned Uranium Mines Impact Investigation," U.S. Department of Agriculture Forest Service, April 18, 2007, 10–11, http://uranium.sdsmt.edu/Downloads/NCHUraniumMinesImpactReport04-18-17.pdf.

60. See, for example, summaries of conditions in many locations in *The Uranium Mining Remediation Exchange Group (UMREG) Selected Papers 1995–2007* (n.p.: International Atomic Energy Agency, 2011).

61. Peter Diehl, "Uranium Mining in Eastern Germany: The Wismut Legacy," World Information Service on Energy Uranium Project, last updated April 17, 2011, http://www.wise-uranium.org/uwis.html.

62. Susan E. Dawson and Gary E. Madsen, "Psychosocial and Health Impacts of Uranium Mining and Milling on Navajo Lands," *Health Physics* 101, no. 5 (November 2011): 618–625.

63. S. Darby, D. Hill, and R. Doll, "Radon: A Likely Carcinogen at All Exposures," Annals of Oncology 12, no. 10 (2005): 1341–1351.

64. Kate Brown, *Manual for Survival: A Chernobyl Guide to the Future* (New York: Knopf, 2019).

65. Rebecca Staudenmaier, "Germany's Nuclear Phaseout Explained," Deutsche Welle, June 15, 2017, https://www.dw.com/en/germanys-nuclear-phase-out-explained/a-39171204.

66. "Materials Transportation," U.S. Nuclear Regulatory Commission, last updated December 2, 2020, https://www.nrc.gov/materials/transportation.html.

67. "Workplace Injury and Fatality Statistics," Electrical Safety Foundation International, accessed May 30, 2021, https://www.esfi.org/workplace-injury-and-fatality-statistics.

68. Stephan Maranian, "Controversial Northern Pass Project Struck Down," *St. Anselm Crier,* February 18, 2018.

69. David Russell Schilling, "Electromagnetic Field Lights Up Field of Florescent [sic] Tubes," Industry Tap into News, February 26, 2013, https://www.industrytap.com/florescent-bulbs-unplugged-and-shinning-tapping-electromagnetic-fields/1763.

70. Peter Elliott, David Wadley, and Jung Hoon Han, "Determinants of Homeowners' Attitudes to the Installation of High-Voltage Overhead Transmission Lines," *Journal of Environmental Planning and Management* 59, no. 4 (2016): 666–686.

71. "Electric and Magnetic Fields in California Public Schools," www.emfrf.com/emf-information/.

72. "School Site Selection and Approval Guide," California Department of Education, 2004, http://www.cde.ca.gov/ls/fa/sf/schoolsiteguide.asp#highvoltage.

Chapter 13

1. Katharina Bouhholz, "Saudi Aramco flirts with $2 Trillion Valuation," Statista, December 13, 2019, https://www.statista.com/chart/20300/market-caps-of-selected-companies-on-dec-12-2019/.

2. One million Btu equals approximately 90 pounds of coal or 125 pounds of kiln-dried wood or 8 gallons of motor gasoline.

3. Commercial energy does not include such energy sources as wood and dung that are gathered by individuals, mostly in less-developed countries.

4. "U.S. Corporate Capex Study: Trends Are Relatively Flat for 2014," Fitch Ratings, September 23, 2013, cited in "Powering America: The Economic and Workforce Contributions of the U.S. Electric Power Industry," 2017, https://mjbradley.com/sites/default/files/poweringamerica.pdf.

5. Estimated based on the electric power industry's impact on sales of all U.S. industry sectors and the ratios of those industries' national sales to GDP contributions.

Modeled estimates and datasets were provided by Emsi. As cited in "Powering America."

6. "2020 U.S. Energy and Employment Report," National Association of State Energy Officials, https://www.usenergyjobs.org/.

7. "Energy Employment by State," National Association of State Energy Officials, 2020, https://www.usenergyjobs.org/2020-state-reports.

8. "Energy Employment by State."

9. Jeffrey Kluger, "Time and Space," *Time*, November 2016, http://world.time.com/ timelapse/ U.S. Department of Energy.

10. "DOE Releases Second Annual National Energy Employment Analysis," U.S. Department of Energy, January 13, 2017, https://energy.gov/articles/ doe-releases-second-annual-national-energy-employment-analysis-0.

11. "Occupational Employment Statistics," U.S. Bureau of Labor Statistics, May 2016, https://www.bls.gov/oes/current/oes450000.htm.

12. "34 States Have Active Oil and Gas Activity in U.S. Based on 2016 Analysis," Fractracker Alliance, March 23, 2017, https://www.fractracker.org/2017/03/34-states-active-drilling-2016/.

13. Macrotrends, https://www.macrotrends.net/1369/crude-oil-price-history-chart.

14. "Oil and Gas Extraction," NAICS 2111, U.S. Bureau of Labor Statistics, accessed July 28, 2020, https://www.bls.gov/iag/tgs/iag211.htm#workforce.

15. Table 1. U.S. coal production, 2014–2020, U.S. Energy Information Administration, https://www.eia.gov/coal/production/quarterly/.

16. "Annual Coal Distribution Report 2016," U.S. Energy Information Administration, November 2017, https://www.eia.gov/coal/distribution/annual/.

17. Details on the fortunes of coal power may be found in a constantly update format here: https://www.carbonbrief.org/mapped-worlds-coal-power-plants.

18. Scott Simon, "Support for Marijuana in Coal Country," *Weekend Edition,* NPR, January 6, 2018, https://www.npr.org/2018/01/06/576197724/ support-for-marijuana-in-coal-country.

19. Aaron Schrank, "Powder River Basin Coal Producers to Increase Exports to Asia," Wyomic Public Media, August 22, 2014, http://wyomingpublicmedia.org/post/ powder-river-basin-coal-producers-increase-exports-asia.

20. "2020 U.S. Energy and Employment Report."

21. IAEA, https://pris.iaea.org/PRIS/WorldStatistics/UnderConstructionReactorsByCo untry.aspx.

22. Markets Insider. 2021, https://markets.businessinsider.com/commodities/uranium-price?op=1.

23. "World Uranium Mining Production," World Nuclear Association, last updated December 2020, http://www.world-nuclear.org/information-library/nuclear-fuel-cycle/mining-of-uranium/world-uranium-mining-production.aspx.

24. See their September 2017 financial statements here: http://www.kazatomprom.kz/en.

25. Colin Thomas, "845 Cameco Workers Receive Temporary Layoff Notices," CTV News, November 8, 2017, http://saskatoon.ctvnews.ca/845-cameco-workers-receive-temporary-layoff-notices-1.3669570.

26. "World Uranium Mining Production."

27. Maitali Ramkumar, "How Refining Margins Are Key Indicators of Refining Profitability," Market Realist, November 14, 2016, http://marketrealist.com/2016/11/refining-margins-key-indicators-refining-profitability/.

28. Ramkumar, "How Refining Margins Are Key Indicators of Refining Profitability."

29. "Natural Gas," Wikipedia, accessed July 28, 2020, https://en.wikipedia.org/wiki/Natural_gas.

30. For a more complete list of the products that derive from the processing of oil and gas, see "Products Made from Oil and Natural Gas," Petroleum Services Association of Canada, accessed July 28, 2020, https://oilandgasinfo.ca/products/.

31. Uranium-235 fissions with low-energy thermal neutrons because the binding energy resulting from the absorption of a neutron is greater than the critical energy required for fission.

32. Heavy water reactors, such as those used in Canada, require little or no enrichment but do require the production of "heavy" water, D_2O. The letter D stands for deuterium, the second isotope of hydrogen.

33. "Economics of Nuclear Power," World Nuclear Associations, last updated March 2020, http://www.world-nuclear.org/information-library/economic-aspects/economics-of-nuclear-power.aspx.

34. "The DOE Laboratories System," U.S. Department of Energy, accessed July 28, 2020, https://science.energy.gov/laboratories/oak-ridge-national-laboratory/.

35. "Coal-Fired Electricity Generation in 2019 Falls to 42-Year Low," U.S. Energy Information Administration, May 11, 2020, https://www.eia.gov/todayinenergy/detail.php?id=43675.

36. Daniel Moore, "As Coal Cools Off, Railroads Close Tracks and Cut Jobs Across the Country," Pittsburgh Gazette, February 15, 2016.

37. Moore, "As Coal Cools Off."

38. Pittsburgh Intermodel Rail Terminal, CSX, https://www.csx.com/index.cfm/about-us/projects-and-partnerships/pittsburgh-intermodal-rail-terminal/.

39. For a real-time piracy map, see "IMB Piracy and Armed Robbery Map," ICC Commercial Crime Services, https://www.icc-ccs.org/piracy-reporting-centre/live-piracy-map.

40. Carolyn Embach, "Oil Spills Impact on the Ocean," Water Encyclopedia, accessed July 28, 2020, http://www.waterencyclopedia.com/Oc-Po/Oil-Spills-Impact-on-the-Ocean.html.

41. U.S. Energy Information Administration, https://www.eia.gov/energyexplained/electricity/electricity-in-the-us-generation-capacity-and-sales.php.

42. "Frequently Asked Questions (FAQs)," U.S. Energy Information Administration, accessed July 28, 2020, https://www.eia.gov/tools/faqs/faq.php?id=65&t=2.

43. "Fact Sheets," Nuclear Energy Institute, accessed July 28, 2020, https://www.nei.org/resources/fact-sheets.

44. "Nuclear Industry Jobs," Nuclear Energy Institute, accessed July 28, 2020, https://www.nei.org/resources/fact-sheets/nuclear-industry-jobs.

45. See https://www.indeed.com/q-Power-Plant-jobs.html.

46. J. D. Dolan, "Anniversary of a Disaster: Edison's Mohave Generating Station Explosion," *Los Angeles Times,* June 9, 2010.

47. Nicole Feldman, "The Steep Costs of Nuclear Waste in the U.S.," *Stanford Earth Matters,* July 3, 2018, https://earth.stanford.edu/news/steep-costs-nuclear-waste-us#gs.8a8x75.

48. Laura Strickler, "Cost to Taxpayers to Clean Up Nuclear Waste Jumps $100 Billion in a Year," NBC News, January 29, 2019.

49. James Burgess, "U.S. Nuclear Waste Disposal—Only One Company Cashing In," OilPrice.com, January 21, 2014, https://oilprice.com/Latest-Energy-News/World-News/U.S.-Nuclear-Waste-Disposal-Only-One-Company-Cashing-In.html.

50. W. Sproat, "Yucca Mountain Cost Estimate Is Increased," United Press International, August 5, 2008.

51. "Statistics," Nuclear Energy Institute, https://www.nei.org/resources/statistics.

52. "Decommissioning Nuclear Facilities," World Nuclear Association, accessed July 28, 2020, https://www.world-nuclear.org/information-library/nuclear-fuel-cycle/nuclear-wastes/decommissioning-nuclear-facilities.aspx.

53. David Duerr, "Transportation of Shippingport Reactor Pressure Vessel," *Journal of Construction Engineering and Management* 117, no. 3 (September 1991): 551–564.

54. According to some authorities, the United States also has the largest, best-trained, and best-educated nuclear community in the world, with over 150,000 professionals and workers spread out among government, academia, and industry. James Conca, "Electricity and Jobs in America," *Forbes,* August 3, 2017.

55. "2020 U.S. Energy and Employment Report," https://www.usenergyjobs.org.

56. "Renewable Energy and Jobs—Annual Review 2018," International Renewable Energy Agency, https://www.irena.org/publications/2018/May/Renewable-Energy-and-Jobs-Annual-Review-2018.

57. "Renewable Energy and Jobs—Annual Review 2017," International Renewable Energy Agency, http://www.irena.org/publications/2017/May/Renewable-Energy-and-Jobs--Annual-Review-2017.

58. "2017 U.S. Energy and Employment Rate," U.S. Department of Energy, January 13, 2017, https://energy.gov/downloads/2017-us-energy-and-employment-report.

Chapter 14

1. Daniel Kahneman, *Thinking Fast and Slow* (New York: Farrar, Straus and Giroux, 2011).

2. Bruce Tonn et al., "Power from Perspective: Potential Future United States Energy Portfolios," *Energy Policy* 37, no. 4 (2009): 1432–1443.

3. Tonn et al., "Power from Perspective." https://doi.org/10.1016/j.enpol.2008.12.019.

4. Auden Schendler and Andrew P. Jones, "Stopping Climate Change Is Hopeless. Let's Do It," *New York Times,* October 6, 2018.

5. Mark Dyson, Alex Engel, and Jamil Farbes, "The Economics of Clean Energy Portfolios," Rocky Mountain Institute, 2018, https://www.rmi.org/insight/the-economics-of-clean-energy-portfolios/.

6. Paul Sullivan, "East Germany's Old Mines Transformed into New Lake District," *The Guardian*, September 17, 2016.

7. UNESCO, "Zollverein Coal Mine Industrial Complex Essen," accessed July 3, 2020, https://worldheritagegermany.com/zollverein-coal-mine-industrial-complex-essen/.

8. Donella Meadows et al., *Limits to Growth: A 30-year Update*, 3rd edition (Vermont, US: Chelsea Green Publishing, 2004); Paul R. Ehrlich and Anne H. Ehrlich, *The Population Explosion* (New York: Simon & Schuster, 1990).

9. From 2010 to 2015, while China's GDP grew at an average annual rate of 7.8 percent, energy use fell by 18.2 percent (exceeding the national goal of a 16 percent reduction) and dropped from 0.617 to 0.505 million tons of oil equivalent per unit of GDP (at 2010 constant price). Such impressive results have led the International Energy Agency to call China a "global efficiency heavyweight." Thibaud Voïta, "The Power of China's Energy Efficiency Policies," IFPRI, September 2018, https://www.ifri.org/en/publications/etudes-de-lifri/power-chinas-energy-efficiency-policies.

10. The world has invested $2.9 trillion in green energy sources since 2004, according to new research, with China leading the way in recent years with its push toward solar power.

11. China is the number one country for solar water heating capacity in the world, with 290 GWt in operation at the end of 2014, accounting for about 70 percent of total world capacity.

12. "China maintained its position as a wind energy powerhouse, installing 19.7 GW, while the European Union added 15.6 GW of capacity. The U.S. installed a little over 7 GW of capacity." Anmar Frangoul, "From China to Brail, These Are the World Titans of Wind Power," CNBC, June 4, 2018.

13. "Masdar City," Wikipedia, accessed July 3, 2020, https://en.wikipedia.org/wiki/Masdar_City.

14. "The Sustainable City," Wikipedia, accessed on June 4, 2020, https://en.wikipedia.org/wiki/The_Sustainable_City.

15. Andrea Lo, "Can India's Amaravati Become the Next Sustainable City?," CNN, October 15, 2018, https://www.cnn.com/style/article/amaravati-india-sustainable-city/index.html.

16. Leanna Garfield, "Saudi Arabia Is Building a $500 Billion Mega-City That's 33 Times the Size of New York City," *Business Insider*, February 22, 2018.

17. Kingdom of Saudi Arabia, "Saudi Vision 2030," https://vision2030.gov.sa/en.

18. Bill Bostock, "Everything We Know About Neom, a 'Mega-City' Project in Saudi Arabia with Plans for Flying Cars and Robot Dinosaurs," *Business Insider*, September 23, 2019.

19. "The Road," Wikipedia, accessed June 4, 2020, https://en.wikipedia.org/wiki/The_Road.

Index

Tables and figures are indicated by *t* and *f* following the page number

Printed in the USA/Agawam, MA
September 12, 2022

798393.035